PHOENIX TRIUMPHANT

PHOENIX TRIUMPHANT

The Rise and Rise of the Luftwaffe

E. R. HOOTON

ARMS AND
ARMOUR

Arms & Armour Press
A Cassell imprint
Villiers House, 41–47 Strand, London WC2N 5JE

Distributed in the USA by Sterling Publishing Co. Inc., 387 Park
Avenue South, New York, NY 10016-8810

Distributed in Australia by Capricorn Link (Australia) Pty Ltd,
2/13 Carrington Road, Castle Hill, New South Wales 2154

British Library Cataloguing-in-Publication data: A catalogue
record for this book is available from the British Library.

ISBN 1 85409 181 6

Edited and designed by Roger Chesneau/DAG Publications Ltd

Printed and bound in Great Britain by
Hartnolls Limited, Bodmin, Cornwall

CONTENTS

Foreword   7
Comparative Ranks   12
Abbreviations   13

1   **A Premature Epitaph**   *November 1918 to November 1923*   17
2   **The Lipetsk Years**   *January 1924 to December 1929*   44
3   **Birth of an Eagle**   *January 1930 to December 1932*   69
4   **The Fledgeling**   *January 1933 to June 1936*   94
5   **The Spanish Trial**   *July 1936 to March 1939*   120
6   **The Steel Façade**   *June 1936 to September 1939*   146
7   **Blitzkrieg/Sitzkrieg**   *September 1939 to May 1940*   175
8   **Eagle Over the Sea**   *September 1939 to June 1940*   212
9   **Sieg im Westen**   *May to June 1940*   239

Appendix 1: German Air Forces, 11 November 1918   273
Appendix 2: Austrian Air Forces, 4 November 1918   275
Appendix 3: Expenditure upon Air Forces, 1933–1939   276
Appendix 4: Pre-War Production and Plans   277
Appendix 5: Aircraft Production, 1933–1939   279
Appendix 6: *Luftwaffe* Expansion, 1935–1938   279
Appendix 7: German Aircraft Deliveries to Spain, 1936–1939   280
Appendix 8: *Österreichische Luftstreitkräfte*, March 1938   280
Appendix 9: *Luftwaffe* Order of Battle, 1 September 1939   281
Appendix 10: *Luftwaffe* Order of Battle in Scandinavia,
     9 April and 10 May 1940   288
Appendix 11: *Luftwaffe* Order of Battle in the West, 10 May
     and 5 June 1940   290
Bibliography   299
Glossary   307
Index   313

FOREWORD

Like a genie from a bottle, the *Reichsluftwaffe* materialized on the European scene in 1935 and grew into a giant which menaced Germany's neighbours. Within five years the *Luftwaffe*, as it was commonly called, was the most powerful air force in Europe and played a vital part in making the Third Reich the master of the continent by June 1940. The victory in 1940 brought to a triumphant conclusion two decades of struggle to overcome the restrictions imposed by the Versailles Treaty. This struggle took place not only in Germany but also in Austria, which faced the equally stringent restrictions of the Treaty of St Germain and whose air force was absorbed by the *Luftwaffe* in 1938. It involved not only the nations of western Europe and Scandinavia but also those of eastern Europe as well as Japan and the distant United States. The way in which Germany, and to a lesser degree Austria, overcame the Versailles restrictions and created an air force capable of subjugating half of Europe is the subject of this book.

I have had two objectives, the first to explain how the Germans and Austrians re-established their air forces from 1930 onwards and the second to show how the *Luftwaffe* was used in both peace and war. The first part must inevitably use the Nazis' rise to power as a turning point. Before this period the Germans acted with great caution and secrecy, and readers will be surprised to discover that the use of the Lipetsk training and evaluation base in Russia was at one time concealed even from the German Government who wished it closed! This is not the only the surprise. It has been generally assumed that the Armistice of 1918 ended the development of German military (and naval) air power until the establishment of Lipetsk in the mid-1920s. In fact it becomes clear that Germany made only a token effort to meet the stipulations of the Armistice, and even at the beginning of 1919 the Army and the Navy had a combined strength of 9,000 aircraft. Before the Versailles Treaty was drafted the German Defence Ministry (*Reichswehrministerium*) was making strenuous efforts to rebuild German air power through the *Freikorps* squadrons, of which some 35 existed. The Ministry intended these squadrons to become the basis of the new air force, but this was prohibited by the Versailles Treaty. However, they continued to exist as *Polizei-Fliegerstaffeln* and saw action as late as the spring of 1920. The *Reichswehrministerium*'s efforts to contravene the Versailles Treaty strengthened the Allies' resolve to emasculate Germany's aircraft industry, which did not recover until the Nazis took power, thus compromising the country's hopes of rebuilding an air force.

I have tried to trace the complex trail followed by Germany towards this latter goal during the 1920s and early 1930s with its numerous domestic and foreign influences. It is clear that the greatest problems, apart from the Versailles restrictions, were those derived from the conservatism of the German military leadership. Yet progress continued at an ever-accelerating rate after Seeckt's fall, and by 1932, when the decision to create an air force was actually taken, the embryo *Luftwaffe* existed. What is astonishing is that, although the Germans made every effort to conceal their intentions until the 1930s, the Allies were never deceived and remained well informed of their progress. Moreover, after 1930 German security became noticeably more lax as steps were taken to complete the breakage of the Versailles shackles.

Readers will not be too surprised that Hermann Göring plays a relatively small, though important, role in the creation of the *Luftwaffe*. The decisive influences in the first three years were Erhard Milch, Walter Wever and Werner von Blomberg. Milch's contribution on the organizational and industrial side has been well chronicled, but what has not been appreciated is his influence on the less glamorous yet vital development of navigation aids: these gave the *Kampfgruppen* a theoretical all-weather, day/night capability which exceeded that of the RAF. Wever shaped the *Luftwaffe*'s operational doctrines and produced advanced concepts made all the more remarkable because he entered the *Luftwaffe* as a reactionary who was not even convinced of the need for an autonomous air force. His willingness to apply the latest technology to operational problems made him unique among senior *Luftwaffe* officers and his premature death deprived Germany of an officer of great vision. Blomberg's contribution was to permit the *Luftwaffe* an autonomous existence out of both military conviction and political sympathy. When his subordinates in the *Reichswehrministerium* consolidated the separate Army and Navy air activity to keep it out of Nazi hands, he first allowed them to complete their work then let it fall, like a ripe apple, into the hands of Göring and Milch to develop.

One of the most insidious influences, and one largely ignored, was the experience gained with the First World War *Luftstreitkräfte*. With the exception of Sachsenberg, Germany produced no air power prophet of note and relied far more upon her wartime experience than the United States and Great Britain to develop a philosophy for the *Luftwaffe*. In particular, German air leaders—with the exceptions of Wever, Wimmer and to some degree Milch—underestimated the importance of technology upon tactics and operations, but by the end of 1936 the first two had been removed from the scene while the influence of the third was declining sharply. The remainder drew upon their wartime experience when technical advance brought only temporary superiority and sought a 'permanent' solution in improved tactics. There were many other examples—in particular relating to the *Luftwaffe* training organization, which was little different from that of the *Luftstreitkräfte*—but the most interesting instance of failing to learn from experience lay in the development, procurement and production organization, where the *Generalluftzeugmeister* created for Udet in 1939 duplicated the failed *Flugzeugmeisterei* created by Siegert in 1918.

It was as a result of wartime experience that Wever shaped the *Luftwaffe*'s role—which, even sixty years later, few aviation historians understand. Traditionally, they have categorized air forces as either 'tactical' or 'strategic', with the *Luftwaffe* classified as the former, but any examination of the *Luftwaffe*'s role shows such a division to be simplistic. Not only did the *Luftwaffe* always have a 'strategic' role (even if it did not have the means of effective strategic bombing), but factories featured as objectives in most of its campaigns. Unfortunately, historians have neglected to explain the obvious anomaly of a 'tactical' force attacking industrial targets. A good definition of strategic bombing—leaving aside, for the moment, its impact on morale—was to be found in *OKW Direktive Nr 9*, which called for 'The destruction of industrial plant whose loss would be of decisive significance for the military conduct of the war'. Such an objective provides a basis for defining a 'strategic' air force, although this was obviously not the *Luftwaffe*'s prime role.

The circle can be squared by interpreting the *Luftwaffe*'s role against the background of contemporary German military thought. This accepted the traditional terms of 'tactics' (the art of fighting battles) and 'strategy' (the art of organizing campaigns), but in the last century the Germans recognized that the growing size of armies made such a division arbitrary and the General Staff introduced an intermediate stage as the operational (*operativer*) art. This applied to operations by forces at Army and Army Group level and is sometimes called 'grand tactics' but will be described in this text as 'operational-level activity'. In German military thought the term 'tactics' now applied to operations within Army Corps while 'strategy' referred to activity at theatre level or above.

The German Army had always thought in 'tactical' air terms, with missions including fighter patrols over the front line, close air support and tactical reconnaissance (*Nahaufklärung*). While not rejecting these requirements, Wever persuaded his former colleagues to accept that the most effective means of using air power was through operational-level activity, which might be defined in today's terms as counter-air, battlefield air interdiction, long-range reconnaissance (*Fernaufklärung*) and even air resupply roles. The *Luftwaffe*'s prime mission was to gain air superiority throughout the operational depth by destroying the infrastructure of the enemy air force and by aerial attrition, and then to support operations on the ground (and to a lesser degree over the sea) by interdicting enemy communications, formations on the march and bases. The operational-level role also included attacks upon industrial targets, a 1938 training manual stating that 'Attacks on power sources and supply lines are carried out only if they serve as preparation for ground or sea operations.' Such attacks were intended to disrupt supplies to the front rather than to destroy the enemy's industrial base, which the Germans often hoped to exploit for themselves.

Readers who have difficulty understanding the concept of the operational level are in good company, for *General* Paul Deichmann asked some of his *Luftwaffe* staff officers to define such operations and received different answers from every one! In fact, tactical and operational-level support of the Army, which had once been part of the *Luftwaffe*'s role, became its primary one, especially on the Eastern Front and in the Mediterranean. Even in the first years of the war, the *Luftwaffe* failed in both Poland and the West to win air superiority from the outset because of poor intelligence but was able to recover and achieve its aim while simultaneously supporting the Army. In the Battle of Britain it again failed to win immediate air superiority and all attempts to recover proved to be in vain.

The mythology of the *Luftwaffe*'s development is as pervasive after six decades as that of the classical world. I have tried to lay many ghosts (including the most famous one—the supposed lack of German interest in heavy bombers) and to indicate that the *Luftwaffe* was not the only air force to demand a dive-bomber capability in a heavy bomber. I have made little attempt to describe the development of individual aircraft because there is a wealth of literature on the subject: for example, there are comprehensive stories in William Green's *Warplanes of the Third Reich*, while the researches of Joachim Dressel and Manfred Griehl on German fighters and German bombers have now been published. Wever's vision has been mentioned, but the successes or failures of the *Luftwaffe* cannot be taken in isolation and in order to provide perspective to the story I have sketched contemporary developments in foreign air forces and the problems the latter encountered. I have tried to describe in some detail the *Luftwaffe*'s operational experience. In Spain the Condor Legion had to overcome numerical and qualitative inferiority and its success owed everything to the new generation of combat aircraft. However, these aircraft also played the leading role at Guernica, and I believe that I have produced new evidence to explain what has hitherto been regarded as an atrocity. Poland saw the *Luftwaffe*'s theories tested and occasionally found wanting, while in Norway a disastrous start was recouped by individual bravery and skill. Norway also demonstrated the vulnerability of navies operating in conditions of enemy air superiority, a lesson which proved as true in 1940 as in 1982.

Two areas which have been largely neglected are both covered in some detail in this book for the first time—the *'Sitzkrieg'* and the Western Campaign. The *'Sitzkrieg'* saw the *Luftwaffe*'s first attempts at strategic air defence, with radar used to support day interception and the re-creation of a night fighter force during the autumn of 1939. Both had a significant influence upon strategic and operational-level reconnaissance, as will be discovered. The task of describing operations in the West during 1940 was believed impossible because of the wholesale destruction of *Luftwaffe* records, but, by using the surviving fragments, Allied records and unit histories written by dedicated German amateur historians, I think that this has now been achieved. I have carried the story to the verge of the Battle of Britain for the sake of continuity.

Inevitably the reader will find some that is familiar, for the overall story has attracted much attention in the past. Yet there will be much which is unfamiliar or surprising, for I have used published sources, documents, diaries and staff publications. I have tried to flesh out the characters of the 'stars' in this story (which may lead to a re-evaluation of some), but there are no heroes and no villains for this is not a melodrama. The centre of attention is, naturally enough, the aviation, but I have tried also to indicate the importance of the groundcrews, the signallers and the supply organization in order to provide a comprehensive understanding of this multi-faceted subject and its leadership.

As this book has been written for English-speaking readers, the names of countries, cities and most towns have been anglicized, for example Brunswick for Braunschweig, Cologne for Köln, Munich for München and Vienna for Wien. However, the names of airfields have been rendered in their original form, to ease the task of the amateur historian using German-language publications. As German military and social ranks are quite distinctive I have left these in the original also, and the same philosophy applies to military ministries, organizations, and formations. Civil establishments such as the Foreign and Transport Ministries have been anglicized for ease of understanding and to avoid confusion. For the *Reichswehr* I have used the terms *'Reichsheer'* for the Army and *'Reichsmarine'* for the Navy, but this has not been considered desirable for the *Wehrmacht* for the terms *'Heer'* (Army) and *'Kriegsmarine'* (Navy) are not that well known. Distances, volumes, areas, speeds and altitudes are in metric terms where the original references were thus, but conversions to Imperial measurements have been provided in most instances.

This book could not have been written without the help of many people. Above all I would like to express my profound thanks to my good friend Alex Vanags-Baginskis for his advice, his assistance and the provision of a gold mine of documents and illustrations, as well as for 'de-bugging' the draft. I would also like to thank *Herr* Hans-Eberhard Krüger for his advice and for his tremendous generosity in providing research material and photographs. Special mention should also be made of Mr Peter Kilduff who gave me much information concerning the aviation and aviation personalities of the First World War and the immediate post-war period as well as invaluable research material. Many of my colleagues at Jane's provided specialist information, and among them I would like to thank Dr Kathy Bunten, Ian Hogg, Ken Munson and John Taylor. I also express my thanks to Will Fowler, Mike Gething, William Green, *Admiral* Gert Jeschonnek, Dr Richard Osborne, John Terraine and Group Captain S. A. Wrigley for their help, while the assistance given by the late *Herr* Steffen Papsdorf must also be acknowledged. The views expressed in this book are, however, my own and do not necessarily represent the views of those who have helped me. I also express my thanks to Peter Burton and Roderick Dymott of Arms and Armour Press for their patience, support and advice.

My primary research for this book was conducted at the *Bundesarchiv*'s *Militärarchiv* in Freiburg, the Imperial War Museum, the Public Record Office, the Ministry of Defence Whitehall Library and the Library of the Royal United Services Institute in London as well as at the Vincennes-based *Service Historique de l'Armée de l'Air* and *Service Historique de l'Armée de Terre*, and I would like to express my gratitude to the staffs of these establishments for their courtesy and assistance. Other organizations which proved very helpful were the *Cuartel General del Ejercito del Aire, Servicio Historico y Cultural Instituto de Historia y Cultura*; Deutsche Luftwaffering e.V.; the Dutch Air Staff Historical Section; the Dutch Army Military History Section; the *Heeresgeschichtliches* Museum, Vienna; the *Instituto de Historia y Cultura Naval* (with particular thanks to *Señor* José Ignacio González-Aller Hierro); the Norwegian Armed Forces Museum; and the US Air Force Historical Research Agency, Maxwell Air Force Base, Alabama. Special tribute should also be paid to the staff of Langley Library, Berkshire, who (to paraphrase the Mounties) always get their book, and to the staff of the Reference Section of Slough Central Library, whose facilities allowed me to read key microfilms.

Lastly, I would like to thank my wife Linda and daughters Caroline and Jennifer for their patience while I have wrestled with this subject.

E. R. Hooton

COMPARATIVE RANKS

Luftwaffe	RAF	Armée de l'Air	USAAC
Generalfeldmarschall	Marshal of the RAF	–	General (five star)
Generaloberst	Air Chief Marshal	–	General (four star)
General der Flieger	Air Marshal	Général	Lieutenant-General
Generalleutnant	Air Vice-Marshal	–	Major-General
Generalmajor	Air Commodore	–	Brigadier-General
Oberst	Group Captain	Colonel	Colonel
Oberstleutnant	Wing Commander	Lieutenant-Colonel	Lieutenant-Colonel
Major	Squadron Leader	Commandant	Major
Hauptmann★	Flight Lieutenant	Capitaine	Captain
Oberleutnant	Flying Officer	Lieutenant	First Lieutenant
Leutnant	Pilot Officer	Sous-Lieutenant	Second Lieutenant
Fahnenjunker	–	Aspirant	–
Oberfeldwebel	Flight Sergeant	Sergent Chef	Master Sergeant
Feldwebel	Sergeant	Sergent	Sergeant
Unteroffizier	Corporal	Caporal	Corporal
Obergefreiter	Leading Aircraftsman	–	Private 1st Class
Gefreiter	Aircraftsman 1st Class	–	Private
Flieger	Aircraftsman 2nd Class	–	Private

★ Also Rittmeister (Cavalry)

ABBREVIATIONS

A number of abbreviations are used in the text, of which the most common are listed here.

A-Day	German equivalent of D-Day, the commencement of military operations	*Feld FA*	*Feld Fliegerabteilung*
		FFA	*Freiwilligen Fliegerabteilungen*
AA	Anti-aircraft	*FJR*	*Fallschirmjägerregiment*
AASF	Advanced Air Striking Force (RAF)	*Flak*	*Fliegerabwehrkanone*
		Flik	*Fliegerkompanie* (Austrian)
Akaflieg	*Akademischen Fliegergruppen an Technische Hochschulen*	*Flivo*	*Fliegerverbindungsoffizier*
		GFA	*Grenzschutz Fliegerabteilungen*
AON	*Aviatsiya Osobovo Naznacheniya* (Russian)	**GHQ**	General Headquarters
		I	*Istrebitel'* (Russian)
AufKlGr	*Aufklärungsgruppe*	**IAACC**	Inter-Allied Aeronautical Commission of Control
BA	*Bundesarchiv*		
BAFF	British Air Forces in France (RAF)	*Idflieg*	*Inspekteur der Flieger*
		IWM	Imperial War Museum
BCR	*Bombardement, Combate et Reconnaissance* (French)	*Jafü*	*Jagdfliegerführer*
		JaGschw	*Jagdgeschwader* (Austrian)
BEF	British Expeditionary Force	*Jasta*	*Jagdstaffel* (German in First World War, Austrian in 1930s)
BFW	Bayerische Flugzeugwerke		
BMfHW	*Bundesministerium für Heerwesen* (Austrian)	*JG*	*Jagdgeschwader*
		JGr	*Jagdgruppe*
BMW	Bayerische Motoren-Werke	*KG*	*Kampfgeschwader*
Bogohl	*Bombengeschwader der Obersten Heeresleitung*	*KGzbV*	*Kampfgeschwader zbV*
		KgrzbV	*Kampfgruppe zbV*
BordFlGr	*Bordfliegergruppe*	*Kofl*	*Kommandeur der Flieger*
DEM	*Détection Eléctromagnetique*	*Kolu*	*Kommando der Luftstreitkräfte* (Austrian)
Deruluft	Deutsch-Russiche Luft-verkehrs GmbH		
		Koluft	*Kommandeur der Luftwaffe*
DLR	*Deutsche Luftreederei*	*KüFlGr*	*Küstfliegergruppe*
DLV	Deutschen Luftfahrt-Verband e.V (to March 1933); Deutsche Luftsport-Verband e.V. (from March 1933)	*KuK*	*Kaiserlich und Königlich*
		LA	*Lehrabteilung*
		LG	*Lehr Geschwader*
		LKS	*Luftkriegschule*
DVL	*Deutsche Versuchsanstalt für Luftfahrt*	*LnAbt*	*Luftnachrichtenabteilung*
		LStrKr	*Österreichische Luftstreitkräfte* (Austrian)
DVS	Deutschen Verkehrsflieger-schule GmbH		
		Lv	*Luftverteidigungszone*
E-stelle	*Erprobungsstelle*	*LVA*	*Luchtvaartafdeling* (Dutch)
FA	*Fliegerabteilung*	*LvKdo*	*Luftverteidigungskommando*
FAA	Fleet Air Arm (British)	*MA*	*Militärarchiv*
FAR	*Flieger-Ausbildungsregiment*	*MLD*	*Marine Luchtvaartdienst* (Dutch)
FBK	*Flughafenbetriebskompanie*		
FdL	*Führer der Luftstreitkräfte*	*NSFK*	*Nationalsozialistischen Fliegerkorps*
FEA	*Flieger-Ersatzabteilung*		

ObdL	Oberbefehlshaber der Luftwaffe	**USAAS**	US Army Air Service
OHL	Oberste Heeresleitung	**USAAC**	US Army Air Corps
OKH	Oberkommando des Heeres	**USAAF**	US Army Air Force
OKM	Oberkommando der Kriegsmarine	**VVS-RKKA**	Voenno-Vozdushnye Sily Raboche-Krestyanskaya Krasnaya Armiya (Russian)
OKW	Oberkommando des Wehrmachts	**WaPrw 8**	Waffenamt Prüfwesen 8
ÖLAG	Österreichische Luftverkehrs Aktiengesellschaft	**Wekusta**	Wetterkundungsstaffel
		Wivupal	Wissenschaftliche Versuchs- und Prüfanstalt für Luftfahrzeuge
PLW	Polskie Lotnictwo Wojskowe (Polish)	**zbV**	zur besonderen Verwendung
PRO	Public Record Office	**ZG**	Zerstörer Geschwader
R	Razvedchik (Russian)	**ZOAE**	Zone d'Opérations Aérienne Est (French)
RAD	Reichs Arbeitsdienst		
RAF	Royal Air Force	**ZOAN**	Zone d'Opérations Aérienne Nord (French)
RDLI	Reichsverband der Deutschen Luftfahrtindustrie	**ZOAS**	Zone d'Opérations Aérienne Sud (French)
RLM	Reichsluftministerium		
RKKVVF	Roboche-Krest'yanski Krasnyvozdushny Flot (USSR)		

German aircraft abbreviations

RWM	Reichswehrministerium	**Ar**	Arado
SA	Sturm Abteilung	**Bf**	Bayerische Flugzeugwerke
SB	Skorostnoi Bombardirovshchik (Russian)	**Bü**	Bücker
		BV	Blohm und Voss
Schlasta	Schlachtstaffel	**Do**	Dornier
SES	Seeflugzeug-Erprobungsstelle	**Fi**	Fieseler
Severa	Seeflugzeug-Versuchs- abteilung GmbH	**FW**	Focke Wulf
		Go	Gotha
SHAA	Service Historique de l'Armée de l'Air	**HD**	Heinkel biplanes (to early 1930s)
SHAT	Service Historique de l'Armée de Terre	**HE**	Heinkel monoplanes (to early 1930s)
Sipo	Sicherheitspolizei	**He**	Heinkel (from early 1930s)
SS	Schutzstaffel	**Ju**	Junkers
StG	Stuka Geschwader	**Kl**	Klemm
Stofl	Stabsoffizier der Flieger	**Ro**	Rohrbach
Stuka	Sturzkampfflugzeug		

Unit designations

T1	Truppenamt Heeresabteilung
T2	Truppenamt Heeres- Organisationsabteilung
T3	Truppenamt Heeres-Statistische (to 1929); Truppenamt Abteilung Fremde Heer (from 1929)
T4	Truppenamt Heeres- Ausbildungsabteilung
TB	Tyazhyoly Bombardirovshchik (Russian)
TrGr	Trägergruppe
TsAGI	Tsentralnii Aero- Gidrodinamicheskii Institut (Russian)

In the *Luftwaffe*, fighters, bombers and dive bombers were organized into *Geschwader*, each with a Headquarters (*Stab*) and a nominal three *Gruppen* which had roman numerals (e.g. *I/KG 1* was the first *Gruppe* in *KG 1*). Each *Gruppe* had three *Staffeln*, numbered consecutively through the Geschwader. Thus *4./KG 1* would be the fourth *Staffel* of the second *Gruppe*. Independent *Gruppen* had arabic numerals (e.g. *JGr 126, KGr 126, KGrzbV 11*.

Geschwader designations changed several times. Until May 1939 a three-figure designation indicating *Luftkreis*, role and lineage was used. In the *Lehr Geschwader* the suffixes *K*, *J*, *St* and *Sch* stood for *Kampf* (Bomber), *Jagd* (Fighter), *Stuka* (Dive Bomber) and *Schlacht* (Combat) respectively.

Reconnaissance and naval *Staffeln* were organized into *Gruppen*, also with arabic numerals and with up to five consecutively numbered *Staffeln*. A suffix *H* (*Heer*) or *F* (*Fern*) was used to indicate the type of *Aufklärungsgruppe*, allowing the designation to be abbreviated. Thus *2.(F)/121* is the second *Staffel* of *Aufklärungsgruppe (Fern) 121*, while *3.(H)/41* is the third *Staffel* of *Aufklärungsgruppe (Heer) 41*. It should be noted that some *Staffeln* in *Aufklärungsgruppe (H)* were long-range units, although most were Corps squadrons. *Küstfliegergruppen Staffeln* followed the same principle but the suffixes were *F* (*Fern*), *M* (*Marine*) and *Mz* (*Mehrzwecke*), for Long-Range, Tactical and General Purpose respectively.

CHAPTER ONE

A PREMATURE EPITAPH

November 1918 to November 1923

T he autumn of 1918 saw Germany's Western Field Army at the nadir of its fortunes as, after four years of war, it reeled towards the Rhine under a succession of Allied hammer blows. With increasing numbers of soldiers preferring ignominious captivity to glorious sacrifice for the Fatherland, Germany's military leaders demanded that the Government seek an armistice, and on the morning of 11 November the guns fell silent.

But the price of armistice was high, with the Field Army forced to surrender half its weapons and all its aircraft. The Allies had originally demanded 2,000 of the latter, but the *Luftstreitkräfte*, the Field Army's air and air defence arm, had only 1,520 first-line aircraft, which were supplemented by 864 naval first-line aircraft (and six airships), most of them in the West (see Appendix 1). With the social fabric of the Empire rapidly unravelling, the German delegation sought to retain 300 aircraft for internal security but the Allies refused. Under Clause IV of the Armistice agreement, Germany had to 'surrender in good condition 1,700 fighting and bombing aeroplanes—in the first place all aircraft of the (Fokker) D.VII type and all night-bombing aeroplanes'. This clause meant that the *Luftstreitkräfte* would be emasculated, for the Fokkers were Germany's most important fighters in terms of both numbers and quality, while all the heavy bombers operated at night.

The Germans had no choice but to accept. During the morning of 11 November telephone bells tinkled at dozens of airfields and shocked commanders were informed that they must surrender their aircraft. However, pride frequently overcame the habit of obedience, and in a final act of defiance many squadron personnel destroyed their aircraft. Some blamed mutinous troops who were looting and rampaging in an outbreak of revolutionary exuberance, but most simply mutilated 'unserviceable' aircraft.[1]

The actions of the *Luftstreitkräfte*'s leading fighter formation, *JG 1 'Richthofen'*, showed the mettle of Germany's airmen when the unit's wartime odyssey ended at Tellancourt, south-west of Longwy. By 11 November the *Geschwader* had claimed 644 enemy aircraft and had lost 56 pilots killed, including the legendary Manfred von Richthofen. Another 52 had been injured, including Manfred's popular younger brother Lothar, leaving their cousin Wolfram as the only scion of the family still with the *Geschwader*. Its square-jawed commander, *Oberleutnant* Hermann Göring, refused tamely to hand over his fighters to the French and decided to obey earlier orders for a move to Darmstadt. The aircraft would fly there while the groundcrews and administrative personnel went by road under the command of his adjutant, *Oberleutnant* Karl Bodenschatz. The pilots, equally disgruntled, were in agreement and one scrawled in chalk across the *Geschwader* status board the premature epitaph 'Born in war, died in war'.

Bodenschatz's convoy took most of the day to reach Darmstadt, passing through dozens of road blocks established by Soldiers' Councils. These latter grew more radical as the convoy neared the Fatherland, to become Workers' and Soldiers' Councils modelled on those which had brought the Bolshevik Revolution to Russia in 1917. The Councils were contemptuous of officers, and when some of Göring's pilots accidentally landed at Mannheim airfield the local Council seized their aircraft and sent on the disarmed pilots by road. A furious Göring reacted by flying to the airfield with the remainder of *JG 1* and threatening to machine-gun it unless the aircraft and the officers' pistols were returned. The terrified Council quickly complied, and the whole *Geschwader* landed at Darmstadt. There a staff officer ordered the unit to fly the aircraft to Strasbourg and surrender them to the French. Led by Göring, the pilots took off once again but instead of flying south-west they circled the airfield and, one by one, landed their aircraft so heavily the undercarriages were smashed. A few days later, on 19 November, the *Geschwader* was officially disbanded at Aschaffenburg, where Göring (soon to become a *Hauptmann* in the Reserve) assured them, 'Our time will come again'. Afterwards he had to fight off a crowd who tried to strip him of his medals.[2]

On 12 December the Germans claimed that they had handed over all serviceable aircraft, and while the Allies publicly accepted this statement they later discovered that only 516 landplanes and 58 seaplanes, together with nearly 4,100 engines, had actually been surrendered. The Allies estimated that another 250 aircraft had been abandoned either in the West or the Baltic States.[3] Yet on 2 March 1919 the German Government was secretly informed that the Services had 9,000 aircraft.[4]

Under the Armistice agreement the Rhineland (Germany west of the Rhine) was to be occupied by the Allies, and German troops marching into the recently proclaimed Republic encountered chaos. The shock of defeat, and starvation caused by the Allies' blockade, heightened social tension. Strikes paralysed many cities as Workers' and Soldiers' Councils seized power and sought to make the Republic into a socialist utopia. Such views were anathema to an Army leadership equally concerned about Poland, which had risen, phoenix-like, from the ashes of three empires. The Poles were trying to seize not only those parts of the former German Empire that had a predominantly Polish population but also 'mixed' areas, notably Posen (Poznan), Silesia and West and East Prussia. The Germans could do nothing, because Eastern Field Army had collapsed, leaving the young Republic helpless in the face of both foreign and domestic foes.

The new Chancellor, Friedrich Ebert, hoped that the Western Field Army would meet both internal and external challenges, but its war-weary men sought only demobilization. With its demise there was no role for the *Luftstreitkräfte*, whose headquarters was disbanded by its commander, *Generalleutnant* Ernst von Hoeppner, on 21 January 1919. In fact some of Hoeppner's men had created their own Soldiers' Council and were authorized by Paul Göhre, the new Under Secretary of State in the War Ministry, to create a *Deutschesluftamt* on 11 November to supervise commercial aviation. Göhre asked the highly experienced *Idflieg*, *Major* Wilhelm Siegert, to head the Directorate, but the *Major* had no sympathy for the Republic. However, the wounded wartime hero *Hauptmann* Ernst Brandenburg was persuaded to become the leading official.

A fortnight later, on 26 November, Brandenburg joined the Interior Ministry when it created the *Reichsluftamt* under 45-year-old August Euler to supervise civil aviation. Euler was not only the first German to hold a pilot's licence but also the country's first aircraft manufacturer, but as an advocate of small business he had vehemently opposed the Army's efforts to control wartime production. Consequently he was soon left on the sidelines, producing only components, but in the new Republic his anti-militarist stance was a political advantage and his appointment may well have owed something to his friendship with the Interior Minister, Heinrich Albert.[5] However, Euler was caught between the Councils and the Army leadership. The former were the immediate problem, for they often occupied airfields and installations as well as instigating strikes at aircraft factories. Yet many factories continued to meet wartime contracts after the Armistice: 200 aircraft each came from the Hannoversche Waggonfabrik and BFW, while Junkers produced 37 CL.Is, Siemens-Schuckert made D.III/IV fighters and Zeppelin (Staaken) produced the six-engine R.VIII. The Soldiers' Councils remained patriotic, and in the Rhineland the Council at Bruchsal organized the flight of 110 trainers from *Bayerische Militärfliegerschule 7* (Germersheim) and *2* (Lachen Speyerdorf) across the Rhine on 5 December rather than let them fall into Allied hands.[6]

Siegert, a pioneer of German military aviation who had lost several fingers in 1913 while stopping a runaway aeroplane, was determined to maintain Germany's military air arm. But he was almost exhausted 'after four and one-half years of duty uninterrupted by vacations or Sundays', as he wrote to Euler.[7] His duties were made more onerous by criticism during the last months of the war of the German aircraft industry's failures.[8] On 13 December he resigned as *Idflieg* and was replaced by the equally experienced *Major* Wilhelm Haehnelt, who had been the staff officer responsible for *5.Armee*'s air support during the Battle of Verdun in 1916. Later in the year he had been transferred to *2.Armee* on the Somme, where he was *Kofl* during the Battle of Amiens in August 1918.

Siegert was not discharged until 4 March 1919 and probably moved to the Prussian War Ministry (which was responsible for the armies of all states except autonomous Bavaria) to join his old commanding officer *Oberst* Hermann von der Lieth-Thomsen, formerly Hoeppner's Chief of Staff and now head of the Ministry's Aviation Department (*A7L*). Thomsen, as he was usually called, was 52 years old and resembled an elderly heavyweight wrestler, but he was an excellent staff officer who had been involved in military aviation since 1908 and in 1915 had become *Feldflugchef*, with Siegert as his Chief of Staff. The following year his close association with *General* Erich Ludendorff, the First Quartermaster General and *de facto* Chief of the General Staff, ensured that he became Hoeppner's Chief of Staff, at which time Siegert became *Idflieg*.[9]

Thomsen had been determined to control civil aviation since 1917, and when he and Haehnelt met Euler on 19 December he made clear that he would not recognize civilian authority.[10] To the officers Euler was an irritating lightweight, and their first concern was to help the War Ministry to restore central authority and secure the eastern borders. The Ministry had been forced to recognize that its regiments were incapable of either task, but during December staff officers led by *Major* Kurt von Schleicher suggested creating reliable units of volunteers, *Freikorps*, to restore Government authority. In the absence of any alternative this was accepted by Chancellor Ebert, but the political situation was so delicate that the War Ministry could not openly create these units, only connive at their formation.

The Spartacist rebellion in Berlin on 5 January spurred the creation of the *Freikorps*, which accelerated on 9 January when the Government officially called for volunteers for the *Grenzschutz Ost* (Eastern Border Patrol) and raised some 50,000 men, mostly demobilized veterans but supplemented by ultra-nationalist students and even schoolboys. Other units were formed to deal with the internal enemies, collectively called the 'Reds' whether they were Liberals, Socialists, Bolsheviks or Anarchists, whom the veterans blamed for Germany's defeat. Some units were based upon wartime formations, such as *Hauptmann* Andrée's which was raised from *Luftschiffer-Battaillon 3* for operations in the Ruhr, while Hoeppner helped to restore order around Frankfurt-am-Main. The War Ministry maintained personnel records as well as arranging payment, supplies and equipment, including tanks. The last *Freikorps* units were disbanded in 1924, many of the first wave having been absorbed during 1919 into the *Reichswehr* and, later, into the police.

Although documentary evidence is lacking, there is abundant circumstantial evidence that Department *A7L* organized air support for the *Freikorps*. Thomsen raised squadrons with numbers in the 400 series and usually designated them *Freiwilligen Fliegerabteilungen* (*FFA*) or *Grenzschutz Fliegerabteilungen* (*GFA*). While the squadrons had to be created from scratch, there were plenty of aircraft in the depots (*Flugparks*) and training establishments, with some 6,000 on the inventories in January 1919 and more coming from the factories.[11] Thomsen equipped his new squadrons with a variety of both old and new aircraft and the *Freikorps* inventory included the Halberstadt C.V and LVG C.V/VI corps (multi-role) aircraft and Halberstadt CL.II/IV, Hannover CL.V and Junkers CL.I close support aircraft, together with the Albatros D.Va, Fokker D.VII/VIII, Junkers D.I and Siemens-Schuckert D.IV fighters.

Haehnelt's organization located equipment and arranged its distribution; *Leutnant* Rudolf Hess, formerly with the Bavarian *Jasta 35*, was one of those involved in delivering aircraft. But the 'Reds', too, were trying to create an air force, and sometimes the *Freikorps* had to take airfields and installations by force of arms, or smuggle supplies past 'Red' road blocks, the highly strung former commander of *Marine Jasta 2*, Theo Osterkamp, being involved in this activity.

Men were easier to find, resentful of defeat and revolutionary chaos but with nothing else to occupy their time. Typical of them was *Vizefeldwebel* Albert Lux, from Alsace, who had claimed seven victories with *Jasta 27* and had no desire to become a French citizen after the war. When he returned home after demobilization he received a letter from Haehnelt inviting him to Berlin, where he learned about the new squadrons. Lux was delighted to have the opportunity to fly again and soon joined a squadron under *Oberleutnant* Schatz at Breslau.[12] However, he later claimed that Schatz was an uninspiring officer, leading a squadron of lazy political debaters, and, disenchanted, he made contact with the Polish enemy, for whom he dropped propaganda leaflets before deserting with a Fokker D.VII. In the 1930s he was arrested for espionage when visiting Germany, though he was eventually released.

Lux must have been extremely unlucky, for the *Freikorps* squadrons were led by the cream of Germany's airmen, such as *Oberleutnant* Oskar *Freiherr* von Boenigk (commander of *JG 2*), *Oberleutnant* Otto Dessloch (*Jasta 17*), *Leutnant* Werner Junck (*Jasta 8*), Göring's close friend *Hauptmann* Bruno Loerzer (*JG 3*), *Hauptmann* Erhard Milch (*JGr 6*), *Leutnant* Carl-August von Schoenebeck (*Jasta 33*) and *Hauptmann* Joachim von Schröder (*Bogohl 7*). From the Navy came the air aces *Kapitänleutnant* Friedrich Christiansen and *Leutnant zur See* Gotthard Sachsenberg. Several commanders, including Boenigk (*FFA 418*), Sachsenberg (*Fliegerkampf-geschwader Sachsenberg*) and *Hauptmann* Krocker (*FFA 423*) used cadres from their old squadrons, such as Krocker's *FA(A) 211*, to form new ones.

Boenigk's squadron, with Junck's *FFA 400* and Milch's *FFA 412*, was among those serving in Silesia under the command of *Major* Hugo Sperrle, who had ended the war as *Kofl 7.Armee*. At least ten other units (*GFA 402–408, 411, 431* and *432*) are known to have operated against the Poles along the new eastern border or in East Prussia, where they were under *Fliegergeschwader von Bredow* at Königsberg and where future *Luftwaffe* Chief of Staff Jeschonnek flew with *GFA 129*. It would appear that these eastern squadrons were the first into action, for on 7 January German aircraft bombed the well-stocked Poznan-Lawica airfield, which had fallen into Polish hands the previous day. Little damage was done and the airfield proved to be a valuable source of equipment for the newly born Polish Air Force. Boenigk was later given command of *Flieger-Jagdstaffel des Armee-Oberkommando Süd* and shot down at least one Polish reconnaissance aircraft with a Fokker D.VII. However, there was little fighting in the air because the Poles had only three squadrons (some 25 aircraft) facing westward: the bulk of their air force was deployed against the Russians and Ukrainians in the east. Fewer than 100 sorties had been flown against the Poles in the west by June, when the Versailles Treaty forced Germany to hand over parts of West Prussia to the young nation. Some 250 German aircraft were captured by the Poles, including twenty machines which had already been sold to them by Germans. Others were copied at the Central Aircraft Workshop, and a French report in January 1920 stated that half Poland's twenty squadrons were equipped with aircraft of German design.

The Berlin-based squadrons were next in action from 11 January, when the *Freikorps*, supported by *FFA 420* and *FFA Hauptmann Krocker* (later *FFA 423*), together with a newly formed *Jasta* and *FA* of a *Freikorps* unit, the *Landeschützenkorps*, attacked the Spartacists. The latter used aircraft from Adlershof-Johannisthal to drop leaflets over Berlin after their uprising, but the *Freikorps* took the airfield on the 11th and within two days the rising had been crushed. Ironically, given the hatred the Right was later to feel for the new Republic, this victory shielded the Constitutional Assembly meeting in Weimar to draft a new constitution. *FFA 420* had been formed at Postdam on 5 January and both its commander, the hard-eyed *Hauptmann* Ulrich Grauert, and one of his officers, Rudolf Meister, were destined to become *Luftwaffe* generals.

Fighting flared up again between 5 and 15 March, both sides using aircraft to bomb and machine-gun strongpoints, the 'Reds'' aircraft being based at Kottbus and

Adlershof-Johannisthal (where a *Marine Landfliegerabteilung* was based). Both sides also used aircraft for reconnaissance, Grauert's squadron again being prominent together with *FFA Loewe* (created by Loewe on his own initiative) and Möhn's *FFA 421*, while an unidentified *FA* supported *Freikorps Hülsen*.

Even before Berlin had been secured the *Freikorps* were marching on 'Red' strongholds in the northern ports, the industrial Ruhr and Saxony. *Hauptmann* Creydt's *FFA 419* supported the assault on Bremen in February, then joined that on the Ruhr, while Grauert and Loewe (whose squadron may have been redesignated *FFA 422* later) supported *Generalmajor* Paul von Lettow-Vorbeck, hero of the East African campaign, who eventually took Hamburg in July. Von Lettow-Vorbeck's units included von Loewenfeld's *III.Marinebrigade*, which is reported as having an *FA* under the naval air ace Christiansen. However, no such unit is shown in orders of battle published in the official histories. Outside Berlin the 'Reds' had few aircraft and ground fire was the greatest threat to airmen. The most distinguished victim was the former commander of *Jasta 13*, Franz Büchner, shot down and killed while on a lone reconnaissance flight near Leipzig.

During April the *Freikorps* concentrated 30,000 men around Bavaria, where a Socialist Republic had been proclaimed. The regime had instituted a 'Red Terror' in which several hundred men, women and children had perished. The main assault was to come from the west, and the *Reichswehrministerium*, or *RWM* (as the War Ministry was renamed on 6 March), assigned the task to *Generalkommando Oven*, which included *Hauptmann* Steffen as *Kofl* controlling *FFA 409, 421* and *422*. But the majority of squadrons were Bavarian (they had 130 airmen on 10 March), including *FFA Fürth, Schmalschläger, Krieglsteiner* and *Dessloch*. The *Freikorps'* spearhead, the *Bayerisches Schützenkorps*, was supported by *FA Krausser (Bayerische FA 1)*, *FA Häfner (Bayerische FA 2)* and *Schlasta Anschütz*.[13] In addition there was an unidentified Württemburg *FA* and a couple of Bavarian courier *Staffeln*.

The attackers were instructed to engage any enemy aircraft they saw but to be careful where they aimed their machine guns for fear of hitting civilians. In fact the defenders had few aircraft and even fewer pilots, most coming from the former *Bayerische FEA 1 (Schleissheim)*, and the aircraft were used for courier roles. The attackers used their aircraft for reconnaissance and ground attack as well as for directing artillery fire. Within a few days Munich was surrounded and the squadrons were instructed to begin an aerial blockade despite poor weather (which improved later). Before the assault leaflets were dropped calling for the surrender of the city and warning the civilian population to stay indoors. At least two aircraft were shot down by small-arms fire, one crew being saved from summary execution by the spirited action of nearby hospital staff. The assault was made in the early hours of May Day, but two days were to elapse before the city was secure.

Surprisingly, most *Freikorps* squadrons served not in Germany but in newly independent Latvia, where the young government offered land and citizenship to German troops who would protect it. *General* Rüdiger Graf von der Goltz was given command of the force, which had German Government backing, and in January 1919 he appealed for volunteers. Among those who responded was Sachsenberg, whose *Marine Jagdgeschwader* was being demobilized in his home town of Dessau, which was also the location of the Junkers factory.

He established his new *Geschwader* at Jüterbog, some 40 miles (60km) south-west of Berlin, and received a variety of aircraft. His connections with Dessau also allowed him to obtain the unique all-metal Junkers aircraft made of steel tubes covered by corrugated duralumin sheets. By mid-February Sachsenberg had three strong squadrons, *FFA 413* with Rumpler C.IV corps aircraft, *FFA 416* (later *Freiwilligen Jasta 416*) with Junkers D.I and Fokker D.VII fighters and the close-support *FFA 417* (later *Freiwilligen Schlasta 417*) with Junkers CL.Is. Army pilots also joined him, of whom the most distinguished was the former *Jasta 7* commander and holder of the *Pour le Mérite* Josef Carl Jacobs (41 victories).

The *Geschwader*, with 80 aircraft including reserves (a third of them Fokker D.VIIs and another third Junkers), landed east of Riga, the Latvian capital, in late February where it joined more squadrons, including Schoenebeck's *FFA 421*, Loerzer's *FFA 427* and *FFA 424, 425, 426* (sometimes described as *Fliegergeschwader Weinschenk*) and *433*. Supported by 120 aircraft, Goltz, with 25,000 men, marched on Riga, which he stormed on 22 May. Sachsenberg, with 32 aircraft, was subordinated to one unit, *1. Garde Reserve Division*, and *Schlasta 417* made numerous low-level attacks on retreating Bolshevik columns, armoured trains and railway stations. Both it and *FFA 413* were often also used to drop supplies to the vanguards who had outrun their supplies, a task repeated by the *Luftwaffe* in the same area 22 years later.

The enemy initially had few aircraft or anti-aircraft guns and in April had fewer still when three Nieuport 17s of the Red Latvian air corps defected and landed at Sachsenberg's airfield at Peterfeld, near Jelgava. The biggest problem lay with the poor quality of the petrol and oil, which caused many forced landings. This was a matter of concern, for, having seen the massacres perpetrated by the enemy, the crews could have no illusion about their fate if captured. The enemy deployed more anti-aircraft guns and machine guns, but Sachsenberg noted that, while conventional aircraft were frequently badly damaged, the all-metal Junkers often escaped with a few bullet holes.

Unfortunately Goltz turned from crusader to conquistador and the German Government was forced to repudiate him. Goltz allied himself with White Russian forces, but British and Latvian military pressure, and the cutting off of supplies from Germany, made his position untenable, although he now had 50,000 men. Wastage had already reduced his air strength (Sachsenberg had only twenty aircraft after Riga had been taken) and on 28 July *FFA 425* and *433* were merged into *Flugpark Kurland*. Under domestic and international pressure, Goltz resigned and was replaced on 18 November by *Generalleutnant* Walter von Eberhardt. Eberhardt had been the first *Idflieg* (1913–14) and now led his bitter troops on an epic march to East Prussia through Lithuania. The squadrons had already begun their withdrawal, some to form a new air force in Germany, and their strength slowly declined, leaving only a handful by the time they reached Germany in December 1919.

In the first half of 1919 some 35 squadrons with about 250–300 single-engine aircraft were created to support the *Freikorps*, of which about a score were in existence at any one time. *Freikorps* operations contributed little to the future development of the *Luftwaffe*, but many of its future leaders fought within their ranks either in the air or on the ground, including Kurt Pflugbeil, Rudolf Bogatsch, Josef Schmid and Wolfgang von Wild. Equally noteworthy are those who did not join the *Freikorps*, including both the *Luftwaffe* commander Göring and the Bavarian air ace Robert, *Ritter* von Greim, together with the air aces *Oberleutnant* Ernst Udet and *Hauptmann* Eduard *Ritter* von Schleich. All these men lived in and around Munich, and the presence of 'Red' militias restricted their activities; indeed, Schleich escaped arrest thanks only to a timely warning by a former member of his *JG 4*. After the *Freikorps* had taken Munich, Göring, Udet and Greim took to barnstorming while Schleich became interested in politics. As with so many in Germany, the post-war chaos polarized his political beliefs, making him fiercely anti-Communist and eventually leading him to join Adolf Hitler's *Nationalsozialistische Deutsche Arbeiter Partie* (NSDAP), popularly known as the 'Nazi' party.

As Government authority was restored, *Freikorps* units were absorbed by the *Reichswehr*, including, during the summer of 1919, the air squadrons. After Bavaria had been secured *FFA 422* transferred to Berlin-Döberitz but in late June it was renamed *Truppen-Fliegerstaffel 15* after moving to Lübeck-Blankensee, from where it helped to suppress a brief 'Red' rising in Hamburg. It was replaced by a *Jagdstaffel* with twelve Fokker D.VII/VIIIs whose pilots included Loerzer (indeed, the unit may have been the old *FFA 427*). *GFA 432* became *Artillerie Fliegerstaffel 103* (later *116*) while Milch's *FFA 412* became *Reichswehrfliegerstaffel 117*. Other wartime air bases, such as Böblingen (south-west of Stuttgart), Königsberg in East Prussia and Schleissheim (Munich), were occupied by mixed units, including *Truppen-Fliegerstaffel 102* and Sachsenberg's *FFA 413*, which latter reached West Prussia in August and was redesignated *Truppen-Fliegerstaffel 34*.[14] The process can be most comprehensively followed in Bavaria (see Table 1), whose forces formally lost the autonomy they enjoyed in the Empire yet retained a considerable degree of independence.

Within the *RWM*, Thomsen's department was expanded and he received an assistant in 39-year-old *Hauptmann* Helmuth Wilberg, one of the most experienced and successful commanders of the *Luftstreitkräfte*, who had distinguished himself as *Kofl 4.Armee* during the Third Battle of Ypres by his aggressive use of close-support squadrons during the day and of heavy bombers during the night. Wilberg had a keen eye for talent and in August 1918 appointed Milch, a non-flying candidate staff officer, to command first *FA(A) 204* and then *Jagdgruppe 6*.

The shape of the peacetime air force had been one of the tasks of Hoeppner's Deputy Chief of Staff, *Major* von Oppeln-Bronikowski, but, with Germany's defeat,

---

TABLE 1: BAVARIAN *FREIKORPS* SQUADRONS ABSORBED BY THE *REICHSHEER* IN 1919

Date	*Freikorps* designation	*Reichsheer* designation
April 21	FA Häfner (Bay. FA 2)	Artillerie-Fliegerstaffel 121
May 14	FA Krausser (Bay. FA 1)	Truppen-Fliegerstaffel 21
June 1	FFA Krieglsteiner and Dessloch	Truppen-Fliegerstaffel 21b
June 25	Schlasta Anschütz	Bayerischen Schlasta
July 1	FFA Fürth and Schmalschläger	Truppen-Fliegerstaffel 21a

Source: Based on Pletschacher, p.152.

and a peace conference scheduled, it was a matter Wilberg needed to re-examine. On 18 May 1919 he submitted a memorandum to *General* Hans von Seeckt, the Army's *de facto* commander, proposing the creation of sixteen air bases (*Fliegerhorste*). Each would have about 100 aircraft, to provide a total strength (including trainers, liaison aircraft and reserves) of 1,800 machines and 8,000 men; the *Reichsmarine* would have another 1,200 personnel. The proposed organization was similar to that of the pre-war Prussian and Bavarian Armies and it also marked the post-war Army's first attempt to control naval aviation. Thomsen had made a similar move in October 1916 but was unable to overcome the Navy's resistance.[15] Wilberg's plan was apparently accepted as part of the force structure of a *Reichsheer* which the General Staff envisaged as comprising 24 divisions (300,000 men) supplemented by a militia.

The Austrians had similar plans to Wilberg's for their new air force, and they put them into effect. At the end of the war the *KuK Luftfahrtruppen* had some 650 first-line aircraft and 46,700 men (including 7,800 in balloon units), while the *KuK Seefliegerkommando* had a first-line strength of some 200 aircraft (for the disposition, see Appendix 2), with 2,450 men. Many aircraft fell into enemy hands while hundreds of others were absorbed by the new air forces formed when the Austro-Hungarian Empire fragmented on ethnic lines. These new nations included the German-Austrian Republic, which was established with its capital at Vienna and created a *Volkswehr* (People's Militia) from returning veterans and volunteers. Three weeks later, on 6 December, the *Deutschösterreichische Fliegertruppe* was established under *Hauptmann* Anton Siebert, a former *Luftfahrtruppe* officer, with his headquarters in the *Fliegerarsenal* at Wiener Neustadt. In theory Siebert had about 1,000 aircraft, but in practice the majority of the aircraft in Austria were trainers and most of the combat machines were in storage.

Siebert organized the *Fliegertruppe* into *Fliegerhorste*, each the equivalent of a wartime *Flik*. *Fliegerhorste 1* and *2* were established at Wiener Neustadt, *Fliegerhorste 3* and *4* at Graz-Thalerhof and *Fliegerhorste 5* and *6* at Wien-Aspern. A total of 100 former *Luftfahrtruppen* single-engine aircraft were assigned to the *Fliegerhorste*, including Aviatik (Berg) D.I, Öfag-Albatros D.III and Phönix KD and D.I/II/III fighters and Lloyd C.III, Löhner C.I, UFAG C.I and UFAG-built Hansa-Brandenburg C.I corps aircraft.

There were few social tensions within the new Republic and the *Fliegertruppe*'s role was defeating external threats. Hungary's revolutionary upheavals threatened her neighbours during the spring of 1919, but Budapest concentrated its forces against Romania and Czechoslovakia. This allowed the *Volkswehr* and *Fliegertruppe* to concentrate on Carinthia, Austria's southernmost province on the borders with Italy and Yugoslavia. This was claimed by the latter, whose forces occupied the provincial capital, Klagenfurt, in December 1918 but were driven out by the *Volkswehr* the following month.

The Austrians were supported by *Fliegerhorst 2a* (*Hauptmann* Julius Yllam), which flew its first sorties on 20 December and was soon based at Klagenfurt-Annabichl with fourteen aircraft and 42 men. The squadron was expanded to 22 aircraft by the end of April, including some fighters, and they were used for reconnaissance and dropping leaflets. In April fighting flared up again with a renewed Yugoslav offensive which culminated in the recapture of Klagenfurt at the end of May. Yllam's men played a more active role, attacking enemy aircraft, bombing enemy positions and carrying out ground attacks, although the level of activity was low, with only 96 sorties flown by 18 June, when a ceasefire was organized. Three of Yllam's men were killed, but their sacrifice proved worthwhile because Allied pressure forced a plebiscite and the Klagenfurt area reverted to Austria.[16]

The Allies also held the fate of both the Austrian and German air forces in their hands and their future was one of the many topics discussed in Paris during the winter and spring of 1919 as peace terms were drafted. The Allies were unlikely to be magnanimous, for the world had been horrified by the brutality of modern war and its persecution by the Germans with scant regard for rules or conventions. (The *Marine Luftschifferabteilung*, for example, had anticipated British area fire-bombing tactics by 27 years.) The world was resolved that the 'Great War' would be the last, and France was especially determined that Germany would never again be in a position to attack her.

The French had suffered more than any of the Allies, with industry and agriculture laid waste and nearly 1.4 million dead. Yet she had been the foundation of the Allies' victory and had out-produced Germany in all fields, including aviation, despite the loss of half her coal and 75 per cent of her steel production (see Table 2). At the end of the war Germany's industrial base remained greater than France's while her population, 66 million compared with 39 million, had grown steadily during the previous fifty years as the French birth rate had declined. The birth rate was generally 2:1 in Germany's favour, but French military planners knew that by the late 1930s the annual intake of men into the Army would be halved to 120,000 while Germany could call upon 400,000.

French leaders feared a rapid German recovery and a war of revenge. They wished to avoid this by reducing Germany's military forces and draining the economy with reparations for wartime damage. To compensate for their weakness in manpower, the French wove a web of alliances with the new nations of Eastern Europe, including Poland and Czechoslovakia, all of which had a vested interest in restraining Germany. The terms of the peace treaty would be hard, although not as harsh as those imposed by the Germans upon the Russians in 1918. Germany lost territory, population and a third of her industrial capacity, while her armed forces were reduced to little more than an armed police force without heavy artillery or tanks. She also faced an open-ended bill for reparations in kind, with an immediate down payment of 20,000 million marks ($2,000 million) in gold.

The British had been willing to allow Germany a small air force, but military advisers recommended clipping German wings and this was reflected in the peace

	France	Germany	Great Britain	USA
TABLE 2: AIRCRAFT INDUSTRY PRODUCTION IN THE FIRST WORLD WAR				
Aircraft	52,146	45,704	55,061	13,840
Engines	93,100	40,499	41,034	28,509

treaty. Article 198 prohibited the *Reichsheer* and *Reichsmarine* from possessing an air force, although the latter was to retain 100 seaplanes for mine detection until 1 October 1919. Article 201 demanded a reduction of the aircraft industry within six months, while under Article 202 all air force equipment was to be delivered to the Allies within three months for supervised destruction. German industry was prohibited from building or importing even civil aircraft for the six months after the treaty had come into effect. To ensure that the Germans complied, an Inter-Allied Aeronautical Commission of Control (IAACC) would be established as part of an international monitoring organization and would report to conferences of Allied ambassadors.

The Germans had anticipated severe terms and hard bargaining when they were summoned to Paris in April 1919, and they were resigned to the loss of Alsace-Lorraine and their mineral deposits (including most of their country's oilfields, which were capable of producing 47,000 tonnes annually). The rest they expected to negotiate, and the delegation, led by Foreign Minister Count Ulrich Brockdorff-Rantzau, included Seeckt, Thomsen and Wilberg. The 75,000-word draft text of the Treaty presented at Versailles on 25 April proved to be a double shock—not only were the terms hard, but none of the 440 articles was negotiable. When the final draft of the Treaty was published on 5 May it was made clear that Germany either had to accept the dictated terms by 23 June or would suffer invasion. The French were determined to occupy Germany if she did not accept, and both Poland and Czechoslovakia were willing to join the occupation. Neither the British nor the Americans were enthusiastic about occupation, although British opinion hardened when the German Fleet was scuttled at Scapa Flow on 21 June.

There was fierce opposition to the Treaty within Germany and the Government refused to sign, but all that knew military resistance would be futile while peace would bring American financial aid. On 21 June a new government was formed under Gustav Bauer and learned that *Generalleutnant* Wilhelm Groener, who had succeeded Ludendorff, advocated capitulation. The Treaty was signed on 28 June, barely two hours before the French occupation was due to begin, and it came into effect on 10 July.

For the German Army the Versailles Treaty was a humiliating blow which forced a fundamental review of its future. The task was assigned to the General Staff, banned by the provisions of the Treaty but camouflaged as the *Truppenamt*, under Seeckt. Seeckt, the son of a Pomeranian nobleman, was 53 years old and his stiff manner, white brush of a moustache and monocle made him appear the epitome of the Prussian staff officer which he claimed as his role model. Yet he was an arrogant and devious man who rejected criticism. He rarely displayed his feelings, except to politicians, whom he treated with undisguised contempt, and within the Army he was known as 'The Sphinx'.[17]

Having spent the war on the Russian Front and in Turkey, Seeckt recognized the value of mobility, and this was reflected in his memorandum 'Basic Ideas for the Reconstruction of our Armed Forces'. He demanded the exploitation of modern technology with the objective of destroying enemy forces using mobile armies whose offensive power would be strengthened by air power. Yet, having encouraged a thorough review of wartime lessons, modern technology and their applications,

Seeckt was reluctant to accept the implications. He was professionally conservative and not only rejected attempts to motorize the cavalry but also insisted upon their retaining lances! He does not appear to have favoured an air force, rather a rebirth of the *Luftstreitkräfte*.

He retained Wilberg initially as his air advisor and from 1 March 1920 as the *Truppenamt*'s aviation expert with the title *Luftschutzreferat*, or *TA (L)*. Wilberg was assisted by a sergeant and a clerk, and his tasks were to monitor foreign military aviation, to examine German aviation requirements and to disseminate information within the Army. On 30 January 1920 he produced a watered-down version of the air force plan he had prepared before Versailles. This called for a *Fliegerhorst* in each of the Army's seven *Wehrkreise*, but, as military planning throughout the 1920s was dominated by a fear of Poland, he proposed forming three *Staffeln*, two in Berlin and one in Königsberg. To some degree this plan was executed, but with the *Reichsheer Staffeln* replaced by *Polizei-Fliegerstaffeln* in order to circumvent the Versailles Treaty—although the Allies were not deceived.

Wilberg's responsibilities were confined to operational aspects while, in accordance with German military practice, specialists supervised procurement, technical evaluation and intelligence. The technical evaluation of foreign aircraft and equipment was the reponsibility of the *Heereswaffenamt*'s *Inspektion für Waffen und Gerät* through *Hauptmann* Kurt Student, while matters relating to aviation industry and procurement were the responsibility of *Hauptmann* Wilhelm Vogt. Intelligence on foreign air forces was acquired and evaluated from foreign publications by a single officer in *T3*. A cadre of staff officers with some aviation background was also created by establishing *Referenten zbV* in each *Wehrkreis*. These had an officer and two assistants who acted as advisors to the infantry division (and, from 1929, cavalry division) staffs on aviation and air defence questions, one assistant acting as a photographic interpreter.

For these and other tasks essential to the establishment of a cadre air force, Seeckt could pick the cream of the wartime officer corps now being demobilized. The air force was important enough for him to select 180 officers, amounting to nearly 5.5 per cent of the 3,800 front-line officers, from the *Luftstreitkräfte* (the Army had another 500 medical or veterinary officers) despite the objections of his personnel directorate, who preferred officers from the traditional arms. Yet, because there was so little air activity, the majority of air officers in both the *Reichsheer* and *Reichsmarine* had little or nothing to do with rebuilding the air arm. Even those who were involved could not escape the bonds of conventional service duties, for all officers had to serve alternate tours of duty in combat units (infantry, cavalry and artillery regiments) as well as in staffs and the *RWM*.

Here, as in the remainder of the Army, Seeckt's policy was to provide the future air force with high-quality leaders. Those selected were usually men who either were staff officers or had demonstrated their ability to command multi-squadron formations. Without these skills the *Reichsheer* had no place even for heroes, and for this reason none of the top dozen air aces alive in mid-1920 (Udet, Bäumer, Jacobs, Loerzer, Lothar von Richthofen, Menckhoff, Buckler, Dörr, Schleich, Veltjens, Könnecke and Bongartz) were selected, although all had been awarded the *Pour le Mérite*, Prussia's premier award, while Loerzer and Schleich had been *Jagdgeschwader* commanders.[18] Seeckt had no need of aircrew after 1921 because he had no aircraft, but the files of the veterans' organization, *Ring Deutscher Flieger*, could be exploited if the situation changed. Significantly, the organization was led by Hoeppner until his death in September 1922, and then by Siegert until his death in 1929.

The *Ring*, with some 2,500 members, could also be called upon by the *Reichsmarine*, which sailed a solitary course in developing air support. Under the Versailles Treaty it was temporarily able to operate 100 seaplanes from Norderney and Holtenau. By some remarkable sleight of hand *Kapitänleutnant* Faber, the *Flottenabteilung Referent für Seeflugwesen* and former staff officer with the wartime *Marineflugchef*, managed to retain six Friedrichshafen FF.49 floatplanes at these bases. They were used throughout the 1920s for second-line duties such as training and target-towing and some of the aircraft remained in service until 1934!

The postwar Navy was allowed only 1,500 officers and selected only 20 (1.5 per cent) with aviation experience. A third were from the *Marine Luftschifferabteilung*, which had a distinguished record in the strategic bombing and 'operational' reconnaissance roles, although the airships proved extremely vulnerable both to wind and to fighters. Even in the reconnaissance role it was a fair-weather force, and in 1918 airships operated on only 17.5 per cent of days and never more than on 34 per cent in 1915.[19] However, the German and US Navies continued to see a long-range reconnaissance role for the airship, and as part of her reparations Germany built the US Navy's most successful airship, the ZR-3 *Los Angeles* (delivered in October 1924). However, nearly a decade elapsed before Germany built a new airship, the successful LZ 127 *Graf Zeppelin*.

Faber worked alone until 1923, when he was given an office (*A11I*) with a staff of four, yet in 1922 he organized his first training course at Stralsund followed by a second in February 1923. The Allies also allowed the *Reichsmarine* a handful of anti-aircraft guns at the Königsberg naval base, which provided a cover for the clandestine development of these weapons throughout the 1920s. Also in 1923 Faber further strengthened naval aviation through the establishment of air staff officers in the *Stationskommandos Nordsee* and *Ostsee* with roles similar to the Army's *Referenten zbV*.

Work on a naval air arm was assisted by the availability of isolated sites, but most Army bases were only too accessible to the Allied monitors who arrived in mid-September 1919. The Germans could not openly defy the Versailles Treaty, but they constantly sought means of circumventing the irksome restrictions. Their delaying tactics ensured that the monitors, including the IAACC under the RAF's Air Commodore E. A. D. Masterman, were unable to begin work until the end of February 1920. In the interval the volatile state of domestic politics provided the excuse to expand the paramilitary *Sicherheitspolizei* (*Sipo*) and to provide it with an air arm under the nominal control of the Interior Ministry.

On 30 September 1919 *Truppen-Fliegerstaffel 15* was renamed *Polizei-Fliegerstaffel Blankensee* and some other *Truppen-Fliegerstaffeln* were also renamed, the Bavarian *Artillerie-Fliegerstaffel 121*, for example, becoming *Polizei-Fliegerstaffel Schleissheim*. In addition, a few police squadrons were activated, including Milch's *Polizei-Fliegerstaffel Seerappen* at Königsberg. A total of ten squadrons, each with between six and ten aircraft, were ostensibly assigned to various police commands, two each in East Prussia and Bavaria (the Bavarian force was later reduced to one squadron at

Schleissheim), one maritime squadron at Kiel and the remainder in Brandenburg, Hamburg, Saxony, Silesia and Westphalia. They were available to the new *Wehrkreise* under Wilberg's plan but were administered by the Interior Ministry because the Versailles Treaty forced *RWM* to disband Thomsen's *A7L*, probably on 11 August when Thomsen himself was demobilized. *Idflieg* followed, and the creation of the *Polizei-Fliegerstaffeln* was Haehnelt's last task before demobilization on 31 October 1919.

The squadron's personnel were somewhat irked to find that, for form's sake, they were sometimes given genuine policing activities. It was a futile effort to deceive the Allies, who knew as much as the German public. Describing the inventory of *Hauptmann* Kastner's *Polizei-Fliegerstaffel Karlshorst*, the aviation magazine *Flugsport* noted that it included two 'very fast fighters' and that 'Two larger aircraft will soon join this base to transport a larger number of detectives . . . to deal with special cases.'[20] The 'larger aircraft' were euphemisms for heavy bombers, which included the Friedrichshafen G.IIIa and the AEG G.V, the latter being used by *Polizei-Fliegerstaffel Hamburg*. These aircraft may have been bombers converted to passenger transports (the airline Deutsche Luftreederei had thirteen) and then re-militarized.

Some squadrons saw action in March 1920 when *Freikorps* leaders attempted a *putsch* (*coup*) under the nominal leadership of Prussian politician *Dr* Wolfgang Kapp. This *putsch* ended in farcical failure, but it caused left-wing uprisings which the *Polizei-Fliegerstaffeln* helped to suppress. During these operations two aircraft from *Polizei-Fliegerstaffel Blankensee* collided over Schwerin on 18 March with the loss of three lives while en route to bomb the coastal town of Wismar.

A Communist rising in the Ruhr during March and April saw the swansong of the *Polizei-Fliegerstaffeln*. *Wehrkreis VI* faced 50,000 rebels and both *Freikorps* units and the *Sipo* reinforced the *Reichsheer* regiments. *Leutnant* Harding's *Fliegerstaffel*, probably from Böblingen, supported Loewenfeld's *III.Marinebrigade* while Häfner's *Polizei-Fliegerstaffel Schleissheim* supported *13.Reichsheerbrigade*. The *Polizei Fliegerhorst* at Paderborn created *Gefechtsstaffel Münster*, initially with six aircraft but later with eleven. The aircraft were used for reconnaissance, dropping leaflets and ground attacks, with especial success against a truck convoy at Unna on 30 March. The scale of the operations may be gauged from the activities of the *Gefechtsstaffel*, which flew 129 sorties from 13 March to 4 April, dropped 1,300,000 leaflets and made ten ground attacks. Two aircraft came down in enemy territory during the operation, one to ground fire and the other as a result of mechanical failure, but the pilots were helped by sympathetic citizens. As a result of the *putsch*, Seeckt became Army Commander as *Chef der Heeresleitung* while *Dr* Otto Gessler became the Defence Minister.[21]

The Allies knew that Germany was not fully complying with the Versailles Treaty; indeed, in mid-1920 Germany admitted to having 200,000 troops and 800,000 paramilitary personnel. During 1920 and 1921 Allied pressure grew for demobilization and payment of reparations (established in 1921 as RM132,000 million, or $13,000 million) and the *Polizei-Fliegerstaffeln* were merely one issue in the campaign. On 6 May 1920 Germany Army aviation officially ended with proclamation *Nr 42/5.20 W 1Fl*, which was drafted by Wilberg and signed by Seeckt. It concluded: 'We shall not abandon the hope of one day seeing the air troops (*Fliegertruppe*) return to life. The fame of the air arm is engraved in the history of the German Army and will never fade. It is not dead—its spirit lives on!' This was too much for the Allied ambassadors, whose Boulogne Conference on 22 June 1920 demanded the disbandment of the *Polizei-Fliegerstaffeln*. When they met in Spa the following month the ambassadors also demanded the disbandment of the *Sipo*.

Although Berlin claimed that the police's wings had been clipped, French Intelligence disagreed. It reported that, apart from *Polizei-Fliegerstaffel Württemburg*, all the *Polizei-Fliegerstaffeln* had simply been renamed *Luftpolizei Abteilungen* on 7 May 1920, each with establishments of 113 men and six aircraft. On 8 November 1920 the Allies again banned the police air arm, and they repeated this ban on 29 January 1921. The Germans prevaricated until, on 5 May that year, the ambassadors' patience ran out and they presented the London Ultimatum. This demanded the disbandment of the police air arm, and the seriousness of the demand was underlined when the Allies again confiscated German civil aircraft. The Germans were forced to grasp the nettle and on 30 June 1921 the *Luftpolizei Abteilungen* were downgraded to flightless *Luftüberwachungs Abteilungen* with genuine policing and air traffic control duties. Among the men who remained in this force were several future *Luftwaffe* generals, including Karl Angerstein, Karl Koller, Alfred Mahnke, Franz Reuss and Herbert Rieckhoff.[22]

Putting pressure on the German civil air market and industry was one method used by the Allies to force compliance with the letter and spirit of the Treaty. German commercial aviation began in January 1918 with the institution, through an AEG subsidiary, Deutsche Luftreederei (DLR), of an aerial postal service which rapidly expanded after the Armistice. There was plenty of competition, for the end of the war saw manufacturers diversifying into passenger services, converting warplanes to civilian use by removing their weapons and fitting seats. Eagerly anticipating the new market, the manufacturers created a commission on 30 January 1919 to examine civil aviation and the acquisition of demobilized aircraft and airfields.

But they faced competition from an unexpected source, as Euler learned on 20 February when he discovered that the Army was organizing a military courier service which might be used by civilians. This service was initiated late in January by Thomsen and Haehnelt, using 24 aircraft (possibly belonging to DLR) at Berlin-Döberitz and Weimar to fly to and from Weimar, Berlin, Dresden, Stuttgart and, later, Munich, official approval for the venture being received on 6 February. Thomsen and Wilberg recognized the value of commercial aviation but believed that it should be under Army direction in the event of mobilization, as was the case with all other means of transport.

The *Reichsheer* rebuffed all efforts by Euler and civil ministries to gain control of the courier service, which later came under the command of the Berlin-based *Generalkommando Lüttwitz* under *General* Walter von Lüttwitz. Because civil aviation had begun with converted military aircraft, it was a common belief in the post-war world that the process could be easily reversed in time of war. This belief had important consequences for German civil aviation—and the arguments had some merit, as the Spanish Civil War was to demonstrate—but ultimately it was recognized that military and civil aircraft requirements were very different.[23]

DLR, with a fleet of about 85 aircraft, began a commercial Berlin–Hamburg service on 1 March, six weeks before the first regular French service and twelve weeks before a British one. The network was gradually extended throughout northern Germany, and it appears that DLR took over the Army courier service. The *Reichsheer* reduced its efforts to direct civil aviation in late April when Wilberg realized that he could not put an air force in mufti, although other air forces (notably the Hungarian) did achieve this aim. Only the disbandment of *A7L* left the way clear for the civilian direction of aviation, but the *RWM* retained considerable influence until the Nazis came to power.

Meanwhile the manufacturers' euphoria evaporated as the Allies excluded Germany from export markets. Even producing new aircraft and converting old ones did not prevent the account books from falling into the red, and by the time of the Versailles Treaty industry was losing up to RM130,000 ($13,000) a month. Civil aviation was permitted by Versailles under Articles 313–320, but some provisions, including the six-month production ban, created great uncertainty. Euler sought in vain to persuade the Allies to ease restrictions and to force the *Reichsheer* to allow its aircraft and bases to be transferred to civilian control. When the *RWM* learned of this, *Hauptmann* Hellmuth Felmy informed Euler that the Ministry would refuse to hand over any military equipment or establishments.

The IAACC permitted up to 140 aircraft, mostly converted warplanes, to operate commercially, and while the majority of these were single-engine there were a few former heavy bombers, including a Zeppelin (Staaken) R.XIVa used by DLR on the Ukraine route.[24] A few optimistic manufacturers continued with the development of advanced all-metal designs such as the single-engine Junkers F 13, which flew in June 1919, while Zeppelin had IAACC permission to develop *Dr* Adolph Rohrbach's revolutionary four-engine E 20 monoplane passenger aircraft. To ensure that all these aircraft were used exclusively for civil purposes, every German aeroplane was confiscated in April 1920 and then released for sale to commercial organizations from 22 May.

But German intransigence and obstruction saw the Allies unilaterally extend their ban on production, and under the London Ultimatum all German aircraft were once again confiscated. This was to stop manufacturers illegally exporting aircraft which the Allies regarded as part of their reparations, a trade which led to the imposition of a RM50 million ($5 million) fine upon the German Government. Some 1,500 aircraft were shipped, or flown, to Scandinavia, the Netherlands or Switzerland, while the redoubtable Anthony Fokker spirited 220 aircraft out of the country in two trains! As late as March 1922 the Allied ambassadors were informed that the IAACC had broken up an organization which had operated from Wandrup since August 1920 and had smuggled out 66 aircraft to Soviet Russia and Denmark, while thirteen aircraft had been discovered in a warehouse. The aircraft sold to Denmark included Rumpler C.IV/VIIs, Halberstadt C.Vs, Albatros C.XVs and LVG C.IVs but appear to have been exported for civil purposes.

Even more serious was the difficulty assembling equipment which, under the Treaty, should have been disclosed to the IAACC for destruction. Understandably, the Germans were unenthusiastic about this, and there was widespread concealment of aircraft, engines and equipment in disused factories, remote barns and even private homes. By 1 April 1920, according to the French *Colonel* Rorand, only 6,730 landplanes, 262 seaplanes and 8,039 engines had been destroyed.[25] As the IAACC hunted for the remainder, an air of desperation could be detected in its activities. The zeal of the disarmers created confusion: for example, they seized civil aircraft, including those being built for reparations, and even put the E 20 to the welder's torch. Within two months, according to the IAACC, the total *matériel* located had risen to 8,939 landplanes, 1,082 seaplanes and 14,839 engines.[26]

The Allies still had not put into effect the six-month production ban which would precede the beginning of civil aviation activity, and during 1920 German manufacturers became increasingly worried. On Christmas Eve their organization urged the immediate delivery of hidden equipment to the Allies to conform with the Versailles Treaty, and by 20 January 1921 the IAACC could report that the total haul had risen to 15,056 aircraft and 25,276 engines (although a *Deuxième Bureau* report ten days earlier had put the figure at 13,237 aircraft and 25,169 engines surrendered, of which 12,695 and 22,171 respectively had been destroyed). The total amount of equipment destroyed is unclear, various figures having been given, but by the end of 1921 it appears to have been 15,714 aircraft and 27,757 engines.

On 3 November 1920 Masterman estimated that only 10 per cent of the aircraft available at the end of the war or built in the immediate post-war period remained hidden. Nevertheless, he concluded,

> The aerial disarmament of Germany may be regarded as virtually accomplished and whatever material may be still concealed cannot be regarded as constituting a menace to the Allies. It is of old design and must have depreciated greatly in value.[27]

The RAF's Director of Plans concurred, writing on 5 November 1921:

> It will be seen that German air power has been broken and that Germany represents no air menace to the Allies.[28]

Yet, simultaneously, some of the Allies were encouraging Germany to produce new military aircraft, since, with the end of the Great War, rivalries in the Pacific had led to a naval arms race between the United States and Japan. To enhance their scouting forces, both these countries' navies sought submarine-borne reconnaissance floatplanes and turned to the German Caspar-Werke. In 1921 each ordered two aircraft, the American U1 and Japanese U2, which were essentially the same design. They came from the drawing board of Ernst Heinkel, who had produced the wartime Hansa-Brandenburg floatplane fighters in which Christiansen had made his name. Four aircraft were delivered (the *Reichsmarine* purchased a fifth), but by that time the Washington Treaty had temporarily ended the race. Nevertheless, Heinkel had established the beginnings of a ten-year business relationship with the Japanese Navy.

Dornier also built for the US Navy the Do A *Libelle* (Dragonfly) and Cs II *Delphin* (Dolphin) I flying boats, together with the Do H *Falke* (Falcon) landplane fighter. Caspar was encouraged by its success and in 1922 sold S1 floatplanes to Sweden, but during the year Heinkel, apparently urged on by Faber and Christiansen, decided to strike out on his own. On 1 December 1922 he created Ernst Heinkel AG at Warnemünde but arranged for the assembly of his naval aircraft to be undertaken in Sweden by Svenksa Aero AB, the first being the S1, designated the HE (Heinkel *Eindecker*) 1.

The Germans' reluctance to comply with the Versailles Treaty and their attempts to evade its provisions produced two self-inflicted wounds on the aircraft industry: the Allies extended the ban which was scheduled to end in mid-July 1920, and they decided to restrict the performance of future German civil aircraft. The new ban was unilaterally imposed by Allied ambassadors at Boulogne and London in June 1920 as another attempt to enforce German disarmament and accelerate reparations. It was to last three months, but it came into effect only after all aircraft and equipment had been deemed surrendered or destroyed. When the Allies and Germans met at Spa in July to resolve their differences over the Treaty, the former had intended to include the new ban in the Spa Protocol. However, because of an oversight this did not occur, and for many months the Germans bitterly opposed its imposition but could do nothing to reverse the decision.

The continuing ban on civil production was a matter of some concern to the IAACC, as Masterman reported to the British Ambassador on 10 November 1920. He referred to it specifically in connection with the E 20, but there were already problems over the Junkers F 13, with disputes between Junkers and the Commission as to whether or not it was a civil aircraft. Junkers had in fact designed it for civilian use, but it could be converted into a warplane; indeed, about 40 were delivered to Soviet Russia for this role a few years later. Although the British Air Council informed the Foreign Office on 19 January 1921 that civil aircraft manufacturing in Germany did not break the provisions of the Versailles Treaty, Allied doubts remained.

The Paris Agreements of 29 January 1921 confirmed plans for the three-month ban, imposed another on police aircraft and served notice that the Allies would limit the performance of German aircraft to ensure that the needs of civil aviation only were met. The limits were not defined, but they were accepted unconditionally by a German Government horrified at the effects of inflation, the dollar/Reichsmark exchange rate sliding from $1:RM75 in early 1921 to $1:RM7,000 by the end of 1922. Berlin needed American loans to survive and was willing to do anything to obtain them.

Only on 9 February 1922 did the Allies finally notify Germany that the three-month production ban had started and that the manufacture of civil aircraft could begin from 5 May. But on 14 April, barely three weeks before the ban expired, the Allies published their civil aircraft performance limitations. The top speed of German aircraft was restricted to 170kph (105mph), their ceiling to 4,000m (13,000ft) and their maximum range to 200km (125 miles). They were to have a maximum payload of 600kg (1,300lb), and their maximum endurance was to be only 2.5 hours. Germany would have to produce annual lists of all civil aircraft, of holders of German pilot licences and of aircraft factories as well as of production.

Wilberg protested that the Allies' limits would immediately render German aircraft obsolete, but the Government ignored him. To save the economy, Berlin was willing to sacrifice the infant German aircraft industry and it signalled its acceptance on 5 May. The same day the IAACC was renamed the Aeronautical Guarantee Committee and agreed to a two-year review of the standards to bring them closer to those of foreign aircraft. In the event the first, minor amendments were made only in June 1925, and the Allies exacted a price of extending control to engine development. Germany could only make a formal objection.[29]

The Allies' action largely finished off Germany's wartime aircraft industry. Famous names such as AEG, Aviatik, Fokker, Halberstadt, Hannover, LVG, Rumpler and Siemens Schuckert had quit aviation by 1921, and while a few sought to break into new business ventures the majority simply disappeared. In July 1922 Berlin paid the aircraft manufacturer's association RM150 million (worth only $2.8 million) as compensation for lost profits and wages.[30] By then Euler had gone, sidelined in peace as in war. His efforts to secure Allied support for German civil aviation had failed, and he became an embarrassment which the bureaucrats slowly neutralized. The *Reichsluftamt* was absorbed into the *Reichsamt für Luftfahrt und Kraftfahrwesen* and power increasingly passed to Ernst Brandenburg, who retained Army support. In December 1920 Euler resigned and retired to a life of obscurity, to die in 1958 in Munich.

Brandenburg, the new man directing Germany's legitimate aviation activities, was a brilliant wartime organizer who had commanded *Bogohl 3 (Englandgeschwader)*. A short, prematurely balding man, he had led his Gotha heavy bombers against London in daring day and night raids which had earned him the *Pour le Mérite* and caused fury in the British Parliament. To deflect public anger, the British Government had undertaken a fundamental review of air power, led by General Jan Smuts, and this had resulted in the creation of the Royal Air Force. By this time Brandenburg's service career had ended prematurely when he was seriously injured in an accident which forced him to use crutches for several years.[31]

When his Directorate joined the Transport Ministry, under Groener until 1923, Brandenburg became Ministerial Advisor. In 1924 a separate aviation department was created in the Ministry under Brandenburg as Ministerial Director, a position which essentially gave him control of civil aviation and allowed him to support clandestine military aviation. His early years were spent salvaging what he could from the wreckage of German hopes and encouraging, through the rationalization of the airlines, a strong civil air transport system. Patronage through the control of postal subsidies was a useful tool during the early 1920s as he laid the foundations of Europe's most efficient airline. The airlines themselves provided employment for hundreds of former air and groundcrew, but the work was rarely secure and, for the pilots, often dangerous. Many an airman who survived wartime dangers fell prey to the capriciousness of the weather or to mechanical failure, the latter causing the death of Lothar von Richthofen in July 1922.[32]

Despite all the obstacles, dozens of young Germans were already beginning to learn to fly with covert *RWM* sponsorship. In 1920 the DLV was established with Wilberg's enthusiastic support, encouraging young people to fly by opening schools which could also be exploited by the Services. Initially the focus of activity was in the Rhön Mountains east of Fulda, where, in the summer of 1920, the editor of *Flugsport*, Oskar Ursinus, started annual glider meetings to stimulate competition and, indirectly, to encourage national interest in aerodynamics and airframe construction. German ascendancy in gliding continued until the outbreak of war and was exploited in 1940 to take the Belgian fort of Eben Emael. Among the young designers who began their careers in gliding were Siegfried and Walter Günter, who were to design the elegant Heinkel He 70 and He 111. Student was an active participant in gliding, often accompanied by his wife, and Seeckt also visited Rhön.

Rhön and the DLV hinted at a bright future, but for Seeckt the present was bleak until an opportunity arose from an unexpected quarter. In the spring of 1919 the Berlin police arrested for agitation Karl Radek, a representative of the Communist International (Comintern), the organization responsible for revolutionary associations outside Russia. Radek was able to turn his cell into a salon for a wide range of visitors, one of whom was an old friend of Seeckt's, the former Turkish War Minister Enver Pasha. The latter became a messenger between the monarchist *General* and the revolutionary who shared a loathing of newly emergent Poland.

The *Reichsheer* lived under the threat of a two-front invasion, from France and Belgium in the west and from Poland in the east. Either reparations or the Poles' desire for territory might provide the pretext for such an invasion, which Germany was too weak to oppose. Germany's only chance was if Poland was threatened from the east, and in 1919 this occurred when war broke out with the Soviet Russia. The following year Warsaw itself was under threat, and Seeckt observed that he would rather Poland be overrun by the Bolsheviks than assist her. It was not a view shared by all Germans; indeed, 70 aircraft were 'requisitioned' after landing at Polish airfields as a result of 'navigation errors'. Half were Albatros D.III and Fokker D.VII fighters and the remainder were Albatros C.XVs, DFW C.V/VIs and LVG C.V/VIs, all of which would prove to be valuable reinforcements to the Polish Air Force at a time when other sources of aircraft had dried up.[33] They helped the Poles to defeat the invader, but by then the Russo–Polish conflict had started a process of Russo–German military co-operation of immense value to the *Luftwaffe*.

The Russians themselves were only too aware of the inadequacy of their equipment, including aircraft, and were seeking a source of modern arms as well as a potential ally against Poland. For their part the Germans were interested in any means of avoiding the restrictions of Versailles, although good relations with Russia had long been a tradition of German foreign policy. Consequently, there was no serious opposition to Seeckt's indirect contact with Radek, as the result of which, in October 1919, Enver Pasha was able to fly to Moscow to begin the delicate process of establishing contact to seek areas of mutual military benefit.

Although bemused by the approach, at the end of 1919 the Russians sent to Berlin an old friend of the Russian Defence Commissar Leon Trotsky, Viktor Kopp, as a semi-official diplomatic representative. He began a prolonged inspection of the remaining German arms factories, which he completed in late 1920. Modern aircraft were a topic of especial concern to the Russians, whose first-line air strength at the beginning of 1921 was 228 aircraft, 72 per cent of their establishment.[34] The Imperial Russian aircraft industry had been a sickly child, most of its output being copies of obsolete French, British or German aircraft, and consequently the aircraft of the Red Air Force (*RKKVVF*) proved inferior to those of Allied interventionist forces, those of the Poles and those of the German *Freikorps* which they encountered.

The uncertainties of the post-Versailles period had already encouraged several German manufacturers to set up factories abroad—Albatros in Lithuania, Dornier in Switzerland and Italy and Junkers and Heinkel in Sweden, all seeking new markets. When Kopp reported to Trotsky in April 1921 he specifically mentioned the interest of the Albatros works. During the early summer of 1921 a German engineering delegation visited the Soviet Union for a reciprocal inspection and in order to continue negotiations. It was led by *Major Dr* Oskar *Ritter* von Niedermayer, whose wartime activities in Persia and Afghanistan had been compared to those of T. E. Lawrence in Arabia and, indeed, had earned him the nickname 'The German Lawrence'.

With both sides better acquainted, serious negotiations began during the autumn of 1921 in the Berlin apartment of Schleicher, the father of the *Freikorps*. The chief German negotiator was *Oberst* Otto Hasse, the future head of the *Truppenamt*, while Leonid Krassin, the Chairman of the Foreign Trade Council, and Kopp led the Russian delegation. In the field of military co-operation the Russians were interested in many things, but they were especially interested in aircraft production. They were already familiar with the all-metal Junkers aircraft, having captured, but been unable to piece together, three disassembled aircraft (two CL.Is and one D.I) in 1920. However, progress was delayed by a lack of German industrial interest. With little spare capital and without Government guarantees, manufacturers were understandably reluctant to invest in Soviet Russia, where all foreign enterprises had been nationalized. The experience of the Junkers works was to show that their caution was justified.

The Junkers works at Dessau, whose aircraft had served with the *Freikorps'* Baltic squadrons, was one of the largest in Germany, but the post-war years had not been kind. The work force had dropped from 2,000 in 1918 to 200 by 1921, and while the Junkers F 13 was a commercial success, Allied restrictions and a generally depressed market made sales difficult. To create work for his plant *Professor* Hugo Junkers had diversified his products, established an airline and set up the Swedish factory, but these ventures drained him of capital and until he lost control in 1934 the company hovered on the edge of bankruptcy. Junkers was naturally very interested when Niedermeyer approached him in July 1921 about co-operation with the *RWM* in Russia, although Niedermeyer's enthusiasm exceeded his authority. Nevertheless, in November 1921 a delegation, which included Hasse as head of the *Truppenamt*, visited Dessau and concluded a verbal agreement with Junkers, who was to build a factory in Russia to produce aircraft and engines in return for Government funds.

The New Year of 1922 saw Russo–German relations growing warmer, partly because both countries were treated as pariahs. On 16 April 1922 the Treaty of Rapallo established normal diplomatic and commercial relations between them and provided a formal framework for secret military and air co-operation. A few months earlier Chancellor Wirth resolved the financial obstacle to Russo-German military co-operation by arranging to provide the Army with RM150 million ($2.8 million). Within the General Staff the co-operation was to be financed and co-ordinated by *Sondergruppe R* under Seeckt's former adjutant *Major* (later *Oberst*) Veit Fischer, who subsequently commanded *Luftgaue*. Even before the Treaty of Rapallo, on 15 March *Sondergruppe R* and Junkers signed a preliminary contract under which the company would receive a RM40 million down payment, with another RM100 million in capital, in return for establishing an aircraft factory in the Soviet Union. It appears that there was also a similar agreement with Albatros, but the details are obscure although it is known that the aircraft manufacturer Friedrich ('Fritz') Siebel represented the Foreign Ministry in projects to set up aircraft factories. The organization of arms production in Russia was to be conducted by the specially established

Gesellschaft zur Forderung gewerblicher Unternehmungen (*GEFU*). *GEFU*, with a working capital of RM75 million, was under the authority of *General* von Borries and *Major* Fritz Tschunke. The German investment was probably in gold and worth a total of $21 million for in script the total value was only $760,000.

Negotiations concerning the Junkers project continued during the winter of 1921–22 and in mid-January 1922 Niedermeyer and Radek returned to Berlin, where the Russians told Seeckt of plans to renew the war against Poland. Radek warned that this could be accomplished only if the Red Army had modern equipment, including aircraft. The tempo of negotiation increased and the following month an initial agreement was signed to set up a Junkers factory in Russia, although the details took nearly a year to complete. The full agreement was signed on 26 November 1922 and eight days later, on 4 December, the Russians placed their first order for 100 aircraft (twenty J 20 floatplanes, fifty J 21 reconnaissance aircraft and thirty J 22 fighters), although it would be another two months before they ratified the agreement.[35]

Significantly it was Seeckt, and not the 45-year-old Defence Minister Gessler, who was involved in all these negotiations, the *General* having his first meeting with Krassin and Kopp on 8 December 1921. The Minister is unlikely to have received more than vague details, for his relationship with Seeckt was a curious one. Gessler, an impassive-faced lawyer from Nuremberg, regarded it as his patriotic duty to rubber-stamp every effort by Seeckt to release Germany from the shackles of the Versailles Treaty. For his part Seeckt repaid this loyalty by treating the Minister like a lackey. Consequently, Seeckt's planning for the future Army and Air Force was conducted in a vacuum which largely ignored the country's industrial and financial problems. But this was not revealed until 1927, when Groener succeeded Gessler as Defence Minister, and by then Air Force planning had been seriously compromised.

Seeckt's attitude was surprising since the Government's economic problems undermined his clandestine military activity. By the end of 1922 $1 was worth RM10,000 and the mark was fading every day, discouraging most German aircraft manufacturers from investment. The crisis grew more acute when a new instalment of reparations was due, and in despair Chancellor Wirth resigned, to be replaced by Wilhelm Cuno. In December Germany informed the Allies that it could not pay the next instalment of reparations and pleaded for a respite so that the country might rebuild its economy. With the United States retreating into isolationism and the British preoccupied with domestic affairs, everything depended upon the French—who were in no mood to listen.

The new Government of Raymond Poincaré took a hard line on reparations and believed, probably rightly, that Germany was trying to default on her obligations. The French Rhineland Army, supported by twenty air squadrons, was as big as Germany's and Paris had often threatened to cross the Rhine and occupy the Ruhr if the Germans defaulted. On 9 January 1923, supported by a Belgian division, the French carried out their threat. Faced by overwhelming force in the west, and the threat of Polish and Czech intervention in the east, Cuno's government resorted to passive resistance and non-cooperation. The workers went on strike and were supported by their bosses through the Ruhr Fund. With Seeckt's covert support, wilder spirits undertook sabotage, a victim of summary French justice being Leo Schlageter, one of Goltz's veterans, whose name would later be carried by the *Luftwaffe*'s *JG 26*.

Fearing further French advances, the *Reichsheer* quietly mobilized, swelling its ranks with volunteers, while paramilitary organizations prepared to conduct guerrilla operations. None of this was observed by the Allied inspectors, for the increased tension forced them to stay in their headquarters. The preparations included the purchase of fighters, which was undertaken with the authority of President Ebert and Hasse but organized by the industrialist Hugo Stinnes using Ruhr Fund money. An order, ostensibly for Argentina, was placed with Fokker for fifty D.XIs and fifty D.XIIIs.

An improved version of the Fokker D.VII had been produced as the DP I since 1922 by the former fighter pilot Richard Dietrich at his Mannheim works, but these aircraft were unarmed. The French advance forced him to transfer the works to Kassel, where he produced 53 DP IIa's for fighter-training at various schools. The Fokker factory was not the only one to benefit during 1923, for BMW received engine orders, and Student ordered a number of new aircraft from Heinkel. The last appeared only between 1924 and 1928, but they included a corps aircraft which became the HD (Heinkel *Doppeldecker*) 17 of 1924, the HD 35 long-range reconnaissance aircraft and the HD 32 trainer. When Junkers learned of the Fokker order he complained and was promised one of a similar size from the Russian factory.

Pilots for the Fokker fighters were provided by the *Ring* and received refresher training at private flying schools, possibly on the DP IIa, while corps aircraft crews received training on Albatros B.II trainers. The training was organized through the *Referenten zbV*, whose role was expanded to include making the public airminded—which provided an excuse for contact with flying clubs to provide emergency air support. However, these preparations stopped short of creating squadrons, which would have been too blatant a defiance of the Versailles Treaty. The *Reichsmarine*, too, acquired its first new combat aircraft at this time, in the form of ten Heinkel HE 1s, which were ostensibly purchased by the Bücker company for Brazil. They were built at Warnemünde but assembled, tested and stored in Stockholm by Svenska Aero.[36]

The Ruhr Crisis rumbled into the summer, damaging the French economy but ruining the German, with hyperinflation making $1 worth RM4,000 million by the end of October. Unable to cope, Cuno resigned on 12 August, to be replaced the next day by the subtle and pragmatic Gustav Stresemann. The new Chancellor was as patriotic as the most stiff-necked aristocrat, but he recognized that a façade of cooperation rather than open confrontation might solve Germany's problems. On 25 September he ordered the Ruhr strikers back to work, and this paved the way for a series of international conferences which produced the Dawes Plan. This economic package led to a French withdrawal from the Ruhr, made reparations easier, stabilized the currency and encouraged American investment, which laid the foundations for economic resurgence during the late 1920s.

A modern industry and a strong economy are the foundations of military strength, but Seeckt bitterly opposed Stresemann's conciliatory policies, even though they were proving so beneficial to him. It was the Chancellor who reined-in Allied snoopers by claiming that German disarmament was complete, and in late January 1925 the Allies ended snap inspections. The Aeronautical Guarantee Committee existed for a few more years, while Allied embassies retained the right to visit suspected factories and sites, but their activities depended upon German co-operation and notice had to be given of visits. Increasingly the Allies were forced to depend upon the slender reed of trust which Berlin preserved by providing data, including lists of pilots and aircraft. The final report of the Aeronautical Guarantee Committee shows that there were 111 pilots and 111 civil aircraft in August 1922, while at the end of 1923 the figures were 186 and 192 respectively.[37]

By restoring the economy to health and effectively blinkering the Allies, Stresemann prepared the way for the renaissance of German military air power although ostensibly conforming to the Versailles Treaty. Such was his success that Europe's chancelleries took no action when the final report of the Allied Control Commission warned that Germany was ignoring, or circumventing, many of the provisions of the Versailles Treaty, including that which forbade the creation of an air arm. Stresemann, who resigned as Chancellor in November 1923 but remained as Foreign Minister in successive governments until 1929, dismissed the Commission's findings as 'petty details', and Germany's neighbours decided that there was no threat to peace.

The country's return to the international stage was completed with the signing of the Treaty of Locarno on 1 December 1925. This guaranteed Germany's western borders and ended the Franco–Belgian threat, in return for a guarantee that the Rhineland would be demilitarized when the French withdrew in 1930. There was no such guarantee for the more controversial eastern borders, and the Treaty allowed the *Reichsheer* to focus on the potentially more manageable threats from the east. Locarno was, therefore, to shape the *Luftwaffe* in its early days as a force aimed initially at Poland and Czechoslovakia but ultimately against France and Belgium. Germany's eastern neighbours would not have been reassured to learn of the establishment in the Soviet Union an air training and testing base at Lipetsk to which were dispatched the Fokker D.XIIIs purchased at the time of the Ruhr Crisis (the D.XIs being sold to Romania, which used them for advanced training).

While Germany began the 'Roaring Twenties' by taking the first steps to prepare for a new air force, Austria's prospects of emulating her were bleak. The peace treaty signed at St Germain-en-Laye on 10 September 1919 had terms similar to those of the Versailles Treaty, including a ban on military aviation, but it aroused no strong opposition. The most disquieting feature of St Germain was the Allies' refusal to sanction any form of union (*Anschluss* in German) between Austria and Germany, in flagrant violation of the principle of self-determination applied elsewhere in Europe. Yet both German-speaking nations faced desperate economic problems because they lacked markets and resources, and such a union would have been popular in Austria.

The end of the *Fliegertruppe* had been anticipated in Vienna, and by late August 1919 *Fliegerhorste 1, 2a, 5* and 6 had been disbanded. The remainder were formally disbanded by a decree of 27 September which also deprived the new Army of the airfields. But, like the Germans, the Austrians also sought a police force, and the *Staatsamt für Heereswesen* created from the *Fliegertruppe* the *Österreichische Flugpolizei*, ostensibly to protect airfields.

The new organization had *Flughafenpolizeistellen* at Wien-Aspern and Wiener Neustadt and *Polizei-Flugstaffeln* at Fischamend, Graz-Thalerhof and Wiener Neustadt. On 10 February 1920 the Allies demanded the dissolution of the *Flugpolizei*, a move rejected by Vienna on 26 May. But Austria was starving: 96 per cent of Vienna's children were officially reported as being undernourished, and many were sent to Norway under the auspices of the Quakers to recover; some of the *Wienerbårn*, as they were called, were to return as conquerors twenty years later. Meanwhile Austria's desperate financial plight and the need for foreign aid forced nominal compliance with Allied demands in December 1920. An attempt was made to circumvent these demands by creating *Sicherheitsabteilung 27* of the *Wiener Sicherheitswache* and providing it with two squadrons. Once again the Allies put pressure upon Vienna, and about March 1921 this unit was also disbanded. By contrast Hungary was able to maintain an air presence under various guises during this period, despite Allied monitors, and by August 1921 had actually re-established a clandestine air force.

In Austria, too, an IAACC was created, initially under the French *General* Poindron and later (from 21 February 1921) under the distinguished French air leader *Colonel* Joseph Barès. It assigned 71 aircraft to the Allies (most went to Belgium and Japan) then supervised the destruction of 1,200 aircraft and 3,400 engines. Like their northern neighbours, some Austrian manufacturers sought to profit from their labours, and 38 aircraft 'disappeared'. Most of the latter were Phönix D.III fighters and C.I corps aircraft sold to Sweden, but a further 76 aircraft were reported by the IAACC to have been sold to Czechoslovakia, 50 to Poland and five to the Ukraine. Of the aircraft sold to Poland, the majority, 38 Öffag Albatros D.IIIs, were purchased not by the Polish Government but by one of the armies on the Eastern Front.[39] A score or so of aircraft and 150 engines were later discovered by Allied investigators, but by 1922 it was accepted there was no more contraband and the Austrians regained control of Wien-Aspern, Graz-Thalerhof and Klagenfurt-Annabichl airfields. An Aviation Affairs Bureau had been created within the Transport Directorate as early as April 1919, but it was not until 1923 that civil aviation really got under way with the creation of a national airline, Österreichische Luftverkehrs Aktiengesellschaft (ÖLAG). Although Austria was permitted to give flying training to a dozen army officers, the country's parlous financial situation prevented such activity for several years.[38]

If Austrians were resigned to accepting the Allied imposition, most Germans were not. Stresemann's policies initially proved unpopular and unrest flared throughout the country. The most significant incident occurred in November 1923 in Munich, which hosted a hive of ultra-nationalist parties, including the Nazis. All exploited the disgust many felt at the Government's craven acceptance of Allied humiliation and hypocrisy, the financial chaos and the general air of decadence. Many airmen were attracted to the Nazis, including Rudolf Hess and Greim, the latter becoming an ardent supporter of Hitler. During the Kapp *putsch*, Greim took Hitler on his first flight to Berlin, the future *Führer* retching throughout the journey and swearing, upon landing, never to fly again. Like many of his decisions, this was not irrevocable.[40]

A newer, and more distinguished, supporter was Göring, who ended his barnstorming and returned to Munich to study political science. This brought him into contact with Hitler, who gave him command of the political shock troops, the *Sturm Abteilung* (*SA*), at a critical time. During late 1923 there was growing support for Bavarian secession from the *Reich*, and a public meeting on the issue was called at the *Bürgerbräukeller* on 9 November. The Nazis hijacked the event to stage the Beer

Hall *Putsch*, which culminated in a march on the centre of Munich led by Hitler and Göring. The plotters included the future *Luftwaffe* generals Andreas Nielsen (last *Kommandeur* of *K/88*), Werner Kreipe (*Kommodore* of *KG 2*) and Günther Korten (Chief of Staff), but it is unclear which of them actually participated in the march. Göring was one of the leaders and was recognized by Udet, who was more interested in managing the Junkers airline business in southern Germany and running his own *Firma Udet-Flugzeugbau*. Some conspirators attempted to take some of his Junkers F 13s, but Udet refused to co-operate.[41]

Meanwhile the marchers were stopped by armed police and in a gun battle Göring was severely injured. The Nazi leaders fled and were soon arrested, but Göring's wife, the former Swedish countess Carin von Kantzow, helped him escape from hospital to Austria. As a result of his injuries Göring became a morphine addict and remained so until he was captured by the Americans in 1945. Enforced idleness and the side-effects of the drugs meant that he put on weight, and by 1933 he weighed some 20 stone (125kg). Consequently, while he preferred the nickname *'Der Eiserne'* ('Iron Man'), he was often called *'Der Dicke'* ('Fat Man', or 'Fatty').

For the Nazis, as for the German Air Force, the events of 1923 were to be but temporary setbacks, and in the coming years both were to go from strength to strength.

NOTES TO CHAPTER 1

1. Kilduff, pp.85, 151.
2. The story of the last days of *JG 1* is based upon Bodenschatz, p.141ff.
3. PRO Air 9/24: Report of 5 November 1921 by the RAF's Director of Planning.
4. Morrow, pp.146, 236/n13.
5. *Ibid.*, pp.56–9, 99–101, 146–8.
6. PRO Air 1/6b: Pletschacher p.150.
7. Morrow, p.150.
8. *Ibid.*, pp.92, 126.
9. Kilduff, pp.11–12; Morrow, pp.5, 65–6, 193–4.
10. Morrow, p.144.
11. *Ibid.*, p.154.
12. Lux, p.134ff.
13. Based upon Pletschacher, p.152, and official histories (*Nachkriegskämpfen*).
14. The section on *Freikorps* squadrons and their successors owes much to the work of *Herr* Hans-Eberhard Krüger, who has proved very generous with his time and advice. See his article 'Jagdstaffeln über Kurland under Oberschlesien', *Jägerblatt*, April–May 1989, pp.18–27; and 'Die Reichswehr- und Polizeiflieger-Staffeln', *Jägerblatt*, August 1989, pp.8–9 (hereafter Krüger, 'Polizei').
15. Morrow, pp.69–70.
16. The section on Austria is based upon Gabriel, pp.220, 229; William Green, 'Aerial Alpenstock' (hereafter Green, 'Alpenstock'); and *SHAT* File 4N104, Dossier 2: 'Plebiscite Commission Report on Carinthia'.
17. For Seeckt see Carsten, pp.104–8, 115, 213–4; and Seaton, pp.6–10.
18. Based upon Robertson.
19. Robinson, p.347.
20. Krüger, 'Polizei', p.10

21. The description of the *Polizei-Fliegerstaffeln* is based upon Krüger, 'Polizei'; and *SHAT* File 7N2611, *Deuxième Bureau*, German Section reports. For Ruhr operations see *Nachkriegskämpfen Band 9*.
22. Information about police air units from May 1920 is based on *SHAT* File 7N2611; and Völker, *Dokumente*, pp.51–2, 55–7.
23. For German commercial aviation see Davies; Morrow, pp.144, 154–7; and Stroud's article 'The Birth of Air Transport'.
24. Kilduff, pp.84–5; Stroud, p.173.
25. *SHAT* File 4N96-I.
26. *SHAT* File 4N94-I.
27. PRO FO 371/4757 C11056.
28. PRO Air 9/24.
29. For the IAACC's activities and progress see *SHAT* Files 4N80-I, 4N94-I, 4N96 Dossier 1 and 7N2611; and PRO FO 371/4752 C12077, FO 371/4757 and FO 371/4872.
30. Morrow, p,165.
31. See Kilduff, pp.69–72.
32. *Ibid.*, p.118.
33. Cynk, pp.39, 45.
34. Tyushkevich, p.128.
35. For Russo–German and Junkers negotiations see Carsten, pp. 67–8, 70–1, 116, 135–8, 142–3, 147, 234, 362; Erickson, pp.92, 109–10, 144–63. Homze, pp.8–11; and Lennart Andersson, 'Junkers Two Seaters'.
36. For the *Reichswehr* and the Ruhr Crisis see Carsten, pp.154–5; Völker, *Entwicklung*, pp.133–4, 138, 200–1; and Lennart Andersson, 'Secret Luftwaffe'.
37. *SHAT* File 4N97, Dossier 2
38. Cynk, p.38.
39. For Austrian aviation see Gabriel, pp.220–1; and *SHAT* Files 4N105 (the IAACC's final report of 5 July 1922) and 4N97, Reports of the Inter-Allied Military Control Commission.
40. See Herlin, pp.103–4.
41. *Ibid.*, pp.108–16.

CHAPTER TWO

THE LIPETSK YEARS

January 1923 to December 1929

A month after the French occupation of the Ruhr, *General* Hasse and a mission including Student travelled to Moscow seeking closer military co-operation between Germany and Russia. The Russians sought access to Western aviation technology, and their team was led by P. P. Lebedev, the Red Army Chief of Staff, and A. P. Rosengol'ts, the head of the *RKKVVF* and also Trotsky's brother-in-law. The meeting led to the establishment of a German liaison office in Moscow, *Zentrale Moskau (ZMo)*, at the Embassy, but under Seeckt (and later his successor, *Generaloberst* Wilhelm Heye). In 1928 it was transferred to *T3*, which established another liaison office in Berlin.

Stresemann's appointment in August 1923 delayed progress, for he opposed closer military relations with Russia which might compromise the policy of reconciliation with the Allies, especially if military equipment such as aircraft were produced. His Finance Minister, Rudolf Hilferding, regarded co-operation with Russia as a drain on scarce funds and on September 25—the day passive resistance to the French ceased—he demanded an end to military expenditure in Russia. Government concern increased after the Soviet Military Attaché gave German Communists weapons purchased in Germany for the Soviet Union.

Seeckt and Gessler refused to end secret military co-operation with Russia and, with Germany still in chaos, the Government had to buy *RWM* loyalty by bowing to their wishes. Negotiations continued, and on 20 September Lieth-Thomsen was appointed head of *ZMo*. He did not set out for Moscow, using the cover name 'von Litz', until 13 November and the project took shape only in 1924. Thomsen was responsible not only for the planned air testing and training base but also other establishments which were created later, including chemical warfare (1927) and armoured vehicle (1930) schools. His task required skilled diplomacy not only with the Russians but also with his own countrymen, for the Foreign Ministry feared that the *RWM* was seeking to dictate foreign policy. Thomsen remained at *ZMo* until deteriorating eyesight forced him to hand over to Niedermeyer in 1929. 'The German Lawrence' was succeeded three years later by *Oberst* Hartmann, who became Military Attaché in Moscow during the mid-1930s.[1] By the mid-1930s Thomsen was blind, but on 1 August 1939 he was officially appointed *General der Flieger*, a rank he held until his death on 5 August 1942.

The core of German activity in Russia was to be an air training and testing centre—a recognition of the importance of air support in modern warfare. The

Russians offered the use of an existing air base at Odessa but the *Reichsmarine*, with facilities in Germany, was uninterested. Moreover, Odessa was too open, and so an inland site was sought instead. Eventually the spa town of Lipetsk, 300 miles (500km) south-west of Moscow on the Voronezh river, was selected. It was sufficiently far inland to be secure from Allied intelligence and close enough to Moscow for activities to be monitored by both sides; it was also served by a railway. The agreement to establish the base, and to share the lessons learned, was signed on 15 April 1925.

A former factory on the outskirts of the town was used both for accommodation and storage but in succeeding years it was expanded. Eventually the base included two runways, a large complex of hangars, production and repair shops, a modern engine-test shop, accommodation blocks and even a fully equipped hospital. The Germans designated it Wivupal and until 1931 it was under the command of *Major* Stahr, who was succeeded by *Major* Müller. The establishment cost Germany RM2 million ($478,000) annually but was important enough for two visits a year by the head of the *Truppenamt*'s aviation organization.

The relaxing of Allied restrictions in the late 1920s made it easier to give pilots flying training and would-be observers classroom training in navigation, photography and radio operation in Germany, but Lipetsk provided essential practical and operational training. From the summer of 1925 this was Wivupal's prime role, initially for fighter pilots but from 1927 also for observers. Each course had an operational squadron (*Jagd-* or *Beobachterstaffel*) and a demonstration squadron (*Jagdlehr-* or *Beobachterlehrstaffel*), these units being officially designated the *4.Eskadrilya VVS-RKKA* (as the *RKKVVF* was renamed in 1924).

For the first two years activity was confined to summer refresher courses for 'Old Eagles' (the term applied to veteran airmen) from the *Jastas*. Some 30 went through the course conducted by *Freiherr* von Scheneback, but from April 1927 the emphasis was placed upon the new generation of airmen, the 'Young Eagles'. Between 1927 and 1933 another 90 fighter pilots were trained at Lipetsk, two-thirds of whom were *Offizieranwärter*. The courses lasted between 20 and 22 weeks in summer, the highlight being a dogfight between two formations of nine aircraft each with a gun camera. Observers came for a six-month course after passing examinations in Berlin. Lipetsk was especially valuable for artillery observation training, for until then German observers could practice only from mountain peaks. The artillery course began at the *VVS-RKKA* base at Voronezh in 1928, supported by Russian batteries with whom new methods of ranging were developed and practised.

Wivupal's training activities reflected Germany Army attitudes to air support. Although Seeckt wrote in 1923 that the air arm should be independent, it is clear that he was thinking in terms of another *Luftstreitkräfte*, for the training courses were only to provide tactical air support. Significantly there were no bomber courses at Lipetsk, although bomber experiments were conducted from 1928. The fighters' role was clearly confined to supporting the Army in the field by intercepting enemy corps aircraft at low and medium altitude while neutralizing enemy fighters in dogfights. The training courses reflected the belief that fighter and observer training were highly skilled and highly specialized roles.

The observer was especially important for he was a junior officer who was expected to make, on his own initiative, important tactical decisions affecting whole Army corps. Wartime observers and pilots worked as a team and in Germany were nicknamed 'Franz' and 'Emil' respectively, with 'Franz' commanding the aircraft. Although relations with 'Emil', usually a NCO, were good he was regarded by the Army essentially as an aerial chauffeur, and training at Lipetsk suggests that this attitude continued in the post-war German Army.[2] The primacy of the observer continued in the *Luftwaffe* and remained for two or three years of war.[3] About 100 observers were trained in the Soviet Union but by 1931 this training could be conducted more cheaply in Brunswick, where it was concentrated.

Wivupal's training was clearly influenced by the British Gosport method. The traditional hit-or-miss techniques were replaced by a systematic programme covering all aspects of aviation, and instructors were encouraged to discuss matters with their pupils. The instructors were nominally civilians on contract either from private flying schools or from the national airline Lufthansa, and for security reasons they were paid in American dollars. The instructors used the winter months for tactical and formation flying exercises and the summer for training the 'Young Eagles'.

Stahr had some 60–70 permanent German staff who were joined during the summer by about 45 students (three-quarters of them for the observation course). Following security leaks in 1926 the Army became more circumspect about the students dispatched to Wivupal. Active-list officers were banned by the Government from attending, so they resigned their commissions and travelled on false documents. After the course they regained their commissions. Many *Offizieranwärter* were selected because, officially, they had not joined the Army. In 1929, for example, six instructors, 36 pupils and ten *Offizieranwärter* were sent to Lipetsk.

In addition to courses for aircrew, Wivupal also offered courses for groundcrew, of whom 450 had been trained by 1933. From 1928 onwards the soldiers were joined by 50–100 scientists, technicians and military observers to bring the average number of German personnel present during the late 1920s to 200, although in 1931 it reached a peak of 300. The Germans tended to be confined to the base and, after working hours, there was relatively little to do and newcomers might be met by a sign saying *'Herzlich Willkommen am Arsch der Welt'* ('A warm welcome to the world's arsehole').

The first aircraft at Lipetsk were the 50 Fokker D.XIII fighters bought during the Ruhr Crisis and powered by British 450hp Napier Lion engines. On 28 May 1925 the fighters were shipped to Leningrad, then sent by rail to Lipetsk and reassembled. One-third were always in storage, and frequent rotation lengthened the lives of the airframes; indeed, Wivupal's safety record was exemplary, with only three fatalities, two of which were caused in a mid-air collision. For overhaul the Napier Lions (also used in the Heinkel HD 17) were forwarded through Fokker to Britain, but eventually the Wivupal engineers were able to carry out this task themselves. When Lipetsk closed in September 1933 up to 40 of the fighters were still airworthy (two replacements had been received), together with a dozen reconnaissance aircraft—a remarkable achievement after eight years' service.

The Fokkers were quickly joined by a small number of Junkers F 13s and A 20s for training and transport and in 1927 the first consignment of German-built combat aircraft, six HD 17s, joined the establishment. They formed the two observer units initially but later one or two were attached to the *Jagdlehrstaffel*. No new combat aircraft are known to have joined Wivupal in 1927, although at some point between 1926 and 1929 a few HD 21 and Albatros L 68a trainers arrived, together with at least two Fokker D.VIIs. The last were from the *VVS-RKKA* inventory, having been bought from Fokker.

It was always intended to use Wivupal to develop equipment, from bomb sights and bomb racks to complete aircraft, for the future German air arm, but financial and diplomatic restrictions prevented this until 1928. For this work, and for general training, some eighteen aircraft were purchased from the Albatros-Werke, including the L 76a and L 78 corps and the L 77v fighter-reconnaissance aircraft, all powered by the BMW VI. The L 76 proved to be a vicious aircraft which was involved in seven fatal accidents, the most prominent victim being Emil Thuy, a *Pour le Mérite* holder and former commander of *Jasta 28*, who was killed at Wivupal on 11 June 1930.

The Heinkel HD 20 twin-engine reconnaissance aircraft was also briefly used at Lipetsk. The HD 20, and other large aircraft, flew to Lipetsk posing as commercial aircraft of Lufthansa or the Russo-German airline Deruluft. They would fly from Königsberg over Lithuania and Latvia, to enter the Soviet Union at Velikye Luki. The smaller single-engine aircraft were flight-tested, unarmed, in Germany then dismantled, crated and shipped to Leningrad, taken by rail to Lipetsk and reassembled. Munitions were usually smuggled from Germany in coastal sailing vessels.

From the autumn of 1928 Wivupal began to evaluate bomber designs, with the emphasis upon developing an auxiliary bomber (*Behelfsbomber*), a passenger or freight aircraft which could be converted to a bomber (also known as a bomber-transport). Single examples of the three-engine Junkers G 24 (from which the K 30 was developed) and Rohrbach Ro XII *Rohrbach Kampfflugzeug* (*Roka*), a derivative of the Ro VIII Roland I airliner, were evaluated in 1928 and a Dornier Do B *Merkur* in 1929. The operational strength gradually rose from 55 aircraft to 66 by 1 October 1929, this latter figure including 43 Fokker D.XIIIs.[4]

The power struggle between Trotsky and Stalin from 1925 saw a cooling of relations between the Germans and Russians. Lebedev was replaced by B. M. Shaposhnikov and Rosengol'ts initially by Pyotr I. Baranov. From 1931 Baranov was also responsible for aircraft production and his place at conferences was often taken by the visionary Yakov I. Alksnis. Alksnis, a Latvian who spoke German fluently, was an immensely able officer who had been Baranov's Chief of Staff since August 1926. *VVS-RKKA* officers visited Germany and participated in conferences while a *Hauptmann* Schöndorff remained on the *VVS-RKKA* staff until 1931, but at Lipetsk the Germans managed to keep the Russians at arm's length. Some Russian military and civilian personnel were employed as technical assistants, but it was the Germans who carried out most of the repair and technical work. However, representatives of the *TsAGI* aeronautical research establishment often participated in the testing of German prototypes and were influenced by what they learned. The *VVS-RKKA* based a few squadrons at Lipetsk some time after Wivupal opened, but they occupied the other side of the airfield and were strictly segregated from the Germans. It was not until 1932 that joint air exercises were held, while plans for the fighter pilots to work together never materialized owing to the frequent Russian postponements.

While Lipetsk bloomed, the Junkers factory established in the former Russo-Baltiysky works at Fili, south of Moscow, in 1923 rapidly withered and died. It had a German staff of 91 and was operated in conjunction with TsAGI representatives under the talented young Russian designer Andrei N. Tupolev. For the Russian aircraft industry, Fili proved the vital training school which helped to hone a generation of Soviet engineers and designers, and Tupolev aircraft such as the ANT-9 transport, R-3 (ANT-3) reconnaissance aircraft and TB-1 (ANT-4) heavy bomber strongly reflected Junkers' influence.[5]

But for Junkers it was a financial disaster. Inflation eroded the German investment and the *RWM* welshed on its promise to order 200 aircraft, while Russian orders failed to make up the difference, possibly because Junkers' deliveries of duralumin were erratic. Apart from the 100 aircraft ordered in December 1922, the Russians placed contracts for only another twenty J 20 (designated Yu 20) floatplanes (produced at Dessau but assembled at Fili), 72 J 21 (Yu 21) reconnaissance aircraft and 23 K 30 (R-42) bombers, the last built in Sweden by AB Flugindustri. Between 1924 and 1926 only 142 aircraft were apparently built at Fili, which had a reported capacity of 600 aircraft a year.[6] For civil aviation the Russians also purchased 49 F 13s (and assembled another five from parts) from 1922 to 1929, and about twenty W 33s (PS-4s) from 1928 to 1931 with a further seventeen assembled by civil aviation repair shops.[7] Two of the F 13s were armed by the Russians and sold to Iran in 1923. The Yu 20s were used by the Naval Air Force and the Yu 21s were largely used in central Asia against Moslem insurgents, while the R-42s were in the 'heavy bomber squadrons'. All had been withdrawn from first-line service by 1931.

Professor Junkers also hoped to produce aero-engines at Fili and was angered to discover in late 1923 that the *RKKVVF* was buying 380 BMW II/III/IV 120–240hp engines. Early in 1924 he was ordered to co-operate with BMW in the production of their engines at Fili and he protested. Seeckt himself had to intervene, writing on 18 August 1924 that German policy in Russia was shaped by diplomatic and strategic rather than commercial factors. The J 20/Yu 20 was powered by the 185hp BMW IIIa, and the BMW VI was produced in the Soviet Union as the M-17, possibly at Fili.

To recoup his losses *Professor* Junkers debited Fili for his group's overheads (including factories in Sweden and Turkey) and by September 1925 these debits accounted for more than half the RM15 million ($3.55 million) lost at the plant. In 1926 the *Heereswaffenamt* conducted an audit and reported on 15 June 1926 that the company was losing a 'grotesque' RM50,000 ($12,000) for every aircraft produced. Junkers could not afford the haemorrhage of funds, and on 1 March 1927 the factory was taken over by the Russians and placed under Tupolev. The well-equipped plant was then prepared for TB-1 production.

Junkers sued both governments, but his agreement had been only a verbal one which, as Samuel Goldwyn observed a few years later, was not worth the paper it was written on. On 12 January 1926, six days after the Junkers airline had been absorbed by Lufthansa, a German court rejected his case. *Professor* Junkers inspired great loyalty among his workers, and when all legal means of compensation disappeared one of his executives sent a memo to the Social Democrat deputy Philipp Scheidemann revealing details of relations between the *Reichsheer*, Junkers and the Russians. Scheidemann raised the matter in Parliament on 16 December 1926.[8]

Scheidemann's revelations were doubly embarrassing for on 3 December a British newspaper, *The Manchester Guardian*, published a well-informed article on

Russo–German military co-operation. This international scoop, although it created no significant Allied response, threatened to undermine Stresemann's foreign policy. The Foreign Ministry demanded a briefing on Russian–German military co-operation and on 24 January 1927 *General* Wilhelm Wetzell (head of the *Truppenamt* from 1926 to 1927) met Under-Secretary of State Carl von Schubert. The *General* admitted that the reports were essentially true but blandly added they were out of date and that the Junkers factory was on the verge of closing. He admitted that some activities were continuing and that Lipetsk was of vital importance to the defence of Germany. However, he assured Schubert that it was organized as a private flying school, with no connection to the Army, and that no officer on the active list was involved with it. He added that Germany was conducting similar activities in Turkey.

Wetzell's explanation lacked candour, while some aspects were downright misleading; for example, officers on the active list visited Lipetsk to monitor specialist activities while the head of Wetzell's own air section (*Hauptmann* Felmy) visited Lipetsk twice a year. It was Stresemann who made the decision, on 9 February 1927, to continue activities at Lipetsk. But it was on condition that no active list officers participate, and that the situation would be reviewed in the autumn. This decision led to more *Offizieranwärtern* being sent to Lipetsk.[9]

Lipetsk was a vital link between the *Luftstreitkräfte* and the *Luftwaffe* and it was the only one of Seeckt's Russian ventures which proved cost-effective. It provided operational training for the new generation of aircrew, including future *Luftwaffe* leaders, many of whom would be involved in the invasion of the Soviet Union twenty years later. But at the time these future leaders received no encouragement to consider the future use of air power outside the tactical field. Moreover, there was no money to provide the continuous training essential for operational efficiency. Lipetsk did allow Germany to complete the development of military aircraft, but industrial weakness meant that they were usually inferior to those of rival air forces. In fact, after 1930 Lipetsk's importance lay primarily in the development of combat aircraft. In this respect Wetzell was correct to play down its importance to Schubert, for training in Germany became easier with every year that passed, largely as a result of Stresemann's policies.

The only lasting Russo–German aviation relationship lay in the field of civil aircraft. The jointly owned airline Deruluft, founded on 11 November 1921, continued to operate until 31 March 1937, by which time it had became too much of a political embarrassment for both sides. In addition, the early 1920s saw a number of German pilots fly for the Russian airline Dobrolet, of whom the most eminent was Friedrich (Fritz) Morzik, who was to become the *Luftwaffe*'s most distinguished air transport officer, *Kommodore* of *KGzbv 1* and later *Lufttransportchef*.[10]

The increase in activity both at home and abroad led the *Truppenamt* to expand its aviation organization steadily during the 1920s. In 1924 Wilberg's organization, *TA (L)*, was raised to Departmental status and included Operations, Personnel, Technical and Administrative desks. Wilberg was promoted to *Major* (and to *Oberstleutnant* in February 1926), and his organization was concerned with preparing for an air arm and training aircrew, Administrative and Air Defence desks being added in 1925.

On 22 January 1925 the organization was absorbed into the Frontier Defence section (*Gruppe III*) of *T2* as *T2 III (L)*. The Allies quickly discovered this and on 4

June demanded its dissolution, but Hasse merely placed it in abeyance and revived it in the autumn under its former identity. At about the same time the renewed development of military aircraft led to a reorganization of the *Heereswaffenamt*'s *Inspektion für Waffen und Geräte*. *Hauptmann* Kurt Student became responsible for developing and testing both aircraft and equipment, while *Hauptmann* Helmuth Volkmann was responsible for acquisition and economic planning.

As the German air effort moved increasingly from the theoretical to the practical, Wilberg concentrated upon administration, while from 1 October 1924 the planning role became the responsibility of 39-year-old *Hauptmann* Helmuth Felmy in *T1*. Felmy had joined the Army in 1904 and the *Fliegertruppe* in 1912 but then transferred to the General Staff. At the outbreak of war he was an observer with, then the Commander of, *Feld FA 51* on the Eastern Front. Between July 1916 and February 1918 he had commanded *FA 300*, the first of the 'Pasha squadrons' supporting the Turkish Army in Palestine, then he went to the Western Front as Commander of *FA(A) 256*. At the end of the war he was planning the post-war colonial air force but soon joined Thomsen's department and in the *Reichsheer* was rapidly involved in aviation activities.

Neither Felmy nor Wilberg could end the division between the operational and supply organizations in the 1920s. This prevented cohesion in the development of equipment, a fatal flaw which was carried over to the *Luftwaffe*. It was not a unique problem: the French also suffered, and it was the foundation of equipment problems which plagued them almost to the outbreak of war. The German problem was aggravated by the low status of aviation where, until 1931, the senior officer was only an *Oberstleutnant*. As he often dealt with men of equal or even superior rank, his opinions could be ignored by other staff officers, and this was especially true when the future of the Air Force came to be studied.

Nevertheless the 'Air Defence' organization was strengthened partly through the intervention of *Oberst* Werner von Blomberg, Head of Operations at the *RWM*. Tall, blue-eyed and strong-chinned, Blomberg was a brilliant planner whose wartime work had earned him the *Pour le Mérite*. Impetuous and intense, but with a sense of humour, he represented a new breed of middle-class officers in an Army still dominated by the aristocracy, who provided 24 per cent of the officers and nearly 36 per cent of the officer cadets in 1932. Professionally and politically, the aristocratic officers were often extremely conservative and their isolation from German industrial society made it difficult for them to appreciate the implications of new military technology such as aviation. However, Blomberg's generation of officers boldly grasped this new technology and its radical implications for warfare, not only on the battlefield but also at home, because its exploitation required a closer relationship with government and industry.[11]

Blomberg was to play a key role in the development of the *Luftwaffe*, although this has been largely ignored by historians. His interest in aviation was strengthened by his son Axel, who gained his pilot's licence on 1 November 1927. Blomberg wished to strengthen Wilberg's organization and on 1 April 1927 he was able to do so when he became head of the *Truppenamt* as a *Generalmajor*. The same day *TA (L)* returned to *T2* as *T2 V (L)* and received the Aviation Intelligence desk from *T3* as well as responsibility for all operational training in Germany and Lipetsk, which Blomberg visited in 1928. However, on 15 August Wilberg was posted to an infantry regiment, an unwelcome promotion which was normal practice at the time. He did not return to military aviation until 1933, but his eight years' tenure had laid a firm foundation and he was to help build on it upon his return.

His successor, *Major* Hugo Sperrle, was the 42-year-old son of a Württemburg brewer. A great bear of a man, Sperrle had joined the Army in 1903 but his only distinction in the next eleven years was to train as an artillery observer. At the outbreak of war he became an observer with *Feld FA 4*, and eighteen months later he commanded *Feld FA 42* but was severely injured in a crash. Subsequently he commanded an observer school and then became *Kofl 7.Armee*. Following *Freikorps* service, including command of Lüttwitz's air units, he held various staff posts in the *Wehrkreise* and the *RWM* before joining *TA(L)* at the Operations desk. A capable rather than brilliant officer, Sperrle affected a monocle but appears to have regarded aristocratic officers with contempt. He had an abrasive personality, and it was probably this which led within eighteen months to his 'promotion' to an infantry battalion commander. He was replaced on 1 February 1929 by Felmy, whose advisory position in *T1* was taken by *Hauptmann* Heinz-Hellmuth von Wühlisch, who was in turn replaced in 1933 by *Hauptmann* Joseph Kammhuber.

Towards the end of his tour of duty at *T2 V (L)*, Sperrle was pressing for an autonomous aviation authority with greater status and Felmy took up the struggle. Both were supported by Blomberg, now nearing the end of his tenure, but the proposal caused dissension within the *Truppenamt*, earning *T1*'s support and *T2*'s opposition. For the latter, *Major* Wilhelm Keitel (later head of *OKW*) warned in a letter of 14 December 1928 that if Germany's neighbours learned of the existence of such an organization it would create an unfavourable reaction. Yet Keitel's words clearly carried as little weight with the General Staff then as they would later with Hitler.

Blomberg and Felmy were supported by the Army's efficiency expert, *Major* Albert Kesselring, who proposed in 1928 an Air Inspectorate (*Fliegerinspektion*) to centralize all military aviation activities at the *RWM*. This was accepted, but as a face-saving gesture to the conservatives the Inspectorate became part of the *Inspekteur der Waffenschule (In 1)* under the 41-year-old *Generalmajor* Hilmar *Ritter* von Mittelberger, a former member of the *Königlich Bayerischen Fliegertruppen* and the ranking officer on the air officers' list.

On 1 October 1929 *Generalmajor* Kurt *Freiherr* von Hammerstein-Equord replaced Blomberg, whose tenure at the *Truppenamt* ended prematurely, apparently through the machinations of Schleicher, a friend of Hammerstein-Equord. On the same day the new aviation organization came into effect; it would remain unchanged until the Nazis came to power. The old *T2 V (L)* now became *In 1(L)*, and Mittelberger's organization officially, but secretly, became *Inspektion der Waffenschulen und der Luftwaffe*—the first official use of the name '*Luftwaffe*'. Felmy remained as Chief of Staff at *In 1(L)*, which was of equal status to the four *Truppenamt* departments and was expanded to include the *Referenten zbV*.

However, it still had no control over aircraft development: this remained under the *Heereswaffenamt*, whose aircraft development and procurement organization expanded during the late 1920s. On 1 February 1928 the *Heereswaffenamt* rationalized the horizontal development and procurement organization (a structure which had been adopted by *Idflieg*) into a vertical structure by merging Student's and Volkmann's departments and placing them under Volkmann. The new section was subdivided into Development, Procurement and Economic Preparation desks but was further expanded on 1 October 1929 to become *Wa Prw 8* with responsibility for all aspects of equipment development and procurement, including operational testing but not the drafting of requirements.

The head of the new department was to have been *Hauptmann* Paul Jeschonnek, Wilberg's operations officer during the Third Battle of Ypres, but on 13 June 1929 he was killed when his Albatros L 76 crashed in Berlin, and Volkmann briefly took the post. He was soon replaced by the former *Wehrkreis VII Referent zbV*, *Oberstleutnant* Wilhelm Wimmer, an 'Old Eagle'. Wimmer had never been a member of the General Staff but had the ability to convert staff requirements into precise plans. Under Wimmer the new organization successfully developed the combat aircraft with which Germany went to war. When the horizontal structure was re-adopted by Udet in 1938, it was to doom the wartime generation of aircraft.[12]

The end of open conflict with the Allies over the implementation of the Versailles Treaty saw a resurgence of German civil aviation. The final report of the Aeronautical Guarantee Committee noted an increase in pilots and civil aircraft from 111 and 111 respectively in August 1922 to 404 and 240 in December 1924 (with 183 student pilots). By March 1926 there were 968 fully qualified pilots (including a few foreigners), 172 on refresher courses and 341 student pilots, as well as 511 civil aircraft. The total numbers of pilots and student pilots included 94 members of the German services and 48 policemen.[13]

This expansion was carefully nurtured by Brandenburg, who encouraged a rationalization of domestic commercial aviation as the basis for a strong industrial base. This industrial base was itself strengthened through skilful diplomacy. Airline business expanded steadily from 7,733 passengers and 35.57 tonnes of freight in 1922 to 120,000 passengers and 607 tonnes in 1925. Amalgamations reduced the dozens of post-war airlines by February 1923 to two major organizations, Deutscher Aero Lloyd AG and Junkers Luftverkehr, the latter under Sachsenberg who had joined Junkers after his Latvian adventures.[14] Traditionally, German business has adopted rationalization to create larger and more efficient organizations, and this policy applied to transport. Two competitive, subsidised, domestic airlines made no sense to the cost-cutting Government, which preferred the solution adopted by the British, who created Imperial Airways Ltd in 1924 as a single domestic and international carrier.

Brandenburg wished to follow the British example, but there was strong opposition from *Professor* Junkers. However, Junkers' legal action against the Government over the Russian débâcle deprived him of official sympathy and Brandenburg struck at the manufacturer's vulnerable finances. He threatened to withdraw subsidies and, reportedly, persuaded banks to end credit to Junkers, who capitulated. On 6 January 1926 Lufthansa[15] was created with a fleet of 162 aircraft including 48 Junkers F 13s. The new airline had a domestic monopoly and received a substantial Government subsidy which averaged RM18 million ($4.3 million) a year—or 38 per cent of all the Transport Ministry's aviation subsidies—between 1926 and 1929.[16]

The Operations Director was a former Junkers man, 33-year-old Erhard Milch, who had worked with Sachsenberg in various Junkers airlines after his police squadron had been disbanded. An extremely able man, Milch eventually became head of the company's management office at Dessau. Many at Junkers resented his Lufthansa appointment, and Sachsenberg in particular became a bitter enemy who, according to Milch's biographer, constantly intrigued against the new airline.[17]

Milch (who also became Lufthansa's Commercial Director in 1929) rationalized the fleet and introduced efficient working and operating practices. In particular, he recognized that profitability depended upon the time airliners spent in the air, so he sought more multi-engine aircraft with greater endurance and, therefore, greater range. In addition, he ensured that his aircraft could operate at night (Deruluft introduced night operations to Europe) and in bad weather through a comprehensive system of visual and radio navigational beacons (eleven of the latter in 1931) and landing aids at an enviable network of well-equipped airports and seaplane terminals. As early as 1929 an instrument rating became compulsory for all Lufthansa pilots, and in October 1934 the Commander of the RAF's Air Defence of Great Britain observed enviously that in a year Lufthansa's night-time Cologne–Croydon service had cancelled only four flights, displaying a capability his squadrons could not match.[18] The RAF was not unique both in desiring twenty-four-hour operations and in lacking the means to achieve the objective. This capability, for both offensive and defensive operations, was one of the features Milch brought to the *Luftwaffe*, together with a profound knowledge of the aviation industry.

Under Milch, Lufthansa quickly became the most important airline in Europe, flying more miles and carrying more passengers in 1927 and 1928 than its British, French and Italian rivals combined. He also established the closest relations with the *RWM*, which received an opportunity to train men, to develop equipment and covertly to create an air force. The airline also offered a means of creating a bomber force using converted airliners which were tested at Lipetsk during the late 1920s.

From 1928 the German economy went into a slow decline, and although the Government increased the civil aviation subsidies by RM11.7 million ($2.77 million) to RM55.5 million ($13.18 million), it announced a reduction to RM42.8 million ($10.16 million) for 1929, including halving the Lufthansa subsidy. Brandenburg advised Milch to fight the move by 'fixing' Parliamentary Deputies, adding, 'They're all wide open to bribery.' Milch lobbied the Deputies and paid RM1,000 ($237) a month to four, including the newly elected Göring. For personal and business reasons (he was the Berlin agent for Heinkel and BMW), Göring supported all forms of aviation and clearly believed that commercial aviation could camouflage military activity. However, although he was eventually receiving RM50,000 ($11,900) a year from Lufthansa, he spoke only once on their behalf, during a debate on the Air Estimates in June 1929, when he said,[19] 'Save the Air Arm: if you don't, you will live to regret it.'[20]

Milch's commercial success was aided by Brandenburg's diplomatic victory, which freed German industry from the restrictions imposed by the Allies in 1922. Brandenburg succeeded because he used the Versailles Treaty, Articles 313 to 315 of which gave the Allies the right to fly over German territory and to use German airfields without permission. Article 316 stated that aircraft would be subject to

German regulations, '. . . but such regulations shall be applied without distinction to German aircraft and to those of the Allied and Associated countries.' Article 320 restored German control of domestic skies (apart from those over the Rhineland) on New Year's Day 1923, and with this authority Brandenburg exploited Article 316 to harass Allied passenger aircraft, which all exceeded the 1922 limits imposed upon German aircraft. These Allied airliners were promptly seized by the *Luftpolizei* when they landed in Germany for whatever reason. The luckless Compagnie Franco-Roumaine, which flew to Warsaw, Prague and Bucharest, had thirteen of its aircraft confiscated.

In the spring of 1926 these actions led to a conference in Paris and the Paris Agreement of 21 May. In return for promises not to manufacture military aircraft and an end to subsidies for flying clubs and flying schools as well as limits on the number of servicemen with pilot's licences, Germany regained control of her domestic aviation everywhere, including the Rhineland. All restrictions upon civil aircraft development were lifted, while the disbandment of the Allied Aviation Guarantee Committee was set for 1 September 1926. The Agreement saw a brief flutter of hope for German aviation, symbolized by the first east–west transatlantic crossing (April 1928), led by *Hauptmann* Hermann Köhl, a former Commander of *Bogohl 7* and a holder of the *Pour le Mérite*,[21] in a Junkers W 33 and the circumnavigation of the world by the LZ 127 *Graf Zeppelin* (August 1929). However, Brandenburg's ultimate objective, to revive the German aviation industry, was not achieved. Manufacturers were in no position to exploit his success—indeed, production actually dropped in 1926 and failed to reach 400 aircraft a month for the rest of the 1920s, as Table 3 shows.[22]

The aero-industry faced the same problems all over the world. Manufacturers were small and under-capitalized during the 1920s, while both domestic and international markets were fiercely competitive, driving down profit margins which, in turn, reduced investment funding. Production was manpower-intensive because it was craft-based, with most parts and assemblies individually and manually produced. The gradual introduction of all-metal aircraft increased unit costs by 48 per cent[23] since materials were more expensive and production took longer because of the need for machining. To maintain financial viability, manufacturers preferred single-engine aircraft, which assured production volumes although unit profit margins were tight.

TABLE 3: GERMAN AIRCRAFT PRODUCTION, 1924–1929

Year	Produced	Exported
1924	248	8
1925	585	63
1926	168	?
1927	301	54
1928	355	95
1929	330	150
Totals	**1,987**	**370**

Many manufacturers, especially in Germany, had to concentrate upon single-engine light and commercial landplanes, although there was a growing demand for multi-engine commercial aircraft. Only two German companies, Junkers Flugzeugwerke AG and Rohrbach Metall-Flugzeugbau GmbH, could address this market, and both produced advanced, all-metal monoplane designs—the Junkers G 24, G 31 and G 38 and Rohrbach's Romar and Roland. By 1930 French Intelligence estimated, with some slight exaggeration, that a fleet of 56 of these advanced aircraft were on the German civil register, the majority of them G 24s and Rolands.[24] The production of such aircraft was extremely expensive. Junkers constantly hovered at the edge of bankruptcy and Rohrbach went over the edge in 1931, while the U 11 *Kondor* four-engine airliner destroyed Udet's company in September 1928, its assets being acquired by BFW.

The greatest handicap for German industry was the ban placed on military aircraft production for, while these aircraft were more expensive to produce, they offered the best profit margins. American figures showed that the average value of a military aircraft was $11,148 in 1926, $15,640 in 1928 and $14,354 in 1930, with the equivalent civil aircraft prices $4,496, $4,800 and $5,547—a difference of between 40 and 70 per cent. Although German industry exported some 20 per cent of its output between 1924 and 1929, it could not overcome the financial hurdle.[25] A few manufacturers, such as Leichtflugzeugbau Klemm GmbH in Germany and De Havilland Ltd in Britain, prospered purely on civil production (although their trainers were often used by air forces), but for most companies the civil market was their bread-and-butter business. Military production for a protected domestic market spread jam on the bread because it was generally based upon a cost-plus-agreed-profit-margin arrangement, with the bonus of potentially lucrative export contracts.

Paradoxically, during the mid-1920s new German military aircraft were found in Chile, China, Denmark, Italy, Japan, the Netherlands, Spain, the Soviet Union, Sweden, Turkey and Yugoslavia. In some cases the unarmed prototypes were built in Germany, but most manufacture took place abroad, either under German control or through licences, neither of which brought the financial rewards of domestic manufacture. Japan proved an especially good market for licence-built aircraft, with production of the Dornier Do N (as the Kawasaki Type 87 Army heavy bomber), the Heinkel HD 25 floatplane (Aichi Type 2 Navy reconnaissance seaplane) and the Junkers K 37 (Mistubishi Type 93 Army heavy bomber). It was no coincidence that Heinkel was the most successful German manufacturer in this period since he had more consistent success with military aircraft. His HD 37 fighter was produced in the Soviet Union as the I-7, to supplement the income from civil aircraft such as trainers.

Yet producing the new generation of aircraft which followed the technological revolution in the United States from the late 1920s would strain even Heinkel's resources. American aviation had tended to lag behind its European rivals through most of the 1920s, although a mixture of military and civil orders had ensured that the industry survived. Charles A. Lindbergh's solo flight across the Atlantic in May 1927 acted as a catalyst to make Americans air-minded, and aircraft deliveries leapt from 789 in 1925 to 6,139 in 1929. Wall Street discovered that investment in the aero-industry brought good returns, and American industry exploited the flow of money to invest in new technology. By the turn of the decade the American aircraft industry was taking the first steps to produce revolutionary civil and military aircraft which were eagerly acquired by domestic customers.

These aircraft were streamlined, all-metal designs and they capitalized on German technology. Traditional construction maintained the 'box-kite' approach of the pioneers, with frameworks of timber or metal tubes strengthened with bulkheads, external struts and bracing wire. The framework was then covered by a skin of doped linen, wood or sheet metal. The Rohrbach civil aircraft used a wing of cantilever construction with a beam rigidly attached to the fuselage structure. Over metal spars and ribs was laid a smooth, metal 'stressed skin' joined to front and rear spars to create a strong but light wing box which behaved as a spar to share the load. A similar concept for fuselages, monocoque construction, removed the reliance on interior structures and bracing by transferring structural strength to the skin and underlying frames and stringers. European air forces took a more conservative view and adopted the new structures several years after the Americans, although the RAF did purchase for evaluation two Rohrbach designs, the Inverness (Ro IV) flying boat and the Beardmore Inflexible (Ro VI) heavy bomber. Nevertheless, the European aero-engine industry largely kept pace with its American counterpart and in some aspects was ahead of it.

In Germany, however, the malaise of the airframe industry extended also to the engine manufacturers. At the war's end Daimler and Benz, the *Luftstreitkräfte*'s prime suppliers, quit aviation. Benz's Rhineland factory was barred by the Allies from aviation work, while the absence of a domestic market meant that Daimler reverted to supporting the automobile industry. The merger of the two companies in 1926 saw a return to aviation, but the new concern had little success. Here, as elsewhere, post-war restrictions and a shortage of capital severely affected the development of aero-engines, especially that of high-performance engines, where Germany was unable to keep pace with her competitors. By 1929 there were still only four aero-engine manufacturers in Germany, and all suffered from a shortage of capital investment. Consequently most German high-performance engines were foreign designs built under licence, for example the Napier Lion and Pratt & Whitney Hornet (by BMW as the BMW 132) and the Bristol Jupiter (by Siemens as the SAM 22B). Several times the Transport Ministry proposed meeting Germany's military needs by importing engines but, understandably, the *RWM* rejected this on grounds of security. The *Heereswaffenamt* drew up plans to expand the industry and to develop new engines by 1933, but the plans failed to survive the Depression.

Junkers and BMW stayed in business only by substantially reorganizing their affairs and reducing their work forces, while Daimler-Benz scaled down aero-engine production and concentrated upon the automobile industry. Only the small Argus company, which continued to produce automobile engines but returned to aviation in 1926, seemed to thrive with small, air-cooled engines for civil aircraft. Indeed, in 1930 it introduced the very successful eight-cylinder, 500hp As 10 which was to be the backbone of its business throughout the 1930s. But the As 10 was unsuitable for combat aircraft, and by 1933 there were only three aero-engines suitable for military use—the domestically produced BMW VI and the licence-built BMW 132 and Siemens SAM 22B.

Brandenburg propped up the aircraft industry with subsidies, but the decline of the German economy saw these slashed to RM9 million ($2.14 million) for the 1929–30 budget, of which RM2.9 million ($690,000) was for aero-engine manufacturers. The Services could offer little relief until Groener rationalized the system of acquisition because each pursued its own aims with limited funds.[26]

Largely through Seeckt's malign influence, the *RWM* remained in sublime ignorance of the industry's woes until the late 1920s, but it drafted grandiose plans. At the war's end the industry had been capable of producing 2,000 aircraft a month from 35 airframe and 26 engine factories but by 1929 there were only eight manufacturers, half of them producing airframes. Although the *Heereswaffenamt* realized that the aircraft industry had contracted, it was unaware either of the manufacturers' problems or of their inability to meet the growing technical sophistication of the market. The increasing complexity of aircraft by 1930 doubled development timetables to four years for a complete aircraft and five to seven years for an engine.

Although the *Reichsheer* ignored industry's problems during the 1920s, it did not neglect training, and throughout the period it was able to provide refresher training for the 'Old Eagles' and to produce a generation of 'Young Eagles'. As early as New Year's Day 1924 the aircraft manufacturer 'Fritz' Siebel created Sportflug GmbH with *RWM* support. Ostensibly a civil organization to encourage flying, Sportflug allowed the *Reichsheer* to provide basic training for pilots and observers. It was no coincidence that its airfields were all close to *Wehrkreise* headquarters, with schools at Königsberg (*Wehrkreis I*), Stettin (*II*), Berlin-Staaken and Warnemünde (*III*), Schkeuditz near Leipzig (*IV*) and Böblingen, Hannover and Osnabrück (*V* and *VI*). An associate of Sportflug, Aerosport GmbH, provided seaplane training at Warnemünde on wartime Friedrichshafen FF 49s.

The continued desire for autonomy in Bavaria led Wimmer, at that time a *Hauptmann* and *Wehrkreise VII Referent zbV*, to insist upon a separate Sportflug for Bavaria with schools at Schleissheim near Munich and at Würzburg. The latter came under the command of Greim, who had gone to China after the Beer Hall *Putsch*. There he helped to train Chiang Kai Shek's air force before returning to Germany, where he directed training at Würzburg until 1933.

The schools operated some 50 single-engine trainers and touring aircraft, of which the most numerous were the Heinkel HD 21, the LVG B.III and the Udet U 12 Flamingo. They provided flying training from A1 to B1 grades (see Table 4) to civilians but received an *RWM* subsidy and each *Wehrkreis Referent zbV* was responsible for liaison between them, the *Wehrkreise* and Sportflug headquarters. The subsidy ensured that basic flying training was also given to officers on both the active and the reserve lists, while 'Old Eagles' received refresher training. The *RWM* also subsidized some half a dozen private flying schools, including Ernst Udet's in Munich, by sending officers and NCOs for flying training.

These activities produced only a trickle of aircrew and during the 1920s the Army's pool of air leaders slowly dried up as a result of age, illness and accident so that by 1926 only 100 were eligible for flying duty. By November 1930, even with Lipetsk graduates, the 'flying officers' list had only 168 names instead of 180, although 77 (45 per cent) went on to become *Luftwaffe* generals.[27] Advanced flying training, C grade,

TABLE 4: GERMAN PILOT LICENCE GRADES

Licence	Passengers	Take-off weight (kg)	Landing distance (m)	Engines
A1	2	500 or less	300	1
A2	3 or fewer	1,000 or less	450	1
B1	3 or fewer	2,500 or less	450	1
B2	6 or fewer	2,500 or less	450	1
C1	6 or more	2,500 or more	450	1
C2	6 or more	2,500 or more	450	2 or more

became available later with the establishment on 1 April 1925 of DVS at Berlin-Staaken. The DVS's objective was to provide a uniform quality of C-standard training for German and foreign pilots and, although it was a civilian organization, *Reichsheer* students were gradually filtered in. The Director was *Major* Alfred Keller, nicknamed *'Bombenkeller'* ('Bomb Shelter'), a pre-war Army aviator who later specialized in night bombing and, as Commander of *Bogohl 1*, had been awarded the *Pour le Mérite*. After the war he had joined Junkers, where he had become a director.[28]

Within a year of its opening DVS had a fleet of sixteen aircraft, including six Junkers A 20s, three Dietrich DP IXs, two Junkers F 13s and two Dornier *Komet III*s. The expansion of civil aviation saw new DVS schools open at Schleissheim, near Munich (A–B1 grade), and Brunswick (B2–C grade). From 1931 the latter was used for *Reichsheer* observer training, 80 men qualifying by the time Hitler came to power. The DVS fleet was also expanded and it proved to be a major customer for German aircraft, of which it received 130 from Junkers (G 24, F 13, A 20/35, W 33/34), Albatros (L 35), Arado (SC I/II) and Udet/BFW (U 12 *Flamingo*). Seaplanes were also required as DVS was extended with bases at Warnemünde (under *Korvettenkapitän* Goltz) and List, on the island of Sylt, providing A–B1 and B2–C grade training respectively. Civil and covert naval pilots used some of the Junkers aircraft converted into floatplanes as well as 40 Heinkel HD 24s and Heinkel HD/He 42s. Students faced a comprehensive course, which included instrument training, to produce some of the best pilots in the world. They were paid RM80 ($19) a month, of which RM35 ($8.33) was pocket money—a substantial sum when it was possible to go to town, have a good meal with beer and then go to the cinema for less than RM5 ($1.19).

Far from impeding German military training, the Paris Agreement actually improved it by allowing the Germans to replace piecemeal programmes with cohesive ones. The Agreement permitted up to 72 German officers to participate in light aviation, but only at their own expense and with only a dozen a year to train for a pilot's licence. The proviso was intended to prevent the creation of a reserve, and the Germans had to provide the Allies with a list of all members of the Services with current pilot's licences.[29]

However, the Allies had the means neither to monitor these provisions nor to enforce them. As an openly subsidized organization Sportsflug was disbanded and its assets were transferred to DVS and *Akaflieg*. The latter were founded at Darmstadt and Hannover on 1 April 1927 to provide advanced training, but their courses were criticized for an overemphasis on theoretical work.

To replace Sportsflug, Wilberg had protracted negotiations with Brandenburg and received permission from the Transport Ministry to train 40 *Offizieranwärter* a year at DVS Schleissheim before they formally entered the *Reichsheer*. In addition, from 1928 DVS schools at Berlin-Staaken and Brunswick were used for observer training, the former providing the basic theory and the latter practical experience, with final training at Lipetsk. By 1931, however, even practical training was conducted in Germany, with an average of sixteen officers graduating each year at Brunswick. The course was to B2 grade and lasted twelve months, with ten members selected for a six-month fighter training course at Lipetsk. After qualifying they joined the *Reichsheer* but every year took a two- to four-week refresher course with a new training organization, Luftfahrt GmbH (later Deutsche Luftfahrt GmbH), which opened schools at Würzburg and Böblingen on 1 April 1927 and sent its first fighter course students to Lipetsk the following year.

The *Reichswehr*'s poor pay made it difficult to find enough officers to meet the legal quota of twelve eligible for flying training as few could afford the lessons; furthermore, only 6 per cent of those officers with a pilot's licence could afford to continue flying. Only six officers (one naval) sought flying training in 1926 and by 1931 the figure was down to three.[30]

The use of DVS by the Navy indicates how much naval aviation expanded during the late 1920s. In June 1925 Faber's organization became a section (*Gruppe*) within the *Allgemeines Marineamt*'s *Seetransportabteilung*, commanded by the remarkable *Kapitän zur See* Günther Lohmann. The naval aviation section, *Gruppe BSx*, was under *Kapitän zur See* Rudolf Lahs and had four desks, Naval Staff, Training, Technical and Administration. The Staff desk was held by *Korvettenkapitän* Hans Geisler and the Training desk by *Oberleutnant zur See* Wolfgang von Gronau (later by *Kapitänleutnant* Ulrich Kessler), while the Technical desk was initially under *Korvettenkapitän* Joachim Coeler and later *Korvettenkapitän* Siburg.

On 30 September 1929 Lahs completed his assignment and left the Navy with the rank of *Admiral* to become the head of *RDLI*. The next day *Gruppe BSx* was renamed *Gruppe LS* under *Kapitän zur See* Konrad Zander and expanded to a staff of 30, initially remaining under the *Allgemeines Marineamt*, the Navy's administrative organization. However, the *Marinekommandoamt* demanded greater control over naval aviation and during the next two years it gradually edged out its rival to assume total control for a brief period from 1932.

Naval aviation had an annual budget of some RM10 million ($3.80 million) and all training was conducted within Germany's borders. A dozen pilots and observers a year were produced, with initial training at a yacht club at Neustadt, north of Lübeck, and flying training conducted by Severa (see below). When Zander took over he was dissatisfied with the quality of the airmen and extended the flying training course to a year and the observers' course to two years at Warnemünde. He also began training other aircrew, including flight engineers and radio operators, as well as increasing the number of courses for groundcrews. In the early 1930s, when the *Reichsmarine* decided to have fighters, it arranged to complete pilot training at Lipetsk.

The key naval training organization was Severa. Founded in 1924, Severa was led by Osterkamp, whose health had collapsed as a result of the rigours of campaigning with *Fliegergeschwader Sachsenberg* in the Baltic and who had spent some considerable time in hospital before entering civil aviation. Ostensibly Severa was a civilian air freight company which also provided towed targets for naval gunnery exercises. Commercially it had little success, especially after the creation of Lufthansa under the redoubtable Erhard Milch, but from the naval viewpoint it was invaluable. Operating from Holtenau, near Kiel, and from Norderney, it initially provided wartime observers with refresher training in an FF 49 before training a new generation of observers from young officer volunteers. Severa's activities gradually extended into training groundcrews and developing equipment, new bases being added at Wangerooge (1925) and Wilhelmshaven-Rüstringen (1929). In 1926 and 1927 its fleet was expanded with an HE 1, Junkers A 35s and Junkers F 13 floatplanes, later supplemented by Junkers W 33/34 floatplanes. The first multi-engine aircraft it received was a Junkers G 24 in 1926, followed by a second in 1927. The latter was fitted out by AB Flygindustri of Sweden and used for trials of new radio and direction-finding equipment, including Germany's first attempt at a transatlantic flight which ended in the Azores.[31]

Lohmann's activities accidentally brought about a significant change in the development of German military aviation for, a year after Seeckt's fall, they led to the disgrace of Defence Minister Gessler. Seeckt regarded the *Reichsheer* as a cadre which could be rapidly expanded from covert sources and volunteers. Plans in 1925 envisaged 65 divisions, with an air force, using secretly stockpiled equipment which included 675 field and heavy guns. Seeckt crushed open opposition to his plans, which included mobilizing Lufthansa aircraft for use against Poland, but a loose alliance of opponents, nicknamed the *Fronde* after the French rebels of the mid-seventeenth century, grew up within the *Reichsheer*.

The *Fronde*, which included Hasse and Blomberg, sought a more pragmatic expansion to 21 divisions as well as closer links with both Government and industry to produce an efficient national defence plan. They disliked Seeckt's political conservatism, yet his monarchist views brought about his downfall for he allowed the *Kaiser*'s grandson to attend *Reichsheer* manoeuvres without informing the Government. When Gessler, who learned of the affair through the Press, confronted him, Seeckt arrogantly denied everything. When the Minister learned the truth he forced the *General*, who had lost the support even of recently elected President Hindenburg, to resign on 8 October 1926.[32]

His successor was the genial *Generaloberst* Wilhelm 'Papa' Heye, who was more sympathetic to *Fronde* views and was a loyal supporter of Gessler. The Minister arranged a Cabinet meeting on 29 November 1926 so that the *General* could provide a comprehensive briefing on secret rearmament, including air activities. The responsibility for authorizing programmes was now transferred to the Cabinet and Gessler ensured that the *RWM* obeyed. In February 1927 Heye gave another briefing and the Cabinet told him that they were willing to fund illegal rearmament provided the budget did not come under Parliamentary scrutiny. This led to the creation of a multi-departmental military–civilian committee to 'launder' the budget, beginning in 1928.[33]

Ironically, it was another attempt at financing clandestine rearmament which brought about Gessler's fall. Within the Navy such finance was the responsibility of Lohmann's *Seetransportabteilung*, which supported a network of nominally commercial organizations such as Severa. Such clandestine activity frequently leads to a temptation to make commercial gambles, and it was such a temptation to which Lohmann fell prey during the mid-1920s.

Some investments proved useful, such as his purchase of the Caspar aircraft works at Travemünde for aircraft testing in 1926; this became the *Seeflugzeug-Erprobungsstelle* (*SES*) *Travemünde* and later *E-Stelle Travemünde des RDLI*. But many ventures were increasingly on the periphery of naval interest and proved disastrous. One was a film company for which Gessler provided guarantees which led, in 1928, to his unmasking and that of Lohmann. Gessler was forced to resign, as was the head of the Naval Command, who was replaced by *Admiral* Erich Raeder. An audit showed the total losses of Lohmann's enterprises to be RM26 million ($6.19 million), and while he made no attempt to profit from the situation his reputation was compromised. He, too, was forced to resign, and he died in poverty a few years later.[34]

The new Defence Minister, appointed on 18 January 1928, was the former *Generalleutnant* Wilhelm Groener, Ludendorff's successor in 1918 and later the Republic's first Transport Minister. Groener had been closely involved with wartime industrial activities and this experience now proved of great value, for it facilitated the process of creating an integrated defence policy. He began by accelerating the process of establishing Government control over the clandestine rearmament programme, which confirmed the opinion of more conservative officers that he was a 'Red'. Yet Groener was to bring the *RWM* greater respect from within and without than it had ever earned previously.[35]

Groener's first task was to clean the Augean stables, and he demanded a full report of clandestine naval expenditure. This was completed on 13 February 1928 and showed, among other items, that Severa's annual cost was RM1.35 million ($320,000). Lohmann's organization was dismantled and in February 1929 Severa became Lufthansa's Coastal Aviation Department (*Küstenflugabteilung*). But Lufthansa was partly owned by the provinces, and the Prussian provincial government, where the new organization was most active, naturally feared financial losses for which it might be liable. It strongly opposed the new arrangements and in July 1929 the organization was renamed Luftdienst GmbH and became autonomous. It continued its training and development activities under this name until it was absorbed by DVS in 1933.

Groener, Heye, Raeder and Blomberg ensured that integrated armament planning was now feasible and on 19 September 1928 a two-stage programme costing RM350 million ($83.13 million) was approved. The programme was to equip, and to provide limited stockpiles for, an army of sixteen divisions as well as improving industrial production in the event of mobilization. It was to be completed by 1932, but no provision was made for aviation.

The programme was possible because Stresemann had secured economic stability through the appearance of conciliation. Consequently the defence budget, which had been RM490 million ($11.61 million) in 1924, rose to a peak of RM827 million ($196.43 million) in 1928–29. The 1929 figure included more than RM27.26 million ($6.47 million) for industrial preparation, of which RM12.44 million ($2.95 million) was allocated to the aircraft industry. From covert funds the *Reichsheer* had been spending RM10 million ($2.38 million) annually since 1925 on training aircrew and developing aircraft. Half was spent supporting flying training, while Lipetsk cost RM2 million ($476,000). The remainder was spent by the *Heereswaffenamt* on the clandestine development of military aircraft and equipment.

Aircraft development was further aided by Brandenburg in the Transport Ministry, who provided subsidies worth RM321.22 million ($76.11 million), or an annual average of RM45.88 million ($10.87 million), between 1926 and 1932. He also diverted funds from civil aircraft and aero-engine development to military use, and in 1928–29 this amounted to RM3.38 million ($802,000). Not only were the Cabinet informed of the military activity but so too were the principal members of the *Reichstag* Budget Committee. Given the clandestine nature of German air rearmament, it is impossible to discover exactly how much money was involved, but it has been estimated that between 1925 and 1933 some RM100 million ($24 million) was spent on Army training and aircraft development and another RM70 million ($16.5 million) on industrial support. By comparison, during the same period the British Air Estimates totalled £134 million ($29 million), while the French spent about FFr12,000 million ($49 million), although both these figures include naval, civil and colonial expenditure.[36]

Financial security enabled the *Reichsheer* to begin planning its future air arm. The signatures on the Paris Agreement were barely dry when Felmy, in May 1926, drafted 'Guidelines for the Leadership of "Operational" Air War' (*'Richtlinien für die Führung des Operativen Luftkrieges'*), but this was an outline study which failed to address the future status of military aviation. It did however, form the basis of the future 'Air War Leadership' (*'Luftkriegführung'*). Other suggestions were made by Sachsenberg, who had become Junkers' Director-General.

In 1928 and 1929 he submitted two memoranda to the *RWM*, 'Air War—Peaceful Coercion' (*'Luftkrieg—Friedenszwang'*) and 'Thoughts and Suggestions on German Aviation Politics under Consideration of Defence Questions' (*'Gedanken und Vorschläge zur Deutschen Luftfahrtpolitik unter Berucksichtigung der Wehrfragen'*). He argued that, since everyone was aware that Germany was evading military restrictions through its civil air fleet, the *RWM* should exploit the situation. He proposed creating a fleet of fast bomber-transports or auxiliary bombers similar to the Junkers G 24 and claimed that a contract for 100 would stimulate production and the modernization of the aircraft industry, encourage domestic support for military aviation and impress foreigners. Although the *Heereswaffenamt* was then testing the auxiliary bomber concept, there was still no agreement either on their role or even on the value of creating a bomber force. There was also a feeling at *RWM* that Sachsenberg's memoranda were crude attempts to win orders for the cash-starved Junkers factory, whose military products had a poor reputation, and consequently no action was taken. However, Sachsenberg's memoranda did anticipate the *Luftwaffe* expansion programme of his enemy Milch and also the concept of a deterrent air fleet (*Risiko Luftflotte*) to impress foreign opinion. In 1932 Milch himself was to make a similar proposal. He suggested that the *RWM* spend RM4 million ($943,000) a year to produce a fleet of Ju 52/3m auxiliary bombers, but this was rejected on financial grounds.[37]

The *Reichsheer*'s conservative view of air power was clearly demonstrated during the summer of 1927. On 25 May Heye approved plans for an Emergency Army, which would be equipped with a few air squadrons, and barely a month later, on 30 June, *T2* produced the Outline Mobilization Plan for Wartime Armed Forces (*Aufstellungsplan einer Kriegswehrmacht*, usually abbreviated to '*A-Plan*'). Although thios was intended only as a consultation document, it was accompanied by the *Heereswaffenamt*'s estimate of industrial costs, which were placed at RM38,478,600 ($9.16 million).

The *A-Plan* covered only Army preparations for a 21-division force but there was detailed provision for air support which repeated wartime experience and provided the Field Army with a *Luftstreitkräfte* and *Kofls* raised from the *Referenten zbV*. The *Luftstreitkräfte* would include reconnaissance and combat formations as well as creating a supporting organization with schools and replacements. Air defence was to be supported by an observation and warning service, which was actually re-formed the following year using police posts with civilian telephone and telegraph facilities. Plans also existed for an anti-aircraft force of 64 motorized 8.8cm guns supplemented by machine guns, searchlights and direction finders.

It was envisaged that by 1929 a permanent force of nine *Fliegerkurierstaffeln* would be created, each with six aircraft and 30 personnel, all ostensibly civilians. There would be one or two squadrons in each *Wehrkreis* and the total strength was envisaged as 247 aircraft. Most of these would be either placed in store or attached to civilian organizations, and all the aircraft would be based upon existing designs for ease of manufacture. Nearly 60 per cent were to be corps aircraft (Heinkel HD 33, Albatros L 65/II, Albatros L 70), there were to be 57 auxiliary bombers (Dornier Do B *Merkur*, Junkers G 24) and 47 Fokker D.XIII fighters were to be transferred from Lipetsk. While none of the corps aircraft designs proved to be successful, the *Truppenamt*'s air organization ordered individual examples of the Albatros designs and a few Heinkels, none of which, again, proved to be a success.[38]

Under Hammerstein-Equord the *A-Plan* was revised in 1929 to be based upon a 21-division force. This envisaged a 22-squadron (*Staffeln*) air force, each squadron composed of six aircraft. While corps squadrons continued to dominate the structure, the plan included six fighter and three night bomber squadrons and showed a total strength, with reserves, of 200 aircraft, while for training and replacements there were to be seven *FEA*s. A more radical proposal was made by *T1* on 15 April 1930 when it proposed an eighteen-squadron force with six fighter, six bomber, two long-range reconnaissance and four corps units.[39]

This was the first move towards an autonomous air force which Sachsenberg supported in lectures and which many aviation officers sought. The British Air Attaché, Group Captain E. L. Gossage, reported on 22 July 1930 that the *RWM*'s *Major* Ulrich Grauert (who would become a *Fliegerkorps* commander during the Western Campaign of 1940) had told him that Germany should have an air defence force and hinted that his country intended to have one.[40] But the junior status of such officers meant that the *Truppenamt*'s prevalent view was generally more conservative, with a desire only to recreate the wartime organization.

The *A-Plan* gave an impetus for a further expansion of the training organization to provide NCOs and other ranks for groundcrew and aircrew duties. At the end of

1929 *2. Fahrabteilung* was created under *Major* Kurt Pflugbeil, an 'Old Eagle' who had fought the Poles in Silesia before re-joining the Army in 1920. He returned from working on auxiliary bombers at Lipetsk to command two squadrons (*Eskadronen*) tasked with training radio operators and air gunners. Other battalions were created to train groundcrews and flight engineers, while *9. Batterie/Artillerie Regiment Nr 1* assumed responsibility for training anti-aircraft gunners in preparation for the creation of six batteries. By August 1932 these organizations had produced 809 NCOs and other ranks, of whom 79 were aircrew, despite changes in the leadership such as Pflugbeil's promotion to *Rittmeister* and his transfer to *11. Kavallerie Regiment*.

The development of the aircraft which these men would operate had already begun. Much of the early work was conducted by Student, who was 32 years old in 1922. He had joined the Army in 1910 as an ensign in a *Jäger* battalion but three years later he had been detached to a military flying school. He did not transfer to the *Fliegertruppe* until after the Great War had broken out and he eventually became a fighter pilot, commanding first *Jasta 9* and then *Jagdgruppe 3*. He joined the *Reichsheer* and was responsible initially for developing new aircraft and then for training. He was succeeded by Volkmann, who had joined the Army in 1908 and later transferred to the *Fliegertruppe*.

With the Paris Agreement the *RWM* laid down the technical requirements for Germany's first post-war generation of military aircraft, which it hoped to acquire by 1930. In 1926, to assist the preparations for this rearmament, the *Heereswaffenamt* established Fertigungs GmbH to inspect and to approve plans and technical drawings, to issue contracts and finally to evaluate and to test aircraft. Only now did the *Reichsheer* begin to recognize the industrial problem it had so blithely ignored, for there were no common industrial techniques or even common engineering standards. Companies were unwilling to share industrial secrets or to invest either in plant or machinery unless contracts were placed—but Germany was in no position to place large contracts. The fragmentation of Service funds meant that only a few prototypes could be ordered at a time from one or two companies, together with small numbers of production aircraft.

As early as 1925 an aircraft research establishment had been re-created at a former testing station at Rechlin am Müritzsee, north-west of Berlin in Mecklenburg, for the evaluation of prototypes. The *E-stelle Rechlin des RDLI* became Germany's equivalent of Farnborough or Wright-Patterson and was used to test both civil and military aircraft. Until 1930 the latter were usually given civil registrations, sometimes using vacant old numbers, then all were registered to *RDLI*. Later, flight testing was also conducted at the DVL.

Following the Paris Agreement, Student defined four types of aircraft to meet Army requirements, these being referred to by the acronyms '*Heitag*' (*Heimatjagdeinsitzer*), '*Erkudista*' (*Erkundungsflugzeug für die Divisionsnahaufklärungsstaffeln*), '*Najuku*' (*Nachtjagd und Erkundungsflugzeug*) and '*Erkunigros*' (*Erkundungsflugzeug für mittlere Höhen und grösste Entfernungen*). The requirement for a 'long-range reconnaissance aircraft' was, despite the innocuous title, actually a requirement for a heavy bomber but, fearing a repeat of the Lipetsk security leak and a consequent catastrophic impact upon the country's foreign relations, the *RWM* did not use the term 'bomber' until 1929. However, in 1927 the *Heereswaffenamt* proposed a requirement for a four-engine *Grossnachtbomber* (*'Gronabo'*) with a maximum speed of 215kph (133mph) and able to fly at 5,000m (16,500ft).

In 1928 the *Heereswaffenamt* invited companies to submit designs, the winner to be awarded a production contract (although the latter was likely to be small). The testing and evaluation programme began in the autumn of 1928, with armament trials at Lipetsk, and lasted two years. Aircraft evaluated were the Arado Ar SD I (*Heitag*), the Albatros L 77v and L 78 (*Erkudista*), the twin-engine BFW M 22 (*Najuku*) and the Heinkel HD 41 (*Erkunigros*). With the exception of the BFW M 22 and HD 41a, which had 500hp Siemens Jupiters, all the contestants were powered by versions of the 600hp BMW VI. The trials were disappointing: all the contenders were under-powered and some had nasty flight characteristics. Albatros won orders for seventeen L 77s and L 78s, another two being built by Heinkel, whose HD 41 was developed into the He 45. The *Heereswaffenamt* had to revise and to refine its requirements, which were issued in 1929 by Volkmann and included maximum speeds of 220kph (136mph) for bombers, 350kph (217mph) for fighters and 250kph (155mph) for reconnaissance aircraft.

The *Reichsmarine*, too, was able to begin the development of combat aircraft following the Paris Agreement. During the summer of 1926 manufacturers were invited to participate in a competition for a new reconnaissance floatplane, with a prize of RM360,000 ($86,000). The competition was won by the Heinkel HE 5 but it was not until 1928 that nine were purchased, followed by ten HE 9s in 1929. From 1927 to 1929 a wide selection of aircraft were evaluated, including the Dornier R *Superwal*, Rohrbach Ro V *Rocco* and Heinkel HD 15 flying boats, the Heinkel HE 7 torpedo bomber and the Heinkel HE 31 and Caspar C 36 reconnaissance floatplanes, as well as the Heinkel HD 38a (a development of the HD 37) and Arado SSD 1 floatplane fighters. The *Superwal* was selected as the long-range reconnaissance aircraft and six were purchased for operation by Severa and Lufthansa. A dozen HD 38s were eventually ordered, but the others failed to meet the naval requirements. However, in 1929 a couple of Rohrbach Ro VII *Robbe I* and Dornier *Wal* flying boats, the former designed by Kurt Tank, were purchased, ostensibly by Lufthansa. The majority of these contestants relied upon imported engines, the domestic BMW VI being used only in the C 36 and the two fighter designs.

All evaluation took place at Travemünde, which had a staff of 60 and became the principal flight testing centre for the *Reichsmarine* from 1928. The aircraft bore civil registrations, initially of Severa and later either of *RDLI* or the *DVL*. Although the *Reichsmarine* had earlier rejected the use of facilities at Odessa from the late 1920s, it too began to use Lipetsk for armament trials.

In addition to developing its own aircraft, the *Reichsmarine* invested in Dornier's ambitious commercial project, the mighty, twelve-engine Do X flying boat. With Christiansen at the controls, it flew a number of route-proving flights to South America. The Navy hoped to exploit its capabilities for long-range reconnaissance, minelaying and even torpedo bombing, but the design proved too large and too complex, although two were sold to the Italian Air Force (*Regia Aeronautica*). The *Reichsmarine* also encouraged the development of diesel aero-engines. The first 800hp MAN and Daimler-Benz engines proved too heavy but were fitted into what became the *Schnellboot* (S-boat, or 'E-boat' to the British) prototypes. Undaunted,

Daimler-Benz persevered and eventually succeeded in developing the 1,200hp DB 602, the first successful diesel aero-engine, and this was fitted to the LZ 129 *Hindenburg* and LZ 130 *Graf Zeppelin II* airships. Naval research and development funds were also invested in a comprehensive programme to develop radios and navigation equipment to support long-range flights over the sea. This programme involved fourteen aircraft, was directed by *Ingenieur* Beck using Lorenz and Telefunken equipment and included flights as far south as the Azores using a Heinkel HE 10 and a Junkers G 24.[41]

Between 1925 and 1930 the *Reichswehr* received about 90 combat aircraft, the majority of them prototypes for evaluation. Small numbers of Albatros L 76 and L 78 corps aircraft and Junkers K 53 (A 35) reconnaissance bombers were ordered for the *Reichsheer*, while the *Reichsmarine* received Heinkel HE 5 and HE 9 floatplanes. Most of the Albatros aircraft, and some of the Junkers, went to Lipetsk but the remainder operated in Germany concealed among a variety of semi-legitimate organizations.

The decline of the German economy from 1928 made the Government extremely cost-conscious. Groener was irked to discover that in aviation the activities of the two Services overlapped—the *Reichsmarine*, for example, was developing landplanes such as the HD 16 torpedo bomber—and in the summer of 1929 he demanded rationalization. The *Reichsmarine* was extremely reluctant but since 1928 it had been seeking funds for the first *Panzerschiff* ('armoured ship'), a new class of capital ships which were to become the famous 'pocket battleships'. To secure such funds it was prepared to drink the poisoned chalice and on 1 October 1929 it concluded an agreement with the *Reichsheer* which defined their two areas of influence and avoided duplication. Essentially, landplanes would be the *Reichsheer*'s responsibility and seaplanes would be developed by the *Reichsmarine*, while the two services would seek to rationalize requirements wherever possible.

But in the long term the agreement began a slow decline in German naval air power which became apparent only during the early stages of the Battle of the Atlantic. Gradually the Army tightened its control on German air power, and the *Luftwaffe* was to be dominated by former Army officers, largely because of the indifference of the Navy. Until the mid-1920s the *Reichsmarine* regarded itself purely as a coastal protection force, but during the 1926 Autumn Manoeuvres it adopted a more aggressive philosophy of attacking enemy maritime supply routes, and this included the use of aircraft. However, Raeder and most of the world's admirals perceived naval strength in terms of capital ships, with aircraft used to locate the enemy battle fleet and then to direct gunfire upon it. Although expenditure on naval aviation rose from RM200,000 ($47,000) in 1924 to RM7.3 million ($1.72 million) in 1932, there was no serious attempt to develop an air striking force capable of attacking enemy ships far from land.[42]

On 3 October 1929 Stresemann died. Behind a façade of co-operation with the victors, he had created the framework for a German military resurgence, including air power. At the time of his death both German Services were on the brink of breaking the final restraints of Versailles, but the status of the air arm was still undecided and its autonomy seemed a dream. Worse still, on 23 October the Wall Street Crash dammed the flow of money from the United States which had fuelled the German economy. While it accelerated the World Depression, the catastrophe was to have a profound influence upon the development of the German air arm during the next three years.

NOTES TO CHAPTER 2

1. For Russo–German co-operation see Carsten, pp.116, 143–7, 163, 232–4; Erickson pp.144–63; and Völker, *Dokumente*, pp.58–92, and *Entwicklung*, p.134.
2. See Cuneo, pp.142–3; and Kilduff, pp.134–51.
3. See Air Ministry, p.32, para.23.
4. For Lipetsk see also Carsten, pp.236–7; Homze, p.9; Schliephake pp.14, 17–18, 21, Appendix B; Völker, *Entwicklung,* pp.134–5, 140–2, 156, 158–9; Lennart Andersson, 'The Secret Luftwaffe'; and Speidel, 'Reichswehr und Rote Armee'.
5. Alexandrov, 'Junkers Planes in Russia'.
6. Andersson, 'Junkers Two Seaters'; Alexandrov, *op. cit.*
7. Alexandrov, *op. cit.*
8. For Junkers see Homze, pp.9–10; Carsten, p.234; Andersson, *op. cit.*; and Alexandrov, *op. cit.*
9. Carsten, pp.234, 237–8, 255, 275–8.
10. Alexandrov, *op. cit.*
11. For German officers and Blomberg, see Carsten pp.214–17, 310–11, 389.
12. For *Reichsheer* air organization in the late 1920s see Völker, *Entwicklung*, pp.161–4; Homze, pp.24, 60; Morrow, pp.74, 126–9.
13. *SHAT* File 4N97, Dossier 2.
14. For the background to German civil aviation and the Lufthansa story to 1932 see Davies as well as Bley, pp.15–31.
15. Strictly Deutsches Luft Hansa AG (DLH). It was renamed Deutsches Lufthansa on 30 June 1933.
16. See Homze, pp.29/t.1, 31.
17. Irving, *Milch*, pp.12–18, 20.
18. PRO Air 2/13889.
19. Manvell and Fraenkel, p.47.
20. For Lufthansa see Homze, pp.12–13, 30–31; Irving, *Göring*, p.94; and Völker, *Entwicklung*, pp.152–3.
21. See Kilduff, pp.80–1.
22. PRO Air 8/138, RAF Confidential Intelligence Summaries.
23. Holley, p.20.
24. *SHAT* File 7N2620.
25. Holley, *loc. cit.*
26. For the German aircraft industry in the 1920s see Homze, pp.23–8, 31.
27. Völker, *Entwicklung*, *Anlage 15*.
28. See Kilduff, pp.77–80.
29. For the Paris Agreement see Homze, pp.3–4; and Völker, *Entwicklung*, pp.148–9.
30. For training in Germany during the 1920s and early 1930s see Andersson, 'Junkers Two Seaters'; Carsten p.273; Homze, pp.13, 20; Schliephake, p.16; Gefolgene Vergangenheit, *75 Jahre Luftfahrt in Schleissheim*, pp.47–54; Völker, *Entwicklung*, pp.137–41, 145–6, 148–50, 154, 168, 222; and *SHAT* File 4N97, Dossier 2, Final Report of the Aeronautical Guarantee Committee, 28 March 1928.
31. For naval aviation see Andersson, 'Junkers Two Seaters'; Carsten, p.363; Homze, p.38; and Völker, *Entwicklung*, pp.139, 145, 157, 164–5.

32. For the *Fronde* and Seeckt's dismissal see Carsten, pp.245–8, 253, 261; Seaton, pp.15–16; and Deist pp.5–6.
33. Deist, pp.9–10.
34. For the Lohmann scandal and its aftermath see Carsten, pp.242, 244–5, 284–7; Homze, p.38; and Mason pp.145–8.
35. Carsten, pp.272, 284–6, 290; Deist, pp.7–8.
36. See Carsten, pp.274–5; Homze p.31; and League of Nations Armaments Handbooks.
37. For the Sachsenberg memoranda see Homze, pp.32–3, 43; Irving, *Milch*, p.56; Völker, *Entwicklung*, p.183.
38. For the 1927 plan see Andersson, *op. cit*; Carsten, p.273; Homze, pp.30–1, 43; and Völker, *Entwicklung*, pp.159–60, 166–7.
39. Völker, *Entwicklung*, pp.168–9.
40. PRO FO371 14539 C6035.
41. For naval equipment programmes concerning aircraft see Andersson, 'Junkers Two Seaters'; Homze, p.39; Schliephake, pp.24–5, 25n; and Völker, *Entwicklung*, pp.156–8.
42. For the Navy see Homze, p.39; and Völker, *Entwicklung*, pp.181–2.

CHAPTER THREE

BIRTH OF AN EAGLE

S tresemann died as the future of the German air arm hung in the balance. The administrative structure was in place, training programmes were under way and aircraft were being evaluated, while the imminent departure of the French from the Rhineland in 1930 removed the last threat of intervention. But the American economic crisis threatened to undermine progress.

The German Army was quick to exploit the improved strategic situation, recalling some instructors from Lipetsk in 1930 and using them to establish three *Reklamestaffeln* ('skywriting squadrons') on 1 October in *Wehrkreise I* (Königsberg), *II* (Berlin) and *VII* (Nuremberg). Ostensibly owned by Luftfahrt GmbH, each had four unarmed Albatros L 75a and Albatros L 82 biplane trainers, similar to the famed De Havilland Moth, and eleven men. They were actually the first of the *Fliegerkurierstaffeln* planned in 1927 to increase air awareness and they represented future squadrons in the reconnaissance role, carrying qualified observers as well as towing targets to help train *Flak* batteries.

Germany was only a step from creating an air force, but before crossing the Rubicon the Army sought the views of *Oberstleutnant* Ferdinand von Bredow, the Head of Counter-Intelligence. He stated that the French had detailed information not only about the camouflaged German General Staff but also about Mittelberger's 'Air Defence' Inspectorate; he added that an unidentified officer in the British War Office had informed him that, while the British were less concerned about German naval and army rearmament, they were paying considerable attention to air activities.[1]

French records do not confirm Bredow's claims, but the British, from the evidence of their own documents, seem to have been well informed about German air activity. From the late 1920s attachés in Berlin noticed Army officers disappearing from the active list then reappearing without loss of seniority, and they associated this with flying training. Sir Horace Rumbold, the British Ambassador, noted on 29 May 1931 that this had happened to 66 officers, and he named the latest seven who had 'returned to the fold' as *Hauptleute* Baier and Niepmann, *Rittmeister* Pflugbeil and Sperling, and *Oberleutnants Freiherr* von Houwald, Krahl and Schulze-Heyn—of whom all but Niepmann appear in a secret 1930 list of *'Fliegeroffiziere'*.

The British Air Attaché, Gossage, reported floatplanes participating in naval air defence exercises but the German Foreign Ministry routinely denied such activity, although it admitted in April 1930 that DVS aircraft from Sylt 'happened to be in the area'.[2] Gossage observed floatplanes towing targets on 3 August 1931 and reported that officers were being trained in Lipetsk, although he underestimated the total number at 54. He warned London that the Germans hoped to create an air force from 1932, although financial problems might cause delays. Diplomatic sources reported growing German truculence: from January 1930 the publication of lists of pilots fell into disuse, while in the summer of 1931 Berlin banned attachés from visiting aircraft factories and made it difficult for them to visit aviation establishments. Less attention appears to have been paid to the published list of pilots, otherwise the British might have wondered why air aces such as Osterkamp and Greim received Class A pilot licences in October 1927.[3]

Spies provided further information. On 29 November 1929 Gossage reported information from 'secret sources' to the effect that tactical training and air exercises were being conducted at Lipetsk, although the Air Council informed the Foreign Office on 11 January 1930 that it had no independent evidence to confirm (accurate) French claims of civil aircraft being converted to bombers there.[4] A comprehensive intelligence picture was presented to the Foreign Office on 5 July 1933 which justified the comment on the file, 'The fact is that . . . up to now we have condoned various infractions . . . of the Treaty known to us from secret sources or press reports . . .'[5] The British Secret Intelligence Service (SIS) listed reports of German contraventions of the Versailles Treaty from 1926 to 1933 and 40 of these refer to air activity. They show that the British knew about the illicit air organizations (October 1926, August 1930, March 1932), their financing (May 1932) and personnel (February 1930, August 1930, June 1932) and even plans to shield behind civil aviation (November 1929), while from August 1932 there were reports on aircraft evaluation. Naval activity included a June 1930 report that the cruiser *Köln* had conducted anti-aircraft exercises against torpedo bombers. Although the British did not possess the most intimate details, they could still gain a comprehensive picture of activities at Lipetsk (November 1929, August 1932) and of co-operation with various organizations, yet no official protest was made, partly because Whitehall had also come to feel that Germany had been unjustly treated at Versailles.[6]

The French *Aviation Militaire* depended upon the Army for intelligence and it was not until February 1932 that an air attaché was appointed to Berlin. The French do not appear to have been as well informed as the SIS, but during 1930 intelligence reports provided general details of German air activity in Lipetsk, which the French incorrectly concluded was being used as a bomber school. The most comprehensive French report, completed on 8 April 1930, noted that students at DVS Staaken and Schleissheim were flying in formations of seven or eight aircraft. This also made alarmist claims about the potential German air threat, probably timed to influence defence policy as the French withdrew from the Rhineland. The report said that on New Year's Day 1930 nearly half (464) of Germany's 980 civil aircraft (including 51 multi-engine machines) were potentially military aircraft.[7] This conclusion reflected a common belief that, as it had been easy to convert military aircraft to civil use after the war, then the process could be reversed. The idea was the basis of the bomber-transport, and in the early days of the Spanish Civil War many civil aircraft were indeed 'converted' to warplanes while in 1939 the French Navy converted three Centre NC 223.4 transatlantic mailplanes into heavy bombers and during the night of 7/8 June 1940 one attacked Berlin. However, experience in the 1930s (including the USAAC's mail delivery service of 1934) conclusively demonstrated that civil and military aircraft had radically different design philosophies. Tacit Allied approval boosted German confidence, leading to a meeting in Berlin on 29 November 1930 attended by Defence Minister Groener, Foreign Minister Julius Curtius and Transport Minister Theodor von Guerard, together with Brandenburg and Mittelberger. They decided to begin producing aircraft and equipment selected at Lipetsk but to store them in the same way as the British Auxiliary Air Force and RAF Reserve squadrons. It was probably at Curtius's suggestion that no decision was taken openly to break the Versailles Treaty by issuing these aircraft to units. Mittelberger presumably acquiesced because this was a step in the right direction.

Within a month, on 20 December, his Inspectorate published a document entitled *'Richtlinien für die Ausbildung in der Reichswehr auf dem Gebiet der Luftwaffe'* ('Training Guidelines for *Luftwaffe* Duties in the Armed Forces') which outlined air power theory, tactics and training. It was also the first document for general distribution which used the term *'Luftwaffe'*, a term apparently not formally proposed by Mittelberger until his memorandum to the *Heeresleitung* on 24 February 1932.[8] Yet until 1933 the air arm continued to be referred to by a variety of terms, including *'Luftstreitkräfte'*, *'Fliegertruppe'* and *'Flugwaffe'*.

Meanwhile the *RWM* monitored progress towards the establishment of an air arm and tried to resolve internal differences over its status. At the beginning of the 1930–31 financial year (1 April 1930) it reviewed progress on the 1927 plan. Mittelberger's Inspectorate had been reorganized into an embryo air headquarters with nine desks, including Operations, Personnel, Administration, Equipment, Organization and Training.[9] The conference learned that by the end of the year the Army would have some 300 qualified aviators, while a list produced on 1 November 1930 showed 168 such officers on the active list. Although some, such as Mittelberger himself and Wilberg (who was now with *Kommandantur Breslau*), were desk-bound, many on the list were to distinguish themselves. Indeed, it reads like a *Who's Who* of wartime *Luftwaffe* generals and included *Majors* Sperrle, Felmy, Grauert, Volkmann, Wimmer and Student; *Rittmeisters* Pflugbeil, Dessloch and Martini; *Hauptleute* Speidel and *Freiherr* von Richthofen; and *Oberleutnants* Kammhuber, Meister, Korten Jeschonnek, Jordan, Hoffmann von Waldau, Deichmann, Plocher, Seidemann, Oscar Dinort and Kreipe.[10]

But what form would the new arm take? On 17 May 1930 *Generalleutnant* Kurt *Freiherr* von Hammerstein-Equord, Chief of the General Staff, sought the views of his departmental heads and provoked a storm of controversy which took 21 months to blow itself out. Many of Mittelberger's officers, including Grauert at the Personnel and Aviation Politics (*Referat II*) desk, were seeking an autonomous air force on the lines of the RAF. The more conservative members of the General Staff, led by *Oberstleutnant* Wilhelm Keitel of *T2* (Organization) and later by *Oberst* H. Geyer of *T1* (Operations), believed that this would encourage the airmen to fight a private war and neglect the Army, a controversial topic which continues to rage into our own time. They also feared that the shallow pool of trained staff officers might be drained; indeed, the lack of qualified staff officers seriously handicapped the *Wehrmacht* during the Second World War. The General Staff was, therefore, anxious to clip the air arm's wings by retaining control of aircraft development and procurement as well as AA artillery. The conservatives would concede only the resurrection of *Kogenluft*, despite Mittelberger's protests that this placed intolerable restrictions upon an extremely flexible new arm.

Mittelberger shrewdly deflected the opposition on 1 April 1931 by assuring his colleagues that an autonomous air force would never act independently but always in co-operation with the Army. This won him the support of the new Chief of the General Staff, *Generalleutnant* Wilhelm Adam, for on 21 April 1931 he stated that because aircraft and AA forces worked in close co-operation they should be within a single organization which would work closely with the other Services. He refused to change his mind when the departmental heads confronted him on 8 September 1931.

Despite its own doubts, the *Heeresleitung* (under Hammerstein-Equord since October 1930) clearly accepted Mittelberger's arguments. It provided him, on 19 December 1931, with the first *'Grundsätze für den Einsatz der Luftstreitkräfte'* ('Basics for Air Corps Operations') under the headings of Reconnaissance Aviation, Fighter Aviation, Bomber Aviation and Communications, Ground Organization and Meteorological Services. Mittelberger made no major objections and his victory over the conservatives was completed on 24 February 1932 when Hammerstein-Equord stated that the new air force would be a separate service equal in status to the Army and the Navy. Nevertheless, during 1932 *T1* and *T2* continued their rearguard action and denied Mittelberger total control over air arm equipment. Aircraft development remained the prerogative of the *Heereswaffenamt*, the Communications Inspectorate (*In 7*) controlled airborne radios and AA artillery was divided between the Infantry (*In 2*) and Artillery (*In 4*) Inspectorates.

Simultaneously, Mittelberger was fighting the conservatives over the force structure. As part of its post-Versailles peacetime army planning, in 1930 the General Staff anticipated a 22-*Staffeln* force, of which thirteen *Staffeln* (59 per cent) were reconnaissance, six (27 per cent) were fighters and the remainder were heavy night bombers. Even *T1* recognized that the plan was reactionary and on 15 April 1930 reduced the requirement to eighteen *Staffeln* but doubled the bomber strength to six *Staffeln* (33 per cent of the total). Although the bombers probably had a secondary reconnaissance role, the proposal reflected a growing recognition within the *RWM* of the strategic implications of the bomber following wartime experience. Yet hopes of establishing a bomber force were restrained by growing public demands for disarmament following the Kellogg–Briand Pact of August 1928 in which all parties, including Germany, renounced aggressive war. Disarmament was popularly seen as a way of cutting expenditure, and worldwide moves for a total ban on bombers were incorporated in a proposal at the World Disarmament Conference which opened at Geneva in February 1932 (Blomberg, accompanied by Brandenburg, led the German delegation). The German Government was anxious not to be seen building a bomber fleet for fear of precipitating an aerial arms race which Germany could not hope to win.

For this reason, perhaps, Felmy wrote to Heye on 19 May 1930 demanding the implementation of the original 22-*Staffeln* plan between 1933 and 1937. In October 1930 Heye was replaced by Hammerstein-Equord (Adam became Chief of the General Staff), who accepted the Felmy proposal which became part of the post-Versailles army plan with Groener's support. But in June 1932 Groener was manoeuvred by Schleicher into a position whereby he had to resign after suppressing the Nazi stormtroopers. Schleicher, a personal friend of Hammerstein-Equord, was a compulsive intriguer (his apartment was used in the first *Reichswehr*–Russian negotiations) who had been the power behind the scenes for some time and now became the new Defence Minister.

On 11 July 1932 Adam provided him with a review of plans and sought permission both to execute the 22-*Staffeln* programme and to raise seven reserve air training units (140 aircraft and 4,600 officers and men) by 1936. Hammerstein-Equord's expansion plan was suitably amended and rubber-stamped by Schleicher when he met Adam again on Bastille Day (14 July) 1932, the decision also being taken to create AA (*Flak*) batteries by 1 October 1933. The General Staff published the fundamental principles (*Grundlage*) and plans for the New Peacetime Army (*Neue Friedensheer*) the following day. In addition to a military aviation arm (*Fliegertruppe*) there were to be 22 *Flak* batteries for field use plus others for home defence, the latter to be created only in the 1935–36 financial year.[11] Detailed planning for the *Luftstreitkräfte des Neuen Friedensheeres* ('Air Corps of the New Peacetime Army') was conducted by Mittelberger and Felmy, who aimed to begin work in the 1933–34 financial year. Their plans reflected wartime experience with operations under a *Kommandeur der Flieger* and training under a *Kommandeur der Fliegerschulen*, while administration and technical development would be the responsibility of departments directly under the *Heeresleitung*. There was a recognition of growing air strength with fighter, bomber and reconnaissance *Staffeln* organized into regional commands (*Fliegergruppen-kommando*) based at Königsberg (East), Berlin (Centre) and Nuremberg (South). Some of the *Fernauflärungsstaffeln* also had a secondary bomber role, while it was planned to supplement the bomber squadrons with converted Lufthansa airliners. The first-line strength of the new force would be 150 aircraft with 50 reserves, while the schools would have 62 trainers with eighteen reserves, giving a total strength of 280. The training organization would be created first, with the *Staffeln* following from the 1934–35 financial year, a third of them fighter units.

Both Mittelberger and Felmy advocated the development of a bomber force to disrupt the enemy's infrastructure and neutralize his bombers. The development of a four-engine heavy bomber, the Do P, was begun to match the French LeO 206 and Farman F.221/222, but both men recognized that twin-engine aircraft were a better industrial and strategic option. There was some opposition to the inclusion of bombers in the plans because of the World Disarmament Conference, but Adam accepted that they were a key element in air power. To avoid upsetting political sensitivities, the bomber *Staffeln* were officially included among the reconnaissance forces. With this proviso Adam approved the plans on 10 August 1932.

Continued concern about an autonomous air force and the problems of manning it led to renewed opposition from *T1* and *T2*. They were especially worried about providing some 2,000 technically trained officers and NCOs (including 320 pilots and observers) and proposed increasing the reservist and civilian element, offering as a sop a force of 23 *Staffeln*. This would include a deterrent of nine bomber units (based upon civil aircraft), but the number of active reconnaissance *Staffeln* would be slashed from thirteen to eight (which would be doubled on mobilization). Mittelberger and Felmy were vigorously opposing these plans when Hitler came to power on 30

January 1933, three days after Hammerstein-Equord gave the training of air and groundcrews for the new air arm absolute priority. The existing training organization provided the framework, with refresher training for experienced civil airmen conducted from 1932 by the paramilitary veterans' organization *Stahlhelm*, which established the *Wehrflugorganization* (Armed Forces Flying Organization) under Wilberg after he retired from the Army.

Even before the Bastille Day meeting, Felmy was considering air force requirements at the end of the decade. On 1 February 1932 he called for the quadrupling of strength to 80 *Staffeln* (720 aircraft) from the financial year 1938–39. More than half (42 *Staffeln*) would form a 300-bomber strike force under GHQ control, while 31 per cent (25 *Staffeln*) would provide tactical air support. Yet *Luftstreitkräfte* experience continued to constrain vision and there was no attempt to organize the bombers or fighters into large strike formations such as the French used in 1918. The reserve strength was respectable at 25 per cent of establishment, but only 96 trainers were sought, perhaps because Felmy anticipated exploiting the DVS organization. Felmy's emphasis upon bombers and his neglect of strategic air defence defined the *Luftwaffe*'s long-term planning, while his demand for a force of 1,056 aircraft influenced Milch's early production plans. Although the 80-*Staffeln* plan was beyond the capability of German industry, it did spur development for in the summer of 1932 outline military requirements for five Army and Navy 'next-generation' aircraft were issued and they included one for a heavy bomber with a range of 2,500km and a bomb load of 1,000kg. More significantly, there was a requirement for a twin-engine, high-performance aircraft for both military and civil applications, a requirement which was to be the genesis of the next generation of *Luftwaffe* bombers.

But vaulting ambition was humbled by financial and industrial problems of which the Army had been previously unaware. Rocketing unemployment and collapsing output following the end of US investment forced Berlin, and every other government, to slash its budget from 1929 onwards and neither national defence nor civil aviation escaped the pruners' shears. Catastrophe was inescapable, for the failure of Austrian banks in 1931 had a 'domino' effect upon the German financial institutions with which they possessed close business links. One failure was of the Darmstädter und Nationalbank (Danatbank), whose Berlin headquarters were to be the first home of the *Reichsluftministerium* (*RLM*).

The *RWM* was already transferring observer training to Germany, and the need to economize accelerated the decline of Lipetsk and the other Russian aviation facilities, which were costing some RM3 million ($714,000) a year. In 1930 the Germans ceased using the Voronezh firing ranges and, with the qualification of the following year's observer class at Lipetsk, Berlin informed Moscow that it intended to concentrate all such training in Germany. Fighter training continued until mid-1933, but Wivupal's primary goal was to evaluate designs which would be developed into the new air force's first generation of combat aircraft. For this task it had some 200 specialists, technicians, engineers and officers—twice the usual number—in 1931.

Meanwhile the *RWM*, anxious to proceed with the establishment of an air force, turned covetous eyes upon the Transport Ministry's coffers. By 1931 the German armed forces were spending RM15 million ($3.57 million) on military aviation, including industrial support, aircraft evaluation and aircraft acquisition, of which RM7 million ($1.67 million) went on training. By contrast the civil aviation budget was RM18 million ($4.28 million), including several million siphoned into military projects. The total expenditure on aviation in the period 1929–32 was RM175 million ($41.67 million), including RM13.25 million ($3.15 million) for aircraft development. Despite, or because of, Brandenburg's generosity, the Army sought to grab his aviation budget. Mittelberger bemoaned the lack of funds in a memorandum of 11 November 1932, pointing out that his RM10 million ($1.80 million) appropriation would not support an expansion of the air arm, especially in respect of the provision of modern machine guns and bombs.

He had opened his campaign after the Bastille Day meeting by proposing on 15 July that Germany emulate Britain and France by establishing an Air Ministry controlling both military and civil aviation. Within a fortnight Schleicher provided him with valuable ammunition when he stated in a broadcast, 'In view of the serious financial situation we should get the greatest value out of every penny spent on the *Reichswehr* for defence.' Quoting these words, *Hauptmann* Hans Jeschonnek, who had 'flown' Mittelberger's Tactics and Training desk (*Referat I*) since October 1931 and was a younger brother of Paul Jeschonnek, advocated an Air Ministry in an August 1932 memorandum. It would, he argued, provide a means of pursuing military programmes under civilian guise while replacing with a central agency the direction of aircraft development and procurement currently split between the two Services and Brandenburg. His memorandum, endorsed by Mittelberger, proposed that either Brandenburg's department come under *RWM* control or that it remain nominally under the Transport Ministry and control military and naval aviation under *RWM* authority.

Jeschonnek's memorandum planted the seeds which grew into the *Reichsluftwaffe* and the *RLM*, for on October 19 the General Staff's *T2* department not only supported Jeschonnek but went further. On October 19 it proposed merging Army and Navy aviation activities under a joint inspectorate. The only objections to both proposals came from the Head of Training (and the *Luftwaffe*'s *de facto* Chief of Staff from 1933), *Oberstleutnant* Walther Wever, who described them as nonsense, but his views were ignored and on 28 October 1932 the memorandum was discussed by the General Staff. Whatever reservations were felt about an autonomous air force, all agreed that there was a need for the central direction of aviation. It was decided to recommend to the Cabinet the creation within the *RWM* of a central Aviation Directorate (*Luftfahramt*), responsible for military and civil aviation as well as the Peacetime Air Force. It was envisaged as having two divisions, Air Defence and Civil Aviation, with the former including a joint-Service inspectorate.

The Navy does not appear to have been consulted, and its approval was only sought by the General Staff upon Adam' instructions on 8 November. Meanwhile Raeder presented Schleicher with a naval reconstruction programme which included a Naval Air Force (*Marineluftwaffe*) of nine *Staffeln* to be created from 1934. *Gruppe LS* had come under *Konteradmiral* Brutzer's *Marinekommandoamt* in 1931 and Brutzer saw the advantages of joint action, no doubt hoping to exploit the Army's aviation budget. Although Organization and Air Defence desks had been added to *Gruppe LS*, the total budget was only RM7.3 million ($1.67 million). For this reason the *Marinekommando-*

amt hastily rubber-stamped the Army's plan and set in motion the slow eclipse of German naval aviation.

It is a measure of the *Marinekommandoamt*'s interest in aviation that a wartime U-boat ace, *Fregattenkapitän* Ralf Wenniger, relieved Zander on 1 October 1932. Wenniger had sunk 106 ships (totalling 132,728grt) to win the *Pour le Mérite*, which his father also won in the Great War. The younger Wenniger survived the loss of his submarine and joined the post-war Navy, in which he served as First Officer in the cruiser *Berlin* before becoming Zander's Chief of Staff on 1 October 1931. His appointment reflected the Service ethos that officers should be capable of performing any task, and he appears to have been selected for reasons of seniority rather than expertise. His only previous experience with aviation was a chance wartime meeting with Udet in Munich.[12]

Transport Ministry opinion regarding the Army's proposals is not recorded, but they are unlikely to have been received with equanimity. In the event, the plans had support in high places, for in December 1932 Schleicher briefly became Chancellor and paved the way for Hitler's appointment as his successor. It was left to fellow intriguer Blomberg, long an air power enthusiast, to implement the centralization of military and civil aviation after he became Defence Minister in January 1933, but the execution of the plan was to surprise his service colleagues.[13] The crisis facing the aero-industry was equally unresolved at the time the Nazis came to power; indeed, it was only slowly recognized by the Army. In planning for the future air arm, the Army anticipated in 1926 receiving 173 aircraft between 1927 and 1932 and 229 over the next five years, many of them converted civilian aircraft. Both the General Staff and the *Heeresleitung* anticipated that in the 22-*Staffeln* force the percentage of dedicated military designs would gradually increase during the 1930s, and it was expected that these would include heavy bombers.

The *Heereswaffenamt*'s first survey of industrial performance in 1928 predicted a theoretical capacity of 7,006 aircraft for the Army and another 1,746 for the Navy by 1929; by 1933 annual production would be 3,043 aircraft, including 750 for the Navy. How these figures were calculated is a mystery, for they bore no relation to reality and can be most charitably described as optimistic. By the autumn of 1931 German industry had, since the mid-1920s, produced only 70 military aircraft, most of them prototypes.[14] Even two years later the total produced was only about 180, including 75 from Heinkel and 40 from Junkers.[15]

The *Heereswaffenamt*'s first doubts occurred in March 1930 when it considered plans to equip the 22-*Staffeln* force and discovered a production shortfall of 120 aircraft. It concluded that wartime industry would take six months to increase monthly production to 300 aircraft for both Services and this could not meet the anticipated 50 per cent loss rate. Two years later, on 4 April 1932, Mittelberger completed a devastating industrial survey for the General Staff at the start of the 1932–33 financial year. He warned that the industrial base, with only 3,800 people in seven airframe and four aero-engine manufacturing concerns, could not meet the minimum mobilization plans. Only Junkers and Heinkel could undertake limited series production; the other firms were little more than cottage industries. The dream of 300 aircraft a month vanished and a bewildered member of *T1* wrote, 'This situation is totally incomprehensible. What is happening?'[16]

Even Junkers and Heinkel provided no cause for optimism: the former had failed to produce any designs which met military requirements while the latter were producing in 1932 only nineteen military and naval aircraft (He 38, He 45, He 46, He 51, He 59, He 63), including trainers. With limited equipment and small stocks of material, even these companies could undertake series production only in batches of six aircraft per month. Although production could be increased by doubling the work force and introducing double shifts, Mittelberger warned that even after nine months the maximum monthly production rate would still be only 100 single-engine aircraft. Reviewing activity in 1932, he said that Arado was producing eight Ar 64/ Ar 65 fighters for the Army and would produce another seventeen by December, Focke Wulf (which had taken over the Albatros facilities) had produced some prototypes and would build thirteen Ar 64s and He 45s under licence, while Dornier had produced only prototypes.

Mittelberger commented on the lack of suitable engines and warned that there was enough fuel only for three months of operations. Although money had been allocated in February 1931 to develop radios, cameras and weapons, little of the new equipment was in production. Industry's problems were aggravated by the icy blast of the Depression, which finished off several companies or brought them to death's door, with Rohrbach and Albatros the most notable victims. Junkers' chronic financial difficulties grew deeper following another dispute with the *RWM* when Sachsenberg informed Volkmann that his company could afford to produce only one type of aircraft for the Army and, in his opinion, it should be the all-metal bomber-transport. The affronted Ministry promptly ceased supporting the country's largest aircraft and engine manufacturer, although a few aircraft were later purchased for evaluation. Dornier, too, had severe problems through the suspension of reparations orders, the cost of building the Do X flying boat and supporting route-proving flights across the ocean. A RM700,000 ($188,000) Transport Ministry subsidy kept Dornier in business and a similar solution prevented the collapse of Junkers, but the industry was living on a knife-edge.

The engine problem was acute, partly because the *RWM* gave priority to Army motorization and were supported by vehicle engine manufacturers who could find commercial applications for their investment. Few were willing to invest in aero-engines, which forced the *RWM* to rely on a handful of small companies dependent upon foreign designs and imported materials, with little spare capacity and no capability for mass production. The Depression forced severe rationalization upon Junkers, BMW and Daimler-Benz in order that they survive—indeed, as already noted, Daimler-Benz abandoned aviation in favour of automobiles. Only the 'mighty midget', Argus, kept going, with a series of light, air-cooled engines for the lower end of the market, and even here Transport Ministry contracts proved necessary. Mittelberger estimated that, even in ideal conditions, with an adequate flow of raw materials and a double working shift for six months, the industry could produce only 160 engines a month, of which 100 would be suitable for combat aircraft.

The Transport Ministry addressed the requirement for high-power engines in the autumn of 1930 with a request for a 1,000hp (30-litre) class liquid-cooled engine with a high power-to-weight ratio. This ultimately led to the Daimler-Benz DB 600 for the Bf 109 and the He 111 in the mid-1930s, but for its first generation of combat aircraft the *Luftwaffe* had to rely upon engines of 750hp or less, including the licence-built Pratt & Whitney Hornet (as the BMW 132) and Bristol Jupiter (Siemens SH 22, later redesignated SAM 22B) or the domestic BMW VI. The *RWM* rejected on security grounds Transport Ministry suggestions to use imported engines in the first-generation combat aircraft. However, small numbers were purchased from 1933 onwards for prototypes, including the Rolls-Royce Kestrel VI for the Ar 67 and Ar 80 and the Kestrel V for the Bf 109, He 112 and Ju 87, as well as the Rolls-Royce Buzzard for the He 118. As the Army became aware of the problem, the *Heereswaffenamt* tried from 1932 to stimulate the aero-engine industry by placing its first major orders for military engines, including twenty SH 22B-2s for the Do 11 and twenty BMW VI 7 3Zs for the He 51 while encouraging both new engines and improvements of existing designs.

The process of selecting Germany's first generation of combat aircraft, based on requirements issued in the late 1920s, continued until 1933 at Lipetsk. In 1930 a selection of promising prototypes arrived in Russia, Arado producing three single-seat fighters, the SD I, SD II and SD III, to compete against the Heinkel HD 38 (later He 38). For the Najaku requirement Junkers provided a couple of two-seat K 47 fighters, derived from the A 48, which proved of greater worth in evaluating the concept of dive bombing. Its rivals were the Dornier C I (later redesignated Do 10) and Albatros L 84. As Felmy commented, Junkers aircraft were always too expensive and the K 47 was eliminated. The HD 38 and SD II/III showed promise, the former ultimately evolving into the He 51 and the latter into the SD IV (Ar 64) and SD V (Ar 65), which all received production contracts; the Do 10 became the Do 22 for the export market. At Lipetsk the SD V/Ar 65 continued fighter-bomber experiments begun in 1931 with Fokker D XIIIs and carried six 10kg bombs.

Two Heinkel reconnaissance designs beat Focke-Wulf to win Army orders: the He 45 biplane (a derivative of the HD 41) beat the S 39 for the *Fernaufklärung* role, while for corps (*Nahaufklärung*) duties the Heinkel HD 46 (later He 46) parasol-wing monoplane beat the S 40. Some bombers were also tested at Lipetsk, including military versions of the Junkers W 34 (K 43) light bomber, but attention focused upon two Dornier designs. The Do P heavy bomber, a land-based version of the *Superwal*, lacked speed and range but the twin-engine Do F, ostensibly developed as a freight carrier, was selected as the Do 11 to be the *Luftwaffe*'s first bomber.

These trials, completed in the summer of 1933, were Lipetsk's swansong. With Left and Right battling in Germany's streets, the *RWM* was becoming more cautious about exposing the capabilities of Germany's new military aircraft to a country increasingly regarded as a potential enemy rather than ally. The level of activity at Lipetsk declined perceptibly during 1932, and in October of that year the Russians sought a showdown: they demanded that the Germans use the base as fully as before and return all the military aircraft which had been evaluated at Lipetsk but then flown to Germany.

Although Felmy proposed on 28 October that Lipetsk be closed, the Germans rejected the demand and in 1932 maintained high-level contacts with the Russians. Mittelberger met Alksnis, now heading the Chief Directorate of the *VVS* (*GU-VVS*), while Wimmer (who replaced Volkmann as Head of Technical Development in 1929) met Alksnis' former Chief of Staff Sergei Mezhenikov. Among the topics discussed was German aero-engine development—the Russians, too, were having difficulty producing high-performance engines. Following these meetings, licensed production of the BMW VI began in the Soviet Union as the M-17.

With German aircraft evaluation completed at Lipetsk in 1933 and the Nazis in power, the base's role was clearly over—and, in any case, the Germans could now use Rechlin. From the late summer they began to deactivate their base and final negotiations were conducted about its facilities and equipment. It was agreed that most of the movable equipment would go to Germany and the Russians would retain the permanent fixtures and some 30 Fokker D XIIIs which could not be flown back. In September 1933 the base was deactivated, the Russians apparently expressing genuine regret at the Germans' departure.[17] The break was symbolically emphasised by the accidental death in an air crash of Baranov, who had been so intimately involved in setting up Lipetsk. He had headed the aircraft industry's Chief Directorate since June 1931, when Alksnis replaced him as head of GU-VVS, and he was killed in a transport prototype.

Some naval armament tests had also been conducted at Lipetsk, seaplanes such as the He 59a and Focke Wulf W 7 replacing their floats with undercarriages, while the Navy was sufficiently impressed with the He 38 to order a floatplane version (He 38c) in 1932. However, Warnemünde remained the centre of floatplane development, and designs evaluated in the early 1930s included the catapult-launched Heinkel HD 30 single-engine reconnaissance biplanes (ostensibly ambulance aircraft), the He 59b and the Junkers A 48. The He 59 was ordered in 1932 to supplement the small flotilla of HE 5/HE 9 floatplanes, essentially wartime designs, and Dornier J II *Wal* flying boats. Allied Intelligence was aware of the activities at Warnemünde, where discretion was occasionally thrown to the winds. For example, on 5 August 1930 Spanish naval officers saw naval aircraft, including the HD/He 38 and HD 30, during a visit in which they were informed that Heinkel had begun the construction of a torpedo bomber (although it is unclear whether this referred to the HD 16L/W landplane/floatplane—two of which were evaluated for the Swedish Navy—or the He 59).

Overseas armament trials were not confined to Lipetsk and orders were placed for machine guns, bomb-release gear and gun rings developed in Sweden and Switzerland with the aid of a Junkers W 33 and a Dornier *Wal* respectively. Airborne minelaying trials began in 1931, and from 1932 a Swedish-registered single-engine Ju 52ce conducted torpedo-bombing experiments in Oslo Fjord using torpedoes from the Norwegian Navy's Horten Arsenal. Following modification, these were adopted by the German Navy as the F5.

The economic crisis which slowed German progress during the early 1930s was triggered by an ailing Austrian economy which had lost virtually all its pre-war markets and sources of raw material. Efforts to resolve the problem through economic union with Germany in March 1931 were thwarted by international opposition, and soon Austrian banks began to fail. The plague of economic collapse swept across Europe but afforded Germany the opportunity to end reparations unilaterally.

This occurred just as Austria began to re-form her air arm, for the severe restrictions on aviation which Vienna also faced during the 1920s were lifted by a separate agreement signed in Paris on 27 October 1927. This permitted a maximum of a dozen soldiers to become pilots and hold civil licences until 1933—which provided the loophole the *Bundesministerium für Heerwesen* (*BMfHW*) sought. The new air arm was shaped by *Oberst* Alexander Löhr, who was 42 years of age at the time of the second Paris Agreement. He had been born in Romania, the son of a Croatian river captain and a Russian mother, and his appearance displayed a Tartar streak. He had joined the *KuK* Army, become a staff officer, then transferred to the *Abteilung Luftfahrwesen* in 1915. During the mid-1920s he had been placed in charge of *Abteilung L* and on 31 January he had established a school at Graz-Thalerhof with the assistance of the national airline ÖLAG (who were, ostensibly, the owners).[18]

The first Austrian servicemen were actually trained as instructors in Switzerland and Germany, where a few enrolled with DVS in 1929. In 1930 cadets entering the Enns Army School were offered the opportunity to transfer to Thalerhof for a year's flying training, and the first course began during the year. Instructors and students alike wore 'mufti' and the school used three vintage Brandenburger C.Is, later supplemented by a Hopfner HS 8-29 and a Phönix L2c. Cash shortages plagued the school, which received only half of its öS500,000 ($70,000) allocation and had to close from June to mid-September. However, Chancellor Johann Schober did authorize the expenditure of öS25,000 ($3,500) for another three Hopfner trainers, and by the end of the year Löhr had five pilots and six observers on his strength. Arrangements were also made to purchase ten Udet U 12 *Flamingos* and another five Hopfners.

Over the next two years Löhr gradually increased his pool of airmen, with 24 qualified and student aircrew in December 1931 and 46 a year later. In 1930 Yllam, now a *Major*, formed with a Junkers A 35b an operational unit which participated in an artillery shoot at Kaisersteinbruch in June 1931, while during the year the first two of six Fiat-Ansaldo A 120 parasol-wing corps aircraft were delivered. With the remainder *Gruppe Yllam* was able to fly 105 sorties in 1932, despite financial problems having grounded two of the Fiats by the end of the year.

The Fiats were purchased by *BMfHW* from Italy, whose pre-war enmity of Austria was converted to post-war patronage, the Italian fascist dictator Benito Mussolini regarding himself as Austria's protector. Yet links with Germany continued: three DVS instructors helped to start the Thalerhof courses, in 1931 Sachsenberg gave lectures on air warfare and in July 1931 Wimmer paid an official visit to Austria for unknown reasons. By the end of 1932 Austria possessed an infant air arm supplemented by a *Polizei-Fliegerstaffel* of a dozen old aircraft, one of them a wartime fighter. The police had four pilots trained at Thalerhof from May 1930. They usually monitored foreign air traffic, for at the end of 1931 the Austrian register showed only ten passenger and 49 light aircraft, together with 78 pilots.

The selection of corps aircraft for the Austrians' first squadrons was natural as they had been the foundation of air power since trench warfare rendered traditional means of reconnaisance impotent. Since all aircraft co-operated with the Army, the British designation describes their function exactly as they provided tactical (i.e. corps-level) air support for ground troops. Primarily they were used for reconnaissance (visual and photographic) and artillery support (observation and direction), but they were in truth maids-of-all-work, conducting ground attacks, some night bombing and even supply drops. They carried two crewmen, a camera, radio, bombs and flares, while their armament consisted of a fixed, forward-firing machine gun with another one or two on a flexible Scarff-ring type mounting in the rear for the observer. The heavy loads made these single-engine aircraft slow and the biplane configuration was commonly used because the double wing surfaces provided greater lift. However, from the late 1920s parasol-wing monoplanes such as the RWD-14 *Czapla*, the ANF Les Mureaux 113/115/117, the Potez 39 and the Westland Lysander, as well as sesquiplanes such as the Breguet Bre 27, were also used.

Because corps aircraft were vital to Army operations, the fighter evolved as a means of blinding the enemy and became the other mainstay of traditional air power. Fighters, too, were biplanes because the double wing surfaces provided a strong structure, a good climb rate and excellent manoeuvrability. These features aided the interception of corps aircraft from standing patrols or even airfields and allowed second attacks to be mounted rapidly. The armament was normally a pair of rifle-calibre (6.5–8mm) machine guns, because wartime heavy (12.7mm) machine guns and cannon were either unreliable or had too low a firing rate.

The bomber actually preceded the fighter and evolved from the corps aircraft; indeed, in 1918 the backbone of both bomber and corps squadrons in French service was the Bre 14. Although resembling corps aircraft, the conventional light bomber flew faster, higher and further to carry a typical 200kg load externally, although it was also used for long-range photo-reconnaissance. The German Army in the Great War took a different route to meeting both requirements. In the latter years operational-level and strategic reconnaissance was assigned to specialized, high-altitude two-seaters such as the Rumpler C.VII equipped with electrically driven film-strip cameras (*Reihenbildner*). Such aircraft pioneered Second World War strategic photo-graphic reconnaissance, with aircraft virtually immune to interception and taking photographs of great swathes of territory with each sortie. For bombing, airships were originally used until they proved too vulnerable to the elements and enemy fire. Corps aircraft were then pressed into the role, but from 1916 they were diverted first into escort duties and then into close air support. In this role they were organized from 1918 into *Schlasta* and used faster and more manoeuvrable CL class aircraft which relied upon machine guns. Multi-engine G and R class heavy bombers, carrying up to a tonne of bombs internally and protected by two or three machine gun positions, replaced the airship from 1916 but the Allies were slower to follow owing to their inferior industrial base.

The speed and defensive firepower of wartime bomber formations made them difficult for fighters to break up until 1918, when both the British and Germans introduced crude, but increasingly effective, visual ground-controlled interception techniques. Until then interception was largely a matter of chance, and fighters consumed most of their limited fuel supply merely establishing and maintaining contact, leaving little for prolonged engagements. With only hand-signal communi-cations, fighter formations could attack only piecemeal in the face of concentrated defensive firepower. The fighters' machine guns could inflict little damage upon multi-engine bombers in particular, and an aerial victory over a bomber generally remained unlikely. The only way of breaking up bomber formations was by means of anti-aircraft fire, which involved a prodigious expenditure of ammunition—5,000 to 8,000 rounds for each 'kill', according to British calculations. Improved AA and fighter defences were steadily increasing strategic bomber loss rates by the end of the war, but this was not generally appreciated and images such as Brandenburg's Gothas cruising unimpeded over London in 1917 dominated the military and civil consciousness.

Post-war views of air power were shaped by the characteristics of these aircraft and also of two long-range reconnaissance aircraft, the British Bristol F.2B Fighter ('Brisfit') and French Caudron R.11. Both were designed to fight their way to operational-level objectives using radically different design philosophies. The Brisfit was a two-seat, single-engine aircraft with a fighter's performance and the advantage of a rear gun position. It was also used for long-range sweeps over the front, while its endurance, superior to that of single-seat fighters, was exploited for home defence patrols. The twin-engine Caudron shared the heavy bomber's defensive firepower, with five machine guns in two positions (one fired downwards). From the summer of 1918, at the request of *Commandant* Joseph Vuillemin and with trained gunners, these Caudrons escorted bomber formations, which they shielded with their formidable firepower.

Multi-role aircraft proved to be a fatal fascination in the post-war world where the theorist *General* Giulio Douhet regarded them as his ideal warplanes. The British used the Hawker Hart light bomber as the basis for the Demon two-seat (night) fighter and Hardy corps aircraft, while there were many attempts to produce high-performance multi-role aircraft which could emulate the 'Brisfit' and Caudron. These included single-engine fighter/ground attack designs such as the American Consolidated P-30 (PB-2)/A-11 and the Russian Kotscherigin DI-6 as well as the multi-engine Tupolev R-6/KR-6 reconnaissance/escort fighter. But it was the French who embraced the multi-role concept most ardently. Their *BCR (Bombardement, Combate et Reconnaissance)* specification of 1928 led to the Amiot 143 and Bloch 200/210 of the 1930s, but the compromises in meeting conflicting requirements produced designs which were as ugly as they were ineffective. They were incapable of fighting or running, but they remained the backbone of the French bomber force in 1939. The *Luftwaffe*'s planners also dabbled with multi-role combat aircraft but quickly opted for specialization, as did Junkers' former protégé Tupolev, whose ANT-21/29/39 multi-role designs led to the elegant ANT-40 bomber and the *VVS-RKKA*'s SB-2M-100 or SB.

The move from multi-role to specialized designs in the 1930s was an outward expression of a design revolution which benefited first the bomber and then the fighter. It was the culmination of developments in both airframe and engine design whose impact upon air warfare may be likened to that of HMS *Dreadnought* on naval warfare 25 years earlier. The wartime design concepts were rendered obsolete, the most prominent victim being the corps aircraft, which could neither outrun nor outfight the new fighters.

Airframe changes were the most obvious aspects of a revolution fuelled literally and figuratively by aero-engine developments. Duralumin as well as alloys based upon nickel, chromium, manganese and tungsten were increasingly incorporated to produce lighter engines with greater power and efficiency, which rapidly approached the British target of one horsepower for every pound of engine weight. At the end of the First World War 150–200hp engines were the norm, but by the early 1930s

500–750hp engines were common while 870–880hp engines were under development.

American-developed fuel additives enhanced engine performance, with tetraethyl lead suppressing detonation and ensuring smoother operation. From 1925 iso-octanes (hydrocarbons of paraffin) were added to petrol to increase efficiency (which was thereby measured in octanes—the higher the octane rating, the more efficient the fuel). By 1943 the octane rating of fuel for primary trainers was 73 and for advanced trainers (and combat aircraft of the 1930s) 80–96, while modern combat aircraft used 100 octane and above. More efficient fuels encouraged improvements in super-chargers—compressors which increase the density of air or air/fuel mixture—to boost power, especially at high altitude.

However, drag inhibited performance, especially with biplanes, in which the pairs of wings created airflow problems known as 'biplane interference'. Only the mono-plane could exploit the new technology fully by providing greater opportunities for streamlining. This also drove the introduction of retractable undercarriages and the enclosure of all crew spaces, including the cockpit, which improved ergonomic performance. The open gun mount was replaced by a gun turret, manually then hydraulically operated, to provide a steady, rapid-reaction gun platform and elimi-nate the need for escort fighters. The French were especially enthusiastic about turrets—and their designs were adopted by the British—but the Germans rejected them because their weight inhibited aircraft performance. The British and Americans believed that turrets enhanced bomber superiority and would allow formations to roam enemy air space with impunity—until the *Luftwaffe* demonstrated otherwise. The RAF reacted by turning to safer night operations, but the USAAF initially revived the bomber-escort concept with the Boeing XB-40/YB-40 before turning to long-range fighters.

The pioneering work on the new bombers was performed in the United States as increased industrial investment in the late 1920s was rapidly exploited despite the Depression. The Boeing Model 247 (later Y1B-9) of 1931 was soon followed by the Martin 123 (later XB-907), which could fly at 207mph (333kph), the same speed as that of the fastest in-service fighter, the Hawker Fury I. From the 123 emerged the Martin 139, which was ordered by the USAAC in January 1933 as the B-10/B-12. Other countries followed the American precedent. The German outline requirement was drafted in July 1932 and a detailed one appeared two years later, the first of the new bombers, the He 111, entering service in 1936. The SB, its Russian equivalent, entered service about the same time, a year ahead of the RAF's Bristol Blenheim.

The 1920s were a period when bombers steadily eroded the fighter's traditional advantage in speed. The new technologies of the 1930s accelerated this trend to give light and medium bombers virtually a fighter's performance.[19] Even if fighters could intercept the bombers, they could make only one or two attacks and they lacked the firepower to inflict more than token damage. Moreover, the new bombers were capable of operating at ever greater heights—up to 25,000–30,000ft (7,500–9,200m), where biplanes were useless. On 12 March 1932 Felmy noted that fighters no longer seemed to be an effective defence against the bomber.

Throughout the 1930s there existed great enthusiasm for the bomber, and designs such as the Ju 88 *Schnellbomber* ('fast bomber') were sought which could outpace the fighter. Yet official conservatism usually compromised these designs (a fourth crew member was added to the Ju 88), which failed to achieve their purpose. The exception was the De Havilland Mosquito, where official interest in a 'speed bomber' led to Specifications B.18/38 and B.1/40 being written around a private-venture design.

The new technologies enhanced not only speed but also range and load, making possible a new generation of powerful heavy bombers. Four-engine landplane airliners and bombers were largely a European phenomenon until the 1930s, France producing the Farman F.222 and Russia the Tupolev TB-3, but these were old-technology aircraft. Once again, the pioneering work in advanced heavy bombers was performed in the United States, following a USAAC requirement in August 1934. Boeing's response was the Model 299 of July 1935 with evaluation models (Y1B-17) delivered from March 1937. The latter could operate at altitudes up to 30,000ft (9,150m), with a maximum level speed of 256mph (412kph), and could carry a ton of bombs 1,500 miles (2,400km) and smaller loads more than 2,000 miles (3,200km). A turbocharged-engine version entered service from July 1939 as the B-17B with a heavier defensive armament and a maximum level speeds of 291mph (468kph). The *Luftwaffe*'s *de facto* Chief of Staff, *Oberst* Walter Wever, was quick to exploit the new technology for heavy bombers with the 'Ural Bomber' requirement, but he was not alone. By 1939 every major European air force had a modern four-engine heavy bomber design under development—the Short Stirling and Handley Page Halifax in Britain, the Bloch 162 B5 and CAO 700 B5 (also the Bloch 135 B4 and Breguet 482 B4 medium bombers) in France, the Petlyakov TB-7 (later Pe-8) in the Soviet Union and the Heinkel He 177 in Germany.

Surprisingly, air staffs proved extremely reluctant to abandon the traditional fighter design values of manoeuvrability, high climb rates and good cockpit visibility in the face of the challenge. They were universally pessimistic about the ability of fighters to destroy the new bombers and felt that the fighter should retain its established roles of neutralizing corps aircraft and conducting point defence. Yet even the latter demanded a radical change in fighter roles in order to destroy bombers, and this in turn demanded high performance and greatly increased firepower. Although the RAF and the *VVS-RKKA* retained biplanes such as the Gloster Gladiator and Polikarpov I-15/152, many air staffs compromised by seeking mono-plane designs based upon traditional values. Fighters such as the braced-wing Boeing P-26, Dewoitine D.371, D.500/501/510, Morane MS 225 and PZL P.7/11 or cantilever-wing Fokker D.XXI had low wing loadings, fixed undercarriages and a light armament. A similar philosophy was adopted not only by the Bf 109's unsuccessful competitors (even the He 112) but also by Supermarine in its initial response to the RAF's F.7/30 requirement which ultimately became the Spitfire.

Such compromise machines had neither the speed nor the endurance to intercept bombers. For example, they required a greater fuel capacity. In 1920 only six per cent of an aircraft's loaded weight was taken up by fuel, but powerful and thirsty engines increased this to as much as fourteen per cent, which in turn drove a demand for higher wing loadings that only the cantilever monoplane could provide. Prejudice continued to affect the air staffs, even where they were willing to exploit features such as cantilever wings, retractable undercarriages and improved firepower. Some adopted a twin-track approach with both conservative and radical designs.

Conservative designs such as the Morane MS 406 and Hawker Hurricane emphasized the traditional values of manoeuvrability and cockpit visibility—indeed, the Hurricane was originally a redesigned Fury. Some had time-honoured tube and fabric airframes while others, such as the Bloch 151/152, the Curtiss Hawk 75 (P-36) and the Polikarpov I-16 were of monocoque construction, the last using a wooden structure (although later versions, like the Hurricane, had stressed-skin wings). They usually had a relatively light armament, the Hurricane and later versions of the I-16 being the exceptions, but this was adequate for destroying bombers. They were complemented by radical designs such as the Spitfire and the Dewoitine D.520, which were characterized by their advanced construction and powerful armament, although they were less manoeuvrable than the conservative designs. However, they were faster and had a better high-altitude performance, while their firepower was more than a match for a bomber.

The steady growth in fighter firepower is shown in Table 5, and the increased use of the 20mm cannon is noteworthy. Such a weapon, rather than the machine gun, could inflict severe damage upon the bomber: *Luftwaffe* records show that an average of twenty such rounds could destroy even the heavily armed and protected B-17.[20]

TABLE 5: ARMAMENT OF 1930s FIGHTERS

Type	Rifle-calibre machine guns	Heavy machine guns	Cannon
Boeing P-26	2/1 × 7.62mm*	0/1 × 12.7mm**	–
Dewoitine D.371	4 × 7.5mm	–	–
Dewoitine D.500	2 × 7.5mm 2 × 7.7mm	–	–
Dewoitine D.501/510	2 × 7.5mm	–	1 × 20mm
Morane MS 255	2 × 7.5mm	–	–
PZL P.7	2 × 7.7mm	–	–
PZL P.11	2–4 × 7.7mm	–	–
Bloch 151	4 × 7.5mm	–	–
Bloch 152	2 × 7.5mm	–	2 × 20mm
Curtiss Hawk 75			
(H75A-1)	4 × 7.5mm	–	–
(H75A-2)	6 × 7.5mm	–	–
(P-36A/B)	1 × 7.62mm*	1 × 12.7mm**	–
Hawker Hurricane	8 × 7.62mm*	–	–
Morane MS 405	2 × 7.5mm	–	1 × 20mm
Polikarpov I-16			
(Types 1, 4, 5, 6)	2 × 7.62mm	–	–
(Types 10, 18)	4 × 7.62mm	–	–
(Types 17, 24)	2 × 7.62mm	–	2 × 20mm
Dewoitine D.520	4 × 7.5mm	–	1 × 20mm
Supermarine Spitfire	8 × 7.62mm*	–	–

*Actually 0.303in.   **Actually 0.5in.

Germany could not afford the luxury of a twin-track approach and for this reason the *Luftwaffe* selected the Bf 109, which was essentially a bomber-destroyer, over the He 112A, which sought vainly to combine advanced construction and traditional values. Early-production Bf 109s had a weak armament, a problem rectified in the Bf 109E—which demonstrated its superiority over conservative designs, especially at higher altitudes, from 1939 onwards.

Despite significant improvements in fighter performance during the 1930s, the bomber was at the heart of inter-war discussions concerning land-based air power. The question revolved around who would control the bomber, the generals, who wished to supplement their firepower, or the airmen, who saw in aerial bombardment a rapier to bring wars to a speedy if devastating end. Air power enthusiasts later portrayed the generals as reactionaries seeking to clip the airmen's wings, but in reality the disputes were about the interpretation of wartime experience.

Warfare is a constantly shifting balance of defence and offence, and during the Great War the former, based on the firepower of artillery, was in the ascendant. As the French 1921 and 1936 Instructions for Major Units stated, 'The attack is the fire which advances, the defence is the fire which halts', but this created serious tactical and insurmountable operational-level problems. The fortifications which shielded the defenders' artillery proved impossible to destroy, although by 1917 both sides knew how to overcome them. This involved secretly assembling an assault force in overwhelming strength at a vulnerable point, then neutralizing the defensive system with a short, hurricane-like bombardment. In Allied attacks the infantry would infiltrate the wreckage in spurts (French 1921 regulations recommended 2–5km), with pauses to bring forward artillery, until the enemy gun line was threatened with occupation. Air power was an essential element of this battering ram. Even before it was applied, fighters had to clear the skies for the corps aircraft to survey the defences and to conduct counter-battery operations. During the assault, corps aircraft would monitor progress, direct fire and provide some close air support, although the task was generally assigned by the Allies to fighters and by the Germans to *Schlasta*.

In 1918 the Allies often reached the gun line within a day, but post-war theorists took a more cautious view and believed in the 'doctrine of the limited offensive', by which the assaulting troops never left the shelter of their own artillery. They envisaged a series of blows which would stretch the defensive system to breaking point, forcing the enemy to abandon it in favour of a new one. The destruction of communications by bombardment shaped this view, for it made the exploitation of tactical success at operational level extremely difficult. As the attacker advanced he outran his supplies, while the defender could exploit an intact road and rail network to bring up reinforcements and establish a new defensive system. Consequently, after breaking through on the Somme, the Lys and the Chemin des Dames in 1918, the Germans were stopped within a week.

Yet the Germans were confident that they had the keys to both problems. Unlike the Allies, they delegated artillery to battle groups, which conducted the assault and maintained constant pressure. This artillery provided the firepower to exploit any momentary enemy weakness and maintain the momentum of the assault to, and through, the enemy gun line. Their 1921 regulations emphasized these tactics to prevent the enemy gaining time to construct a new defensive system. Although the mechanized element, which was the true solution to operational-level exploitation, was lacking until the late 1930s, the 1933 regulations were sufficiently confident of successful exploitation to consider encircling large enemy formations. While tactical air support was part of *Reichsheer* military thought, this potential also drove a requirement for greater operational-level air support.

The victors also foresaw such a requirement as a result of their experience on the Western Front; indeed, it was emphasized when *Général* Philippe Pétain took command of the French Army in May 1917. His Air Chief, *Colonel* Maurice Duval, replaced strategic with operational-level bombing to support the Army, and when Duval learned that the British planned to create the Independent Air Force for strategic attacks upon Germany, he commented, 'Independent of what? Of God?' Bombers had routinely interdicted communications and attacked supply dumps and reserves, but under Duval they did so in bigger formations, sometimes of up to 200 aircraft. As the *Kagohl* could operate only at night, the *Schlasta* were assigned the daylight operational-level role of interdicting road communications from 24 June 1918 in imitation of both the British and French.[21]

Wartime experience showed that air power was an essential tactical tool with operational-level capabilities and strategic pretensions. Post-war generals maintained their belief in the supremacy of massed artillery but recognized the value of bombers to strike beyond artillery range (this was one reason why the *Reichsheer* desired a bomber force) as well as striking enemy defences. To ensure the integration of air support, military commands were assigned their own squadrons while GHQ would retain a central strike force of fighters and bombers. The typical role of such air forces was outlined by the French Army's 1936 regulations, which stated that 'Air forces will reconnoitre, liaise, protect and fight' (Paragraph 50) and that 'Air reconnaissance for the artillery will take priority' (Paragraph 297).[22] Fighters and bombers were to be used to gain temporary air superiority by attacking the enemy air forces (Paragraph 298), while every aircraft was to be used to attack the enemy throughout his operational-level depth at the direction of GHQ (Paragraph 299). Clearly, the generals feared the loss of the tactical air support so vital for their operations and wished to retain firm control, even when aircraft were used in operational-level attacks. *General der Artillerie* Ludwig Beck (Chief of the German General Staff from 1935 to 1938) commented, after a wargame in 1938, 'Make sure the *Luftwaffe* doesn't fight an operational-level war somewhere in the depth of enemy territory while our infantry remains frozen in a war of attrition'.[23] The generals recognized the bomber's growing destructive power: for example, US Army wargames umpires in the late 1920s calculated that if an encamped division were surprised by three attack squadrons it would lose one per cent of its men and ten per cent of its mules and horses,[24] yet they believed that such effects would be momentary. They remained sceptical of air power's ability to affect events on the battlefield and failed to anticipate the cumulative effects of sustained air attacks upon an army attacked throughout its operational-level depth.

Their fear that autonomous air forces would conduct private wars against enemy air power or even purse the will-o'-the-wisp of strategic bombing was a very real one, as the Holy Day War of 1973 indicates. In 1927 RAF corps squadrons under Wing Commander Trafford Leigh-Mallory provided valuable support for the experimental mechanized force, but the Air Staff reprimanded the War Office for encouraging pilots to conduct low-level attacks during exercises. It was Wever's oft-stated commitment to supporting the Army and his demonstration that this was best achieved at the operational level which reconciled most German generals to an independent air force. Yet the Army still hankered for tactical air support, and many of its officers were transferred to the *Luftwaffe*. With the ageing of its strike force from 1942, the *Luftwaffe* increasingly reverted to a tactical role as an aerial fire brigade.

Wever did not ignore strategic bombing but he regarded it as part of the overall programme of supporting the Army. Similar views were held during the early 1930s by both French and Soviet air leaders, who had long-range bombers, although the Russians made the more substantial commitment by creating the *AON* in 1936. Although the *AON* was an operational-level as well as a strategic bombing force, the prophets approved of it as strengthening the case of air power as a war-winning weapon. They opposed the generals' dilution or fragmentation of air power which, they believed, could paralyse land movement, sink the most powerful warships and smash the heart of enemy power through strategic air attack. Similar prophesies were made before the Great War but only after Brandenburg's raids on London was the concept seriously considered by policy makers when General Jan Smuts investigated British air power. His enthusiasm led to the creation of the Independent Air Force and the RAF, but he never advocated the abandonment of traditional air support.

Nevertheless, the intoxicating vision of air power as the ultimate weapon was the theme of post-war crusaders. These included Brigadier-General William Mitchell of the US Army Air Service and Marshal of the Royal Air Force Sir Hugh Trenchard, although during the war the latter shared Duval's views. The voice of the movement belonged to the Italian Douhet and his writings, collectively published posthumously in 1930 as *The Command of the Air*. A gunner and administrator with close links to the Italian heavy bomber pioneer Gianni Caproni, Douhet originally proposed a bomber-based strike force to complement a separate air force for the traditional services. Criticism drove him to advocate air power as the sole means of modern warfare, using multi-role aircraft which, like the *BCR*, emphasized bombing. He had little air experience, so his dreams were untroubled by the nightmares of creating, supporting and operating a bomber fleet. Such ignorance did not excuse either Trenchard or Mitchell, especially the former, who had actually commanded a strategic bombing force. They and the other prophets ignored the well-documented problems of strategic air operations, exaggerated the vulnerability of civilian morale to such attacks and dismissed strategic air defence. Strangely, Douhet's last work, *The War of 19—*, was overlooked, yet it foresaw a Franco–German conflict won by the latter, whose aircraft first wrecked the French Air Force and then paralyzed troop movements—a scenario Wever himself followed in the *Reichswehr* wargames of 1934.

Mitchell and Trenchard appear to have espoused Douhetism independently of Douhet, whose views they learned more by hearsay than through translation. Although prophets rallied around Douhet's name, with English-language versions of some articles available in the mid-1920s, *The Command of the Air* was published abroad only from the mid-1930s and in Germany only in 1937. It underlined the great public fear of air attack, which was fed by writers who vied to produce apocalyptic visions of future war in which cities were instantly reduced to ashes by a rain of high-explosive, incendiary and gas bombs. In the days before radar the threat was serious—indeed, British air defences during the 1931 and 1932 exercises failed to intercept 25 per cent of attacks during the day and 40 per cent of those at night[25]—and consequently the only effective form of defence appeared to be deterrence based upon one's own bombers.

In his 12 March 1932 memorandum Felmy concluded that bombers were the only means of neutralizing air threats to Germany, while three weeks earlier, on 18 February, Wimmer (an advocate of strategic bombing) wrote that the future belonged to those nations which could create a powerful air fleet that could strike fear into the hearts of the enemy population by day or night.[26] Although German planning before Hitler did not neglect air defence, the emphasis was upon AA guns rather than fighters, and similar views were held by the *Luftwaffe*'s four major enemies in the Second World War.

Mitchell was Assistant Chief of the Air Service in the early 1920s, and his advocacy of an independent air force aroused considerable sympathy among his superiors if not in the War Department. His views automatically challenged the status of the US Navy as the country's first shield, especially when his bombers sank decommissioned warships in 1921 and 1923 (although these were less demonstrations of air power than acts of showmanship). Eventually his intemperate criticism of the Army and Navy led to his court-martial in 1925, and he resigned his commission in January 1926, six months before the USAAC was created.

The new Corps began a five-year expansion which nominally doubled its strength to 45 squadrons at a cost of $147 million. In practice the expansion faced severe economic restraints, but in 1931 the US Army Chief of Staff, General Douglas MacArthur (who attended Mitchell's court-martial), concluded an agreement with the Chief of Naval Operations which gave the USAAC responsibility for continental air and coastal defence. Although the Navy later reneged on the deal, the Army adopted the long-range reconnaissance role in 1933 and these decisions provided the impetus for the purchase of advanced bombers such as the B-10 and, ultimately, the B-17.

While the Corps was committed to supporting the Army, the Air Corps Tactical School increasingly advocated Douhetist views and by 1934 proposed that strategic bombing should be the USAAC's prime role. The Mitchell-influenced faculty was extremely potent, for it trained most of the unit commanders and provided the key forum for discussions about air power. The most prominent dissenter was the fighter instructor Captain (later Major) Claire L. Chennault, who believed in an effective defence based upon efficient detection and high-performance aircraft. Disgruntled at his failure to influence events and suffering ill health, Chennault resigned his commission in 1937 and successfully demonstrated his theories in China. Ironically, his friend and supporter Captain George C. Kenney later became the first head of the US Air Force's Strategic Air Command.[27]

Neither doctrine nor financial restrictions prevented the US Army from addressing and largely solving the human and mechanical problems of bad weather and night flying in the decade after 1925. By contrast the RAF, where Douhetism reigned supreme, was barely conscious of these problems, and while it, too, faced financial stringency, the fault lay with an Air Staff generally indifferent to technical and scientific matters. Operations were seen largely as the triumph of the human spirit over adversity, and pilot training was emphasized to the detriment of other aircrew.

The trend was set by Trenchard, the RAF's first post-war commander, who embraced Douhetism with the zeal of a convert, partly to increase his Service's status, which closely followed that of the Royal Navy (whose air power he also provided). The concept of deterrence gained support in Whitehall because it offered a cheaper alternative to a conscript army in times of conflict, while the RAF's successful battle to maintain its autonomy owed much to its demonstrations of air power's value in replacing the Army in imperial policing. Opinionated but technically ignorant, Trenchard failed to improve the RAF's effectiveness, although its strength in the British Isles expanded from nine squadrons in 1920 to 34 in 1931. However, a lack of aids and poor training meant that it was largely a fair-weather force, and one lacking any credibility, since sorties rarely lasted more than an hour or two. The situation improved only gradually, and as late as 1938 only 14,000 out of 164,000 hours' flying were at night;[28] in the same year the *Armée de l'Air* flew only 75 hours at night compared with 17,600 during the day.[29] The RAF also acted as a cadre for future expansion, and to prevent an accumulation of obsolete aircraft most bomber squadrons until about 1935 had single-engine two-seaters such as the elegant Hawker Hart. Despite the fact that these bombers were devoid of an effective range or payload, the philosophy was perpetuated by the Air Ministry in 1933 with Specification P.27/32 and the grossly inadequate Fairey Battle. Second thoughts by both the Ministry and the designer notwithstanding, these bombers spearheaded Allied attacks in 1940 and suffered appalling losses. Trenchard's obsession with the bomber was maintained by his successors, and the RAF made little provision either for fighters or Army support, the latter being regarded as a waste of resources.

Yet Trenchard's undoubted administrative skills created a support and personnel infrastructure as the basis for future expansion while a system of educational establishments, including an Air Staff College, ensured enthusiasm for, and interest in, air power. He skilfully avoided the problems of 'geriocracy' that he encountered in the British Army through the careful selection of personnel after the war, the introduction of short-service commissions and the encouragement of auxiliary and reserve forces.[30]

French air power was unable to avoid the problem of 'geriocracy' because of the nation's chronic shortage of men. Wartime leaders maintained positions of power and influence: for example, the Air Chief of Staff on three occasions between 1931 and 1934 was *General* Joseph Barès, who had held a similar position from 1914 to early 1917. In 1920 the *Aviation Militaire* had only half the number of officers it required, and most junior officers (*capitaine, lieutenant*) had little formal military education, having been commissioned from the ranks, although, despite official discouragement, by 1928 64 per cent of commissioned newcomers (*sous-lieutenant*) came from major military schools.[31] To conserve manpower the French restricted themselves to short, fair-weather flights and only 65 aircraft were written off in accidents between 1931 and 1938; eight of these accidents were fatal, with ten deaths.[32]

The shortage of manpower dictated a defence policy which was based upon fighting a prolonged war. While covering forces existed to prevent a surprise attack, the peacetime Army and Air Force merely provided a framework for mobilization, and to become effective they required large drafts of reservists and substantial reorganization. From 1930 French military air power possessed only a small strike force, with the remainder mostly concentrated at five regional bases (Le Bourget, Dijon, Lyons, Nancy and Pau) where operational, training and logistics functions were combined—a policy which made it impossible to respond rapidly to situations.

This was the situation when the *Armée de l'Air* was established in August 1933. Its first-line strength in Metropolitan France was maintained between 1920 and 1933 at about 100 squadrons (*escadrilles*) and 900 aircraft, a similar size to that in 1920. But efficiency was hamstrung by a multiplicity of aircraft types which created a severe maintenance problem aggravated by a complex stocking and supply system and a shortage of mechanics. Aircraft development and procurement was arthritic: for example, the first BCR aircraft did not enter service until mid-1935, although the specification was issued in 1928. The procurement organization preferred to meet the Air Staff's requirements with a multiplicity of types—1,858 aircraft of seventeen models in February 1926 alone.[33] Moreover, production contracts were distributed among as many manufacturers as possible, in order to maintain the industrial infrastructure, with the result that most of the FFr1,200 million ($52 million) spent between 1920 and 1928 was wasted.

The BCR requirement occurred not so much because the Air Staff shared the Douhetism of the first Air Minister and wartime airman, *M* Laurent-Eynac, but rather in response from the demand by *Général* Maxime Weygand, the Army Chief of Staff, for greater tactical air support. The Air Staff wished to strengthen the bomber force in the face of growing financial restrictions, and by ostensibly meeting the Army's needs they squared the circle.[34]

The Soviet Union faced neither financial restrictions nor shortages of manpower, and between 1924 and 1929 the *VVS-RKKA* expanded from 341 combat aircraft to 1,300 (and 110 naval aircraft), organized into 39 squadrons and 52 corps detachments. Stalin's drive to make the Soviet Union an industrialized nation beginning with the First Five-Year Plan (1929–33) further boosted the country's strength, for it provided substantial capital investment. Not only was strength increased but also quality, with early products including Tupolev's all-metal, twin-engine TB-1 and four-engine TB-3, while in the early 1930s development began of the advanced I-16 and SB.

Tactical air support was the Soviet Air Force's only role until the 1930s: in 1929 81 per cent of the *VVS-RKKA*'s strength was assigned to 'reconnaissance', 10 per cent of its aircraft were fighters and 8 per cent were bombers. But that year saw the creation of multi-role strike forces (air brigades), and by 1933 there were 22 of these, with some 100 squadrons, to provide operational-level support and some twenty corps squadrons for tactical employment. By 1932 Alksnis had begun concentrating his TB-3 heavy bombers into special units, which heralded the formation of *AON*.[35]

The Depression, which so restricted air force development during the early 1930s, saw a polarization of political opinion on the Left and Right as despair gave way to anger. In Germany the Nazis, legitimized in February 1925, increased in strength, and Göring's fortunes turned for the better. After leaving Austria in 1924, he and his wife went to Italy, where he vainly sought a subsidy from Mussolini for the Nazi Party.

The failure of this mission increased his dependency upon morphine, and when the couple went to Sweden in 1925 his addiction proved so great that he had twice to be admitted to hospital. He was weaned off the drug and returned to Germany in 1927, following a Presidential amnesty, ostensibly to sell Swedish parachutes. He received a cool reception even from Hitler, while the Richthofen Veterans' Association blackballed him because of allegations over his war record. However, Hitler needed someone to strengthen Nazi influence with the great and the good in Berlin, and Göring was the only person with the right social connections. His success was facilitated by his election as one of the handful of Nazi *Reichstag* deputies in 1928. In fact he had to coerce Hitler into nominating him, and he exploited his position to obtain bribes.

As the Nazis gained ground in the 1930 and 1932 elections, Göring became a social lion, despite the death of his beloved wife. As the acceptable face of the Party in Berlin salons, Göring engineered an unsuccessful meeting between Hitler and President Hindenburg in 1931 and he was clearly in the ascendant within the Party. Gradually he became Hitler's right-hand man, then he engineered, with Schleicher and Blomberg, a situation by which Hindenburg offered the Chancellorship to Hitler. With Hitler's acceptance the scene was set for the *Luftwaffe*'s birth.[36]

NOTES TO CHAPTER 3

1. Carsten, p.359.
2. FO 14359 C3232.
3. PRO Air 8/138, No 4, p.34; FO 371/15218, C 707, C 6371, C 6722; FO 371 14359, C 941, C 3232; Völker, *Entwicklung*, *Anlage 15*.
4. PRO, FO 371 14359, C293, C1768.
5. PRO Air 2/1353.
6. PRO Air 2/1353: Report 0160/1825 of 5 July 1933, Appendices A–F.
7. *SHAT* File 7N2620.
8. See Völkers, *Entwicklung*, *Anlage 19*.
9. For the responsibilities of each desk see Völker, *Entwicklung*, *Anlage 17*.
10. For the list see Völker, *Entwicklung*, *Anlage 15*.
11. For these meetings and the plans see Völker, *Entwicklung*, pp.188–9; Homze, pp.46–7; and Seaton, p.30.
12. For Wenniger see *Cross & Cockade Journal* 1982, Vol. 23 No 2, p.189. I would like to thank Peter Kilduff for his kindness in bringing my attention to this information.
13. For the Jeschonnek memorandum and background see Völker, *Entwicklung*, pp.184, 194–200, *Anlage 23*; and Homze, pp.47–9, 55.
14. Carsten, pp.358–9.
15. Andersson, 'The Secret Luftwaffe', p.47.
16. Quoted by Völker, *Entwicklung*, p.184/n.170.
17. Völker, *Dokumente*, pp.74–92.
18. For Löhr see Othmar, p.112; and Mitcham, p.133.
19. On at least one occasion during the Spanish Civil War SBs chased Fiat CR 32 biplane fighters.
20. Price, *Fighter Conflict*, p.82.
21. PRO Air 1 1189/204/5/2595, RAF War Diary, Intelligence, Summary 11 August 1918.

22. Michel Forget's essay 'Co-operation between Air Force and Army in the French and German Air Forces during the Second World War' in Boog, *Conduct*, pp.419–20.
23. Boog, *Luftwaffenführung*, p.174.
24. Maurer, p.243,
25. Jones, p.66.
26. Homze, pp.33–4.
27. McFarland and Newton, pp.25–32; Maurer, pp.331–2, 364, 414.
28. Harvey, p.56.
29. *SHAA* File 2B133.
30. For the RAF in the 1920s see Bond, p.144, 322; Harvey, pp.53, 56–8; and Jones pp.27–66.
31. See Christienne and Lissarague, p.238.
32. *SHAA* File 2B133.
33. Christienne and Lissarague, p.222.
34. For the French Air Force see Christienne and Lissarague, pp.207–309; and Forget in Boog, *Conduct*, pp.415–22.
35. For the *VVS-RKKA* see Boyd, pp.18–45; and Tyushkevich, pp.165, 191.
36. For Göring after the Beer Hall *Putsch* see Irving, *Göring*, pp.68–108.

CHAPTER FOUR

THE FLEDGELING

January 1933 to June 1936

W ithin four years of the Nazis' acquiring power, Germany experienced a radical change in her fortunes. There was a dramatic recovery in the economy and the country forced its way back into the first rank of European nations by breaking most of the shackles of Versailles. In both these achievements the *Luftwaffe* played an important role, symbolized by the changes in its headquarters, the *Reichsluftministerium (RLM)*. Created in March 1933 as the *Reichsluftfahrtkommissariat*, the *RLM* rapidly found its first headquarters, the former Danatbank offices in Behrenstrasse, too restrictive. In January 1935, three months before the *Luftwaffe* flew openly, the foundations of a new building were laid on the corner of the Wilhelm and Leipziger Strassen. Progress was rapid and the old Prussian War Ministry, where Thomsen had had his office, was razed to the ground overnight to make way for a 4,500-room edifice of utilitarian design. By October 1935 a quarter of the offices were occupied, and the building was to remain the *Luftwaffe*'s heart until wrecked by British bombs in 1943.

The *RLM*'s development reflected a universal interest in aviation during the 1930s which no political party could ignore. Even Hitler overcame his dislike of flying to exploit the new medium during his unsuccessful campaign for the Presidency in 1932. With Milch's aid he used a Lufthansa airliner to visit 96 cities in two fortnightly tours. The Deputy *Führer*, Rudolf Hess, was an 'Old Eagle' and *Freikorps* pilot who flew extensively, but it was Göring who was the Party's aviation spokesman. He made a public commitment to a substantial expansion of civil aviation under an organization similar to the British Air Ministry. The latter also administered the RAF, and the Nazis also intended to emulate the British in this area, although Hitler recognized that the establishment of a Party-sponsored autonomous air force might bring him into conflict with the *Reichswehr*. This he wished to avoid while he consolidated power and so, within hours of becoming Chancellor, he appointed Göring *Reichskommissar für die Luftfahrt* as part of Brandenburg's organization within the Transport Ministry.

Göring himself was no intellectual and tended to regurgitate the air power theories of others, but the concept of an independent air force and a Douhet-like policy appealed to his megalomanic instincts. Within three days of becoming Chancellor, on 3 February 1933, Hitler outlined his foreign policy to his military leaders, who were delighted to learn of his intention to break the restrictions of the Versailles Treaty and then launch a campaign of conquest in eastern Europe. Such a policy required a larger army based upon conscription and, by implication, a large air force.

On 9 February one of the few Nazi Cabinet meetings approved plans to create an air force rapidly and allocated RM40 million ($9.52 million). The more astute officers recognized which way the wind was blowing, and this was confirmed during Göring's visit to Rechlin in March 1933 when he demanded greater progress in military aviation. Göring's appointment not only thwarted the *RWM*'s attempts to control all aviation but also brought with it the prospect of the Nazis' control of military aviation. German officers disliked any form of civilian control, and the conservatives at the *RWM* acted quickly to strengthen the Services' grip on military aviation. Mittelberger proposed merging all *RWM* air activities into a *Luftschutz Amt* (*LS-Amt*) by the beginning of the next financial year, i.e. 1 April, and this was eagerly accepted. It was quickly agreed that the *LS-Amt* would be responsible for the creation, operational training and administration of air and *Flak* units. Basic training would remain with a separate inspectorship, while the *Heereswaffenamt* would continue developing and producing equipment. As a sop to the sailors, each department would have a separate *Reichsmarine* section and Wenniger would be Chief of Staff.

There was one factor which the conservatives had overlooked, and that was the character of Blomberg, who was appointed *Reichsverteidigungsminister* a few hours before Hitler became Chancellor. The impressionable Blomberg became a Nazi sympathizer after commanding the East Prussian *Wehrkreis I*. His willingness to meet the aspirations of his political masters led to the promotion of Göring to *General der Infanterie* on 31 August 1933 and Milch to *Oberst* on 19 October, while in turn Hitler later appointed him *Oberbefehlshaber der Gesamten Wehrmacht*. He was not only sympathetic to the Nazis but also to the concept of an autonomous air force, and had been since he was head of the *Truppenamt*. It is clear that Blomberg sabotaged *RWM* efforts to retain control of military aviation so that it might come under Göring's authority. Since the new Air Ministry would come under Blomberg and the autonomous air force would depend upon Army officers, he probably believed that he would retain ultimate control.[1] Certainly Blomberg met Göring and his nominated deputy Milch on 6 February 1933, although it was not until 25 April that Milch could write in his diary 'We get the lot.'[2]

The new *RWM* organization was created on schedule and, with Mittelberger on the verge of retirement, Felmy seemed to be its natural leader. Instead, he was transferred to an infantry regiment and Milch, an old friend of his, affected astonishment at the decision.[3] The new leader was a monocled Army officer, *Oberst* Eberhardt Bohnstedt, who was only five months off retirement and whose only relevant experience had been wartime service as a *Flak* officer. He came from *6. Infantrie Regiment* staff, where he had been an *Oberstleutnant* since December 1929. Hammerstein-Equord was alleged by *Oberst* Walther von Reichenau, head of Blomberg's ministerial office, to have described Bohnstedt as 'the stupidest clot I could find in my General Staff'.[4]

As Bohnstedt settled in, the *Reichsluftfahrtkommissariat* arranged on 25 March for DLV to absorb all civil aviation organizations, including those of the Party's *Fliegersturm*, manned by members of the *SS* and the *SA*. The DLV was renamed Deutschen Luftsportverband e.V. under the presidency of Göring's friend Bruno Loerzer. Its dozen (later sixteen) *Luftsportlandesgruppen* continued to encourage air awareness, with members as young as ten years old, but they also increasingly provided paramilitary training, a function officially assumed only in 1935.

The Berlin-born Loerzer had been friends with Göring since before the war when they were junior officers in the *112.Infanterie Regiment*. Loerzer soon became a pilot in the *Fliegertruppe* and encouraged his friend to join him, initially as an observer. Both later became ace fighter pilots and Loerzer (44 victories) ended the war as Commander of *JG 3*. After service in the *Freikorps* under Goltz, he worked as a cigar salesman, became involved in light aviation and after leading an *SA Fliegersturm* became President of Sportflug. He appears to have been a heavy drinker after the war, and although brave he had few pretensions to competence, his subsequent advancement being due exclusively to Göring's friendship.

Göring now had total responsibility for commercial and light aviation, and he received ministerial status on 27 April as *Reichsminister für die Luftfahrt* at the head of the *RLM*. The Ministry came under Blomberg's authority, although Göring retained access to Hitler. A fortnight later, on 10 May, Blomberg peremptorily informed Bohnstedt that the *LS-Amt* would be transferred to the *RLM* in five days' time. This occurred on schedule, and 15 May became the *Luftwaffe*'s official birthday.

Göring himself was to play little part in the *Luftwaffe*'s development, although he was its Commander and was quick to claim the prerogatives and kudos. Until 1936 he was absorbed with domestic politics (he became Hitler's successor in December 1934) and visited the Behrenstrasse about four times a year. But his political influence was vital, to counterbalance that of the traditional Services, which he regarded as fossils, and ensured that until 1937 the *Luftwaffe* obtained an ever larger slice of the defence budget. This rose from 10 per cent in 1933 to 38 per cent in 1936—a total of nearly RM4,000 million ($1,360 million) in four financial years (see Appendix 3). The German economy failed to keep pace, and while Göring talked blithely of ignoring the problem, the nation entered an inflationary spiral which was not brought under control until 1948.

To supervise the creation of the *Luftwaffe*, Göring relied upon his deputy, and his first choice was *Admiral* Lahs. Lahs, President of the *RDLI*, had directed the Navy's secret air programme in the 1920s but was unwilling to join the *RLM*. Several people proposed Brandenburg, who proved to be too conservative and too stubborn for Göring, and the latter eventually selected the Lufthansa director Milch, long a friend of the Party although not a member until 1933. He may have been recommended by other right-wing politicians, including former Chancellor Franz von Papen, whom he had previously cultivated, but by his own account Milch was reluctant to leave Lufthansa, partly because he knew Göring had once been a drug addict. When the former air ace assured him, untruthfully, this was no longer so, and after Hitler had appealed to his not inconsiderable vanity, Milch changed his mind. He entered the *RLM* on 9 February, although the formal appointment was not made until 22 February.

Milch's cherubic appearance concealed a ruthless egoist with a justified belief in his own abilities, for he was an organizer with technical expertise. The industrious Milch had strong political instincts, a vital shield for a man who was a natural intriguer with a strong streak of outspoken bravado. His relationships were vulnerable to his own insensitivity towards opponents and over-sensitivity towards slights, both real

Above: *Grenschutz Fliegerabteilung 431* pictured at Breslau-Hundsfeld during operations in support of *Freikorps* fighting the Poles in Silesia. Seen here are a Junkers J.I and a Halberstadt CL.IV; the unit also had a Siemens-Schuckert D.IV. (Hans E. Krüger)
Below: High jinks in front of a Pfalz D.III serving with *Fliegerabteilung 429*. (Hans E. Krüger)

Above: Three pilots serving with *FA 431*: (left to right) *Leutnant* Kaubler, *Leutnant* Liebig and *Pour le Mérite* holder *Leutnant* Griebsch. (Hans E. Krüger)
Right, upper: *Leutnant* Lenz flew a Siemens-Schuckert D.IV with *Fliegerabteilung 431* and is pictured here with his mechanic and rigger. (Hans E. Krüger)
Right, lower: Members of *Freiwilliger Fliegerabteilung 422*, which operated in northern Germany and later Bavaria under *Hauptmann* Joachim von Schröder (fifth from the left in the second row). (Hans E. Krüger)

Above: A *Fliegerabteilung 431* Junkers CL.I flown by *Leutnant* Martin and *Hauptmann* Eberstein. (Hans E. Krüger)
Below: *Leutnant zur See* Gotthard Sachsenberg (second from right) in front of a Fokker D.VIII pictured as *Fliegergeschwader Sachsenberg* was assembling at Berlin-Jüterbog. The civilians include a representative of the Halberstadt company, *Herr* Theiss (left). (Hans E. Krüger)

Above: Pilots and observers of *Fliegerabteilung 422* with their Commander, *Hauptmann* Joachim von Schröder (in the lighter-coloured jacket). (Hans E. Krüger)
Below: A Junkers CL.I photographed somewhere over Courland in Latvia. (Hans E. Krüger)

Above: Aircraft destroyed by the Allies at the Austrian Öffag airfield, Wiener-Neustadt. (*Herresgeschichtliches Museum*)
Below: Pictured here during the plebiscite in Upper Silesia in 1920 are a collection of 'civilian' airmen who include *Pour le Mérite* bomber leader *Major* Alfred Keller (fourth from the left in the front row). No information is available to explain their presence, although they were probably providing some support for *Freikorps*. (Hans E. Krüger)

Above: A Dornier *Wal* I flying boat under Italian civil registration. The *Wal* was one of the most successful German aircraft sold abroad in the 1920s, although many were manufactured in Switzerland. *Wal*s later formed the first equipment of the *Küstfliegerstaffeln*. (Dornier)

Below: The first Austrian military flying training course in 1930 at Graz-Thalerhof. The *Offizieranwärtern* (later *Leutnants*) sitting and squatting in the foreground are Roland Borth, Erich Gerlitz, Josef Gutmann, Johann Kogler, Alois Maculan and Roman Steszyn. All except Kogler were killed serving with the *Luftwaffe* during the Second World War. (*Heeresgeschichtliches Museum*)

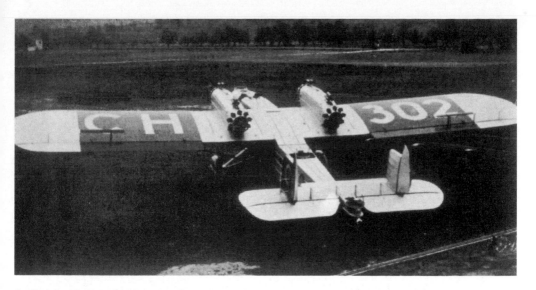

Left, top: A Heinkel HD 55 catapult-launched flying boat. Heinkel was the most successful exporter of German military aircraft before 1933, and 40 HD 55s were purchased by the Soviet Union as the KR-1. (Alex Vanags-Baginskis)

Left, centre: The Junkers A 48/K 47 was a two-seat fighter which appeared in 1928. Three were delivered to Lipetsk in May 1930 but, like most Junkers designs, they failed to impress the German armed forces and were used to train fighter pilots to attack two-seaters. A few were sold to China, but the type's greatest distinction was to be used in dive-bombing trials by the Swedish Air Force from 1932 onwards. (Alex Vanags-Baginskis)

Left, bottom: This Junkers A 48Dy (later D–IRES) was one of two used for 'civil sports and training' in Germany. (Alex Vanags-Baginskis)

Above: The Dornier Do P was designed to meet a *Reichswehrministerium* specification for a four-engine heavy bomber in the early 1930s but proved to be underpowered. (Alex Vanags-Baginskis)

Below: Germany's first post-Great War combat aircraft were Fokker D.XIII sesquiplane fighters. In 1925 they were sent to Lipetsk, where this example was photographed during ther winter with skis to take off in snow. Fifty were purchased (and another two transferred to Lipetsk), and by careful management about 40 remained airworthy when Lipetsk closed in 1933. (Greenborough Associates)

Above: A humble end for a historic aircraft. This Junkers K 47, D-2284, ended its days as a trainer with the *NSFK* but in August 1932 it participated in dive-bomber experiments in Sweden. After a crash, it was converted to a BMW 132 (licence-built Hornet) powered A 48da and resumed testing before being returned to Germany, probably in 1932. (Greenborough Associates)

Below: The Dornier Do 11C was selected to be the backbone of the *Kampfgruppen* and was the first European heavy bomber with a retractable undercarriage and built-in direction-finding equipment. The aircraft displayed poor stability and the wings vibrated, with the result that a major redesign was necessary as the Do 13. This aircraft proved to be even worse, and Sperrle complained bitterly about what the crews called 'The Flying Coffin' in a conference with Richthofen and Keller in 1934. (Alex Vanags-Baginskis)

Above: The Do 13 was substantially redesigned and became the Do 23, with a fixed undercarriage. It equipped *KG 153*, *154*, *155* and *253* and *I/KG 254* but proved to have a disappointingly poor performance and from 1937 was rapidly phased out of first-line service. (Alex Vanags-Baginskis)

Below left: Claude Dornier was at first ambivalent towards the Nazis but as they developed the aircraft industry he supported them. (Alex Vanags-Baginskis)

Below right: Ernst Heinkel was the most successful German aircraft manufacturer of the Weimar Republic period. He supported the Nazis but they 'hi-jacked' his name for *RLM*-owned factories such as that at Oranienburg in which he had only a nominal interest although they produced aircraft designed by his company. (Alex Vanags-Baginskis)

Left: Hermann Göring pictured in full splendour on his 48th birthday in 1941 as *Reichsmarschall*. His previous birthday was overshadowed by the Mechelen Incident, but on 19 July 1940 he was promoted as a reward for the Luftwaffe's success. However, his jumbo-size baton would end up in the Pentagon. (Alex Vanags-Baginskis)

Above: Hitler and Göring review a parade before the war. As Hitler's right-hand man, Göring possessed the political influence to ensure the *Luftwaffe*'s autonomy, but he secured it by a deal with the *Reichswehrminister*, Blomberg. (Alex Vanags-Baginskis)

Right: Erhard Milch, pictured here as a *Generalfeldmarschall* after his promotion on 19 July 1940, wished to replace Göring as Commander-in-Chief of the *Luftwaffe*. His talent for organization and his industrial knowledge made possible the rapid expansion of the *Luftwaffe*, but his ambition brought him into conflict with Göring, although there appears to have been a form of reconciliation in 1939. His only front-line experience after 1919 was a brief excursion commanding *Luftflotte 5* for a few weeks in 1940. (Alex Vanags-Baginskis)

Above: Hans-Jürgen Stumpff was the first personnel director of the *Luftwaffe* and was responsible for Wever and Kesselring joining the Service. He succeeded Kesselring as Chief of Staff in 1937 but appears to have preferred administrative positions, although he commanded *Luftflotte 5* until 1944. (Alex Vanags-Baginskis)

Above right: Albert Kesselring joined the *Luftwaffe* reluctantly in 1933. He became Chief of Staff in 1936 but resigned in exasperation at Milch's intrigues. However, before he did so he ensured that Milch no longer controlled the *Luftwaffe* General Staff and his protégé Jeschonnek ultimately replaced him. Known as 'Smiling Albert' because of his perennial optimism, he commanded *Luftflotte 1* until the Mechelen Incident and then replaced Felmy at *Luftflotte 2*. (Alex Vanags-Baginskis)

Right: Before he joined the *Luftwaffe* Walter Wever opposed almost everything he was subsequently to support as its *de facto* Chief of Staff. His change of heart owed much to his decision to study the new Service and its capabilities thoroughly. He had a better understanding of the importance of technology in air force development than most of his successors. (Alex Vanags-Baginskis)

Above left: Hugo Sperrle briefly directed clandestine military aviation within the *Reichsheer* but was transferred to an infantry regiment, possibly because of his abrasive personality. He was the first Commander of the Condor Legion in Spain and later commanded *Luftflotte 3*. (Alex Vanags-Baginskis)

Above: Hans Jeschonnek was *Luftwaffe* Chief of Staff from 1939 to 1943. While his appointment may well have been a political one because of his enmity with Milch, he had been a champion of an independent *Luftwaffe* before the Nazis came to power. (Alex Vanags-Baginskis)

Left: Wolfram von Richthofen was a cousin of the famous 'Red Baron' and had been with him on his last mission. Richthofen was responsible for aircraft development until 1936 and to Sperrle's complaints about the Do 11/13/23 he adopted a 'take it or leave it' attitude. Later he became Sperrle's Chief of Staff in the Condor Legion and was ultimately its last Commander. He made his reputation leading formations of dive bombers, yet originally he had opposed their development. (Alex Vanags-Baginskis)

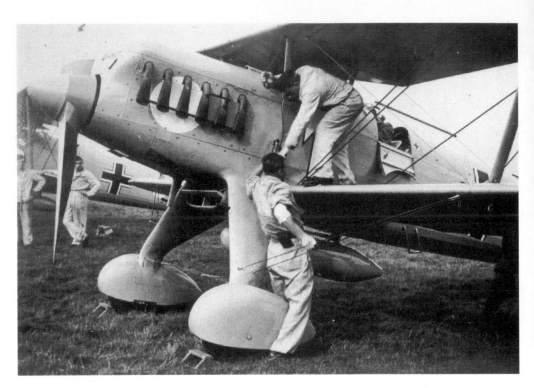

Above: Mechanics prepare to crank up the engine on an He 51B-1 fighter of *3.JG 134 'Horst Wessel'* at Werl, near Dortmund. Developed at Lipetsk, the He 51 proved to be woefully inadequate in Spain and when on escort duty it was sometimes forced to seek shelter behind the accompanying bombers' machine guns! (Greenborough Associates)
Below: The He 70F-1 was a reconnaissance bomber which proved inadequate in both roles but was used by many units, including *3.(F)/123* at Grossenhain. Wever was killed in an He 70 when he failed to carry out pre-flight checks. (Greenborough Associates)

and imagined. But in the first three years of the Nazi regime it was he who equipped the *Luftwaffe* and ensured that it met Hitler's desires. His contribution was unique, not only in organizing production but also in applying the Lufthansa philosophy of operating in almost all conditions—which led to the introduction of all-weather flying training and blind-flying aids.

The only fly in the ointment was that *Herr* Milch *senior* was Jewish, although his wife was Christian. Göring was not rabidly anti-Semitic and once observed, 'I'll decide who is, and who is not, a Jew'; indeed, he shielded from persecution many Jewish friends of his second wife Emmy. The infamous Aryan regulations of April 1933 banned Jews, or anyone of Jewish descent, from public life but made theoretical exemptions for those with distinguished wartime service. Wilberg, who also had a Jewish parent, consequently served in the *Luftwaffe* and was a *General der Flieger* in November 1941 when he was killed in an air crash, while anecdotal evidence suggests that some Jewish pilots remained in training schools. During the summer of 1933 Milch was threatened by a whispering campaign conducted by BFW, with whom he had long had a feud and whose military designs he ignored.

Dr Willi Messerschmitt, BFW's chief designer, and the directors had the ear of high Party officials, including Hess, but fortunately Milch's mother, at his request, provided a document stating that all her children were fathered by her Christian lover. With written confirmation from the terminally ill Anton Milch, Göring's deputy had documentary proof that he was a kosher Aryan and this was accepted by Hitler on 1 November. Interestingly, Milch's personal assistant, Carl Bolle, the former head of DVS, had a Jewish wife and soon left the *RLM* for industry, while in 1935 Milch purged the air defence warning organization of the racially impure.[5]

As *Staatssekretär der Luftfahrt*, Milch held the reins of power and controlled all *RLM* departments and directorates until the autumn of 1937. Bohnstedt's organization was renamed the *LA* while the rump of the old commissariat became the *Allgemeines Luftamt* (*LB*), with responsibilities for civil aviation. Brandenburg's involvement in aviation ceased and he became a minor transport bureaucrat—a sad fate for a man who had done so much to nurture German aviation through the lean years. Milch also received the aircraft development organization from the *Heereswaffenamt* with Wimmer, whose comprehensive knowledge was to prove invaluable. On 1 October 1933 the organization was expanded to directorate status as the *Technisches Amt* (*LC*), while *LA* was renamed the *Luftkommandoamt* and reorganized to amalgamate the Army and Navy elements. Wenniger and the second-ranking naval officer, *Korvettenkapitän* Ulrich Kessler (an 'Old Eagle' and a friend of Göring's), were pushed into minor administrative posts. In 1935 they became, respectively, Air Attaché and Deputy Air Attaché in London/Brussels/The Hague. The organization was unchanged until 1 April 1936, when the *Luftkommandoamt* was expanded with the addition of Quartermaster and Intelligence elements together with four inspectorates, to be renamed the *Führungstab* in the summer. The inspectorates were responsible for developing tactics, training, manuals and equipment for their arms. For evaluating historic experience, a Military History Section was established under Haehnelt, who ultimately became a *Generalmajor*.

Milch established not only the administrative framework but also an infrastructure of bases and facilities without equal in Europe until the early 1940s. By the end of

1933 he had ordered work to begin on fourteen new airfields (and five seaplane stations), and within three years there were 36 fully equipped air bases. Each cost the equivalent of an infantry division but had modern radio and land-line communications, radio and visual navigation aids and meteorological facilities. By comparison, the French (who had only two civil radio beacons in 1933) delayed the development of signal and navigational systems until 1936, while as late as March 1938 the Annual Report of RAF Bomber Command noted that no base had a blind-landing system and only half had a Meteorological Section.[6] The *Luftwaffe*'s bases were supported by an extensive signal network (one company per *Luftkreis* by 1936), the Signal Service having become a separate entity in December 1933 at the suggestion of *Oberstleutnant* Wolfgang Martini, who was to remain its leader until 1945.

Blomberg provided the new Ministry with 224 officers (including 42 from the Navy) between 1933 and 1934, partly to give the new Service a professional foundation and partly to ensure that the Army retained influence. However, he had no intention of sacrificing the expansion of the other Services upon the altar of Göring's ambition, and of the 224 only 49 were on the active list.[7]

One of the first arrivals, on 1 July, was *Oberst* Hans-Jurgen Stumpff, a congenial, efficient, yet curiously anonymous staff officer from Pomerania who celebrated his 42nd birthday a fortnight later. He was an infantryman and wartime staff officer who, according to Milch,[8] had in his personal file a comment that he was suitable material for Army Commander-in-Chief. However, he appears to have been too self-effacing and anonymous a personality ever to have truly qualified for such rank.[9]

Within months Stumpff selected two key figures in *Luftwaffe* history, *Oberst* Albert Kesselring as Head of Administration from October and *Oberst* Walter Wever as Bohnstedt's replacement from September. Kesselring, the son of a schoolmaster, was a month short of his 48th birthday when he entered the *RLM*. He had joined the Bavarian artillery in 1904 and before the war was a qualified observer and the adjutant of a balloon battalion, yet he never became an 'Old Eagle'. He became a staff officer after rallying units during the Battle of Arras in April 1917 and reluctantly joined the *Reichsheer* after *Freikorps* service in Bavaria. Once he was the efficiency expert at *RWM*, but at the time of his transfer to *RLM* he commanded *4.Artillerie Regiment*.

A stocky, balding man, Kesselring was an immensely capable staff officer, well suited to creating order out of chaos and providing the human framework of the new Service. His façade of cheerful confidence, which earned him the nickname 'Smiling Albert', concealed a loveless marriage which his Catholic background prevented him from ending. Perhaps the interest he frequently displayed in his subordinates' welfare was an attempt to create a substitute family.[10]

As Bohnstedt's replacement, Blomberg offered two of his most promising staff officers, *Oberstleutnant* Erich von Manstein and Wever. Both were strong personalities but Stumpff recognized that the charming Wever was the more malleable, and probably the more manipulative, of the two and could work with the *RLM*'s leadership more effectively. On 1 September 1933 Wever was appointed Head of the *LA-Amt* while Manstein went on to a distinguished career in the Army. Wever was 46 years old and barely a year earlier, as Head of Training, he had opposed the concept of an autonomous air force. He was a stocky man with an open face, and when Stumpff selected him Blomberg jokingly complained, 'You're taking my best man.'

Indeed, his file also carried the endorsement that he was a potential Army commander-in-chief. Yet it had not always been so, for the career of this Posen-born officer of middle-class background had shown, like Trenchard's, little promise before the war, which had allowed him to display his clarity of thought and organizational talent. A staff officer from 1915, he had joined *OHL* early in 1917, where he worked closely with Hindenburg and Ludendorff. He was an advocate of 'elastic' defence, and his work with *7.Armee* in the Chemin des Dames had helped thwart the Nivelle Offensive in April 1917; he later became Ludendorff's adjutant. In the *Reichsheer* he advanced rapidly from *Major* in 1926 to *Oberst* in 1932. His nationalism was strengthened by the loss of his province to Poland and, like many middle- and low-rank officers, he supported the Nazis. Nevertheless, there is no evidence that he said, 'Our officer corps will either be National Socialist or nothing at all.'

Wever's great contribution was his ability to create trust inside and outside the *Luftwaffe*. Within it he was content to remain the *éminence grise*, like a good staff officer. The egos of his masters and the capriciousness of their decisions did not faze him, although he cannot have been devoid of ambition. Perhaps the satisfaction of his unique position itself made him willing to accommodate Göring and Milch, who gave him complete freedom of action because they did not regard him as a threat.

When his staff officers sought endorsement for plans, he would play devil's advocate to test the depth of their arguments. *Hauptmann* Paul Deichmann, who joined Wever's command in 1935, complained about this approach but Wever retorted, 'You brood for weeks over something while I have to decide within a minute.'[11] Even if not fully convinced, Wever might still accept a well-argued view after a night's reflection, but he demanded total commitment. To those who complained about a lack of time, he sarcastically commented, 'A day has 24 hours, and if that is not enough use the night.'[12] He established a rapport with the first-line units, which endured critical flying visits. Invariably he would draw the sting by bringing a cake and over coffee he would listen to the views of all the officers. Wever's military talents combined with Milch's industrial and organizational capability and Göring's political support to create the most powerful triumvirate in the *Wehrmacht*—of which Wever was the linchpin.[13]

The conservatism he had shown in 1932 was initially maintained and he sought only fighters and reconnaissance aircraft. But he was open-minded and conscious of his own ignorance and not only learned to fly but also voraciously read everything available on aviation. Within a year he had radically changed his views, supporting the *Luftwaffe*'s autonomy and sharing Wimmer's views on a strategic bombing force sufficiently to authorize the development of the 'Ural Bomber'. Unlike his British and American contemporaries, Wever regarded strategic bombers as part of an integrated force to support the Army and the Navy. He knew that the Nazis believed in a bomber deterrent, and in operational-level bombing he found a rationale which allowed him to meet all demands while retaining the new Service's autonomy.[14]

While Wever's views of air power changed, there had been no change of heart among senior Army officers, especially *Generalleutnant* Ludwig Beck, head of the *Truppenamt* since October 1933. To convince them that air power should be released from the straitjacket of tactical employment, Wever decided to use as a forum the *Reichswehr* Winter War Games which began on 6 November 1934 in the Behrenstrasse and ended with a debriefing on 11 December. They were a *tour de force* which displayed Wever's great confidence in the *Luftwaffe*'s future capability and his ability to harness it. The wargames envisaged a French advance into south-western Germany. Wever took personal command of *Luftwaffe* 'operations', which were largely confined to tactical air support on the first two days. However, bombers 'attacked' French bridges across the Rhine while long-range reconnaissance aircraft 'flew' night intruder operations over French airfields.

On the third day Wever unleashed his 'forces', which not only 'attacked' the French air and ground forces throughout their tactical-operational depth but also 'struck' Paris and industrial centres. It was concluded that the French would sue for peace because of the destruction of their aircraft industry, their air force infrastructure and the attacks on Paris. Yet there were anomalies, especially over German losses and, somewhat petulantly, Wever refused to accept an 80 per cent loss rate for fear of compromising his own argument and brow-beat the referees into accepting a lower one.[15] However, the exercises did display the need for better co-operation with the Army, and there was considerable work during the winter to improve the situation.

At the following year's wargames, which began on 18 November, Wever was once again confident enough to 'fight' with one hand tied behind his back. This exercise involved a two-front scenario (France and Czechoslovakia/Lithuania) and initially the *Luftwaffe* operated as the Army desired, allocating most of its resources, including 25 per cent of its fighters, to tactical air support. The 'results' of the first days showed the complete failure of the *Luftwaffe* to destroy French air power, but on the third day Wever concentrated his assets to destroy the enemy air forces in detail. A feature of the second day was a study of dive bomber operations against French operational-level communications. An analysis of these exercises continued until April 1936, and the lessons were absorbed in *LdV 16*, which was published in May 1936 (see below).[16]

Beck conceded that Wever had won his point, but personally he remained openly sceptical of air power and during the summer of 1938 drove *Luftwaffe* staff officers to the point of apoplexy. As a sop to the Army's fears, the *Luftwaffe* gave the generals operational control of the *Aufklärungsgruppen* (*Heer*) corps squadrons (*Nahaufklärungs-staffeln*) and long-range reconnaissance squadrons (*Fernaufklärungsstaffeln*). Executive control would be the responsibility of each Army and Army Group Headquarters' *Koluft*, who would also co-ordinate fighter and bomber operations with the appropriate *Luftwaffe* command. During the early years of the war the Army sought to have *Kampf-* and *Stukagruppen* assigned to *Koluft*, but every attempt was vigorously opposed by the *Luftwaffe*. Yet the Army still hankered for direct air support, and with so many *Luftwaffe* officers having exchanged *feld grau* for blue-grey there was a considerable degree of sympathy for such views. This was demonstrated during the prolonged and debilitating conflict with the Soviet Union, when the *Luftwaffe* began to abandon Wever's hard-won victories and increasingly committed its squadrons to close air support. The nadir occurred in 1944 when He 177 heavy bombers were used for ground attack operations.

The struggle between the two views was also reflected in the bickering over force structure, which continued within the Army as the Nazis took power and lasted into the spring. To end the argument Felmy proposed to a *Truppenamt* conference on 9 March 1933 a plan for 32 *Staffeln*, to be completed between 1934 and 1935 when the training infrastructure was in place. It was a compromise which included seven cadre reconnaissance units and nine each of fighters and bombers. Discussion continued after Felmy's transfer, and a version was published by Hammerstein-Equord in his 10 April 1933 directive *Vorbereitungen zur Schaffung einer Friedensfliegerwaffe* (Preparations for the Creation of a Peacetime Air Arm). This envisaged 37 *Staffeln* (200 aircraft) by October 1937, with the emphasis on reconnaissance units (see Appendix 4). Half the *Jagdstaffeln* would meet an *In 1* requirement of 10 August 1932 for two-seat fighters to support the bombers. Three *Fliegergruppenkommandos* would administer the air force from October 1934, while operational control would be vested in a *Brigadekommandeur der Flieger* from 1935. The training infrastructure was to be created, under civilian disguise, from 1 October 1933. While the Hammerstein-Equord directive was swiftly overtaken by events, the *Reichswehr*'s emphasis upon Army support, the creation of a strong bomber arm and the corresponding weakening of the fighter forces (especially in the field of strategic defence) continued under the new regime.

Although Hitler and Göring demanded completion of the plans by April 1936, they were not impressed by the Army's caution. Milch used the framework of Felmy's plan to produce an interim one more attuned to his masters' wishes while simultaneously preparing a more grandiose scheme. He was influenced by a former colleague, Lufthansa's traffic manager, *Dr* Robert Knauss, an 'Old Eagle' and part-time aviation writer who joined the *Luftwaffe* at the end of 1933 as *Kommandeur* (later *Kommodore*) of the Lufthansa-based *KG 172* auxiliary bombing unit. In October 1937 he succeeded Jeschonnek as *Kommodore* of the *Lehr Geschwader*. In early 1940 he participated in the planning for *Unternehmen 'Weserübung'*, after which he became Head of the *Luftkriegsakademie*. He died in 1955.

Early in May 1933 he sent his former employer proposals for the future air force. Although they had overtones of Douhet, it is uncertain whether Knauss was familiar with the Italian's theories. He advocated an autonomous air force because he saw it as a means of deterring the threat of a two-front war during the early stages of rearmament when Germany was weakest. He proposed a low-cost deterrent of heavy bombers costing RM80 million ($11 million), the equivalent of five divisions, which could hold the ring until the conventional forces were expanded. He compared the force to the deterrent fleet (*Risiko Flotte*) created by Tirpitz in the 1890s to hold the European naval balance of power. It was an ominous comparison, for the appearance of Tirpitz's fleet spurred an arms race with the British, who interpreted the German action as a direct threat.

The *RLM* recognized the merits of Knauss's case but also realized that industry could not produce heavy bombers. Industrial weakness therefore forced both Milch and Wever to concentrate upon twin-engine bombers which could strike enemy forces and industry within 500km (300 miles) of the border, thus meeting strategic and diplomatic requirements.[17] It was left to Milch to square the circle, and on 6 May, four days before Blomberg informed Bohnstedt of the imminent transfer to the *RLM*, he commissioned plans for an air force based upon Knauss's principles. It would have 51 *Staffeln* (a 45 per cent increase) and these would have double the strength of those proposed by Hammerstein-Equord, to give a first-line strength of 600 aircraft by 1 October 1935 (see Appendix 4). All would be camouflaged as elements of legitimate civilian aviation organizations such as Lufthansa, while the absence of suitable designs meant the temporary abandonment of two-seat fighters. Operational and organizational control of the force would be split between *Fliegergruppenkommandos*, of which two would be created in 1934. The so-called '1,000-Aircraft Plan' was completed by 13 May but was not published until 22 May and the *RLM* was not enthusiastic. As Bohnstedt's Operations Officer, *Oberstleutnant* Kühl, pointed out, industry did not have the capacity to support Milch's programme.

But Milch was already addressing that problem and in a technological *tour de force* lasting three days and nights he exploited departmental responses and his own knowledge to draft a detailed production plan costing RM170 million ($40.5 million). It used off-the-shelf designs, not only those developed at Lipetsk but also civilian aircraft which could be militarized, and was flexible enough to be adapted to the industrial situation.[18]

Milch had the pleasure of seeing his force structure plans approved without demur by Blomberg, who signed a formal agreement with Göring on 27 June. Neither the *Heeresleitung* (Hammerstein-Equord) nor the *Marineleitung* (Raeder) were consulted; they received details only on 12 July, but their protests about the lack of tactical air support were ignored. Although Milch's plan remained the basis of the *Luftwaffe*'s organization for the next two years it was modified to include a provision for dive bombers. The greatest changes were organizational, with the creation, on 1 April 1934, of administrative commands (*Luftkreiskommandos*) and a separate operational command (*Fliegerdivision 1*) similar to that proposed by Hammerstein-Equord. For most *Luftkreise* the Army provided senior officers who were able organizers and administrators but were devoid of aviation experience. Most, such as *Generalleutnant* Leonhard Kaupisch, formerly *Artillerie Führer V*, were elderly and had been transferred to the reserve list in 1931–32. Sperrle became Commander of *Fliegerdivision 1* and was also *Kommandeur der Heeresflieger*. The naval *Luftkreise VI* was under Zander, now a *Konteradmiral*, who received a tactical subordinate on 1 July 1934 when the 'Old Eagle' *Fregattenkapitän* Hermann Bruch was appointed *FdL*.

On 29 June Milch's instruction *LA Nr 1305/33 g. Kdos. A 2 I* began the colossal task of building the *Luftwaffe*, superseding the programme Bohnstedt had barely begun. If the obstacles to a 200-aircraft force were formidable, they appeared insurmountable for a 600-aircraft requirement. In May 1933 the aircraft industry employed only 3,500 people at eight factories evenly divided between airframes and aero-engines. The industry's net production was worth RM37.5 million ($8.9 million), 0.2 per cent of German industrial production, while exports were down from RM20 million ($4.76 million) in 1930 to less than RM4 million ($952,000).

Industry was split into factions (its internecine quarrels continued throughout the Second World War), with the regime's strongest supporters being Adolph Rohrbach (whose firm had collapsed and whose attempts to revive his flagging fortunes ended in a take-over by the Krupp organization which renamed the company Weser Flugzeugbau and later Blohm und Voss), Ernst Heinkel and the BFW designer Willi Messerschmitt. They eagerly anticipated a rich, protected, domestic market and this persuaded Claude Dornier to come off the fence. The only major opponent to Milch's plans was his former employer *Professor* Hugo Junkers, and the action taken against him revealed the regime's darker side and reflect little credit on either Milch or

Kesselring. The old professor was persecuted—the financial juggling of the Fili project was used as the prime weapon—and bullied into handing over control of his factory, the aero-industry's largest, to the state. The strain of the persecution hurried him to his grave, to which Milch hypocritically sent a wreath.

However, the *RWM* had been unhappy with the Junkers management for many years as it had failed to produce any worthwhile military designs. Modernization was vital, but Milch's efforts to prise Junkers' fingers from the company certainly demonstrated the English saying about patriotism being the last refuge of the scoundrel. Yet these actions reflected the attitudes of a regime which prided itself upon ruthless dynamism. Milch and plenipotentiaries later involved in the aircraft industry glibly talked of shooting those who opposed various projects since it was results which mattered. Indeed, Milch's own downfall occurred when he ceased to meet his masters' demands.

Junkers' successor, Heinrich Koppenberg, demonstrated the industrial dynamism of the new regime and made the company a model for the aircraft industry. Koppenberg was the technical director of the Mitteldeutsche Stahlwerke AG steel works, whose Leipzig-based ATG (Allgemeine Transportmaschinen GmbH) was one of six aircraft companies created in April 1933 (the others being Henschel in Berlin, Siebel in Halle, Blohm und Voss in Hamburg, Gotha in Gotha and MIAG in Brunswick). A choleric, bull-necked technocrat with no aviation background, Koppenberg introduced modern mass-production techniques to Dessau and swiftly consigned the 'cottage industry' approach to the dustbin.

Commercial and state funds were available to create new aircraft manufacturers and to expand established ones. The latter usually became the centre of industrial complexes, using factories from unrelated industries, or those of the new companies, to manufacture components and sub-assemblies. For the Ju 52 Koppenberg acquired a locomotive repair shop and provided licences both to ATG and a Blohm und Voss factory. To assist newcomers to the industry, Junkers, aided by Fertigungs GmbH, instituted the 'ABC' programme which produced master rigs that were then copied and supplied to the sub-contractors. This extremely flexible process, based upon American technology, soon became standard with most *RDLI* members except for Dornier.

The Achilles' heel of Milch's plan lay with the aero-engine plants. Co-operation between the prime contractors (Junkers, Daimler-Benz and BMW) and sub-contractors was poor, the latter having great difficulty adjusting to mass-production techniques. The aero-engine programme was plagued by poor production decisions on the part of both the *RLM* and the manufacturers (especially the small ones) and by the lack of skilled labour, as well as by shortages of raw materials and capital. Siemens und Halske, for example, failed to maintain the agreed schedules for the Do 11 bomber's SH 22B-2: only a handful of pre-production engines reached Dornier in 1933, and by 1 March 1934 the *Behelfskampfgeschwader* had received 24 Ju 52/3m bombers but only three Do 11Cs.[19]

The *RLM* encouraged and coerced manufacturers into establishing complexes in central Germany. Factories, each of which had a duplicate, were built on a massive scale with substantial over-capacity so that in wartime they were capable of two 10-hour shifts and could produce 25 per cent more than their ostensible output. By the end of 1933 the industry employed nearly 17,000 people, a year later 53,800 and by mid-1936 124,800, organized into some fifteen companies.[20] For manufacturers, profits were guaranteed with 'cost-plus' contracts—even the humble Klemm 25 had a 61 per cent profit margin—and the *RLM* was willing to support even the most exotic of the manufacturers' fancies. But, despite this, by September 1934, as Göring admitted to Blomberg, industrial expansion had outpaced the capacity of commercial finance to support it. For strategic reasons, therefore, the state increasingly intervened in the aircraft industry. New factories were built and, while often named after well-established companies, they were run by the state just as Junkers was. A typical arrangement was that concerning the building in Oranienburg of He 111 bombers by the state-financed associate company, Heinkel-Werke GmbH. Ernst Heinkel's only connection with this plant which bore his name was his chairmanship of the firm's advisory committee, for which he received a handful of shares worth RM150,000 ($60,500). By the outbreak of war industrial investment amounted to RM2.35 billion ($943.75 million), of which 47 per cent represented Government investment in one form or another.

The *RLM* attempted to rationalize the industry in order to curb the ferocious rivalries. Development work was generally restricted to the larger established companies, leaving fabrication to the smaller concerns and the newcomers. This policy provided Milch with a legitimate reason for neglecting BFW, although the real reason was a grudge against Messerschmitt following an accident with the BFW M20 airliner prototype.[21] Consequently, the designer's proposals were ignored and BFW was left to produce the Dornier Do 11 bomber.

A kaleidoscope of production schedules added to the problems. Hitler and Göring demanded an ever larger air force, and by September 1933 industry had a production target of 2,000 aircraft, but the frenzy of industrial expansion forced constant changes and by the end of the year the *Luftwaffe* had received only 197 military aircraft. In an attempt to create order out of chaos a new schedule, the *Flugzeug-Beschaffungsprogramm 1934*, was published on 19 January 1934 but for obscure reasons was nicknamed the *Rheinlandprogramm* ('Rhineland Programme'). As demands increased or problems developed, other plans followed—an extension of the *Rheinlandprogramm* in January 1935, then the ironically designated *Lieferplan Nr 1* (Production Plan No 1) in October 1935. The latter remained the basis for German aircraft production until the autumn of 1936 (see Appendix 5). It was nominally replaced by *Lieferplänen Nr 2* and *Nr 3* in March and July 1936, but, essentially, these merely incorporated adjustments to the original. These and all the other plans which followed seemed to embody the legendary efficiency of the Germans for they were detailed and comprehensive, covering first- and second-line aircraft as well as prototypes and pre-production aircraft. The schedules showed how many airframes would be produced in each month, and at which factory, and they were as intricate as a Swiss watch. Unfortunately, given the capriciousness of the Nazi leadership and the numerous industrial and economic problems which developed during the 1930s, the plans proved as fleeting as a harlot's smile.

Nevertheless, by December 1934 the factories were producing 160 aircraft a month and had delivered to the *Luftwaffe* a total of more than 1,900 (only 6 per cent below the target), including nearly 1,300 trainers and communications aircraft.

Within a year monthly production had doubled to 300 and the Extended Rhineland Programme was envisaging a qualitative, as well as a numerical, improvement with the phasing in of a new generation of aircraft such as the Do 17, He 111 and Ju 86 as well as provision for prototype four-engine heavy bombers.

However, as *Lieferplan Nr 1* got under way in 1936, a growing shortage of raw materials and unexpected difficulties developing the complex new aircraft forced Milch to take an unpleasant decision. In order to maintain production rates in *Lieferplan Nr 2*, he substituted obsolete aircraft for modern ones in the knowledge that they could initially be used to form new squadrons and later be relegated to second-line duties. The British Air Staff adopted the same philosophy with the Fairey Battle, which was authorized for production in 1935 to facilitate RAF expansion although the design concept was recognized as being totally inadequate. This brought little long-term operational benefit for the RAF—in contrast to Milch's decision: by increasing production of the Ju 52 to replace the inadequate Dornier Do 11 family, Milch provided the *Luftwaffe* with a surplus of aircraft which accidentally laid the foundation of the transport force. The rugged reliability of the Ju 52 earned it the nickname *'Tante Ju'* ('Aunty Ju') and ensured that it remained the backbone of the *Transportgruppen* until 1945.

The sudden change in official attitudes and the flood of money left Wimmer's directorate in the position of rich, greedy schoolboys in a sweet shop. It led to an especially indulgent attitude to prototypes, with manufacturers given every opportunity to prove their most optimistic claims and projects continuing when they should have been curtailed. Some manufacturers were pursuing a dozen projects simultaneously, many duplicating the work of other companies, and scarce resources were wasted. Wimmer's desire to get the best aircraft for the *Luftwaffe* was understandable, especially as concern grew about the performance of the first generation of combat aircraft. Not only were they inferior to that of their potential French opponents, but some of the machines were extremely dangerous to fly. The lack of suitable engines was one cause of the problem, but design faults raised doubts about the thoroughness of the Lipetsk development work. The worst example was the Do 11C bomber, scheduled to be the backbone of the deterrent but dubbed 'The Flying Coffin' by its unfortunate crews. Its wings vibrated so alarmingly that the risk of structural failure led the *Luftwaffe* to restrict angles of bank to 45 degrees. A significant redesign produced the Do 11D and Do 13, but the latter proved even less endearing than its predecessor and several were lost to structural failure after joining the *Kampfgruppen*.

Four months after assuming command of *Fliegerdivision 1*, Sperrle vigorously criticized the Dornier bombers on 18 July 1934. Five months later there was a crisis conference involving Sperrle, *Hauptmann* Wolfram von Richthofen, the head of aircraft development, and *Major* Alfred Keller, the famous *'Bombenkeller'* wartime bomber commander who had recently joined the *Luftwaffe* and commanded *Luftkreis IV*. Despite another redesign to become the Do 13e (Do 23 from early 1935), the bomber remained slow and had a sluggish performance. Wimmer considered ending production and waiting for the new generation of bombers, but the prototype Ju 86ab1 made its maiden flight on 4 November 1934 and the He 111a did not fly until February 1935 and, clearly, production of these aircraft could not begin until late 1935 or early 1936. Reluctantly, the conference agreed to continue with the building

TABLE 6: COMPARATIVE PERFORMANCES OF BOMBERS, CIRCA 1934

Aircraft	Max speed (kph)	Combat range(km)	Ceiling (m)	Bomb load (kg)
Do 11D	259	959	4,099	1,000
Do 23	259	1,351	4,200	1,000
Ju 52	277	997	5,900	1,500
H.P. Heyford	228	1,480	6,400	1,587
Bloch 200	230	997	8,016	1,653
Martin B-10	342	1,995	7,376	1,025

of the inferior Do 13e because delays were operationally and politically unacceptable. Production of the Ju 52 took up the slack, while the Do 23 reached the *Kampfgruppen* during the late summer of 1935.[22]

Despite the Dorniers' structural problems, German bombers compared favourably with their traditionally built contemporaries, as Table 6 shows. Thanks to Milch, all the German bombers had direction finding (D/F) equipment, giving them the capability of operating on cloudy days or in bad weather. Few of their European contemporaries had this equipment—indeed, RAF Bomber Command's Annual Report of 1938 stated that no British bombers had D/F loops.[23]

Richthofen recognized the problems with his aircraft but, in the absence of anything better, his only response to complaints was to comment that 'It is better to have second-rate equipment than none at all.' He was well qualified to steer the development of the *Luftwaffe*'s next generation of aircraft. As a child on his parents' estate he had developed an interest in mechanics by dismantling and re-assembling farmyard machinery. Of vaguely Asiatic appearance, he joined the cavalry and served on the Eastern Front until September 1917, when he transferred to the *Luftstreitkräfte*, serving in his cousin Manfred's *JG 1* and accompanying him on his last flight. After the war, in which he scored nine victories, he became a diploma engineer before joining the *Reichsheer*. He was transferred to the *Luftwaffe* in 1933 and began a steady ascent up the ladder, eventually winning promotion to *Generalfeldmarschall*. He possessed the instinct to 'think on his feet' and proved very popular with both superiors and subordinates, although he could be impatient to the point of rudeness. His frustrations he confined to his voluminous diaries, while in his off-duty hours he played the flute or played cards.

Richthofen and Wimmer had not only to bring into service the *Luftwaffe*'s first generation of post-war combat aircraft but also to complete a technically ambitious aircraft programme defined in the summer of 1932. The most important element was the high-speed medium bomber, for which an outline specification was issued by the *Heereswaffenamt* in July 1932 and modified in March 1933 (possibly as a result of the evaluation of the Martin 139). The new aircraft were also to meet a Lufthansa requirement for a high-speed airliner on prestige services, although the first prototype was to be a bomber. This was emphasized by Milch on 23 May 1933 when he wrote to Dornier about the proposed Do 17 and noted that even the civil version should be capable of rapid conversion to a military role.[24]

The elegant Do 17, later dubbed 'The Flying Pencil', was to require a major redesign of the tail and it was the fastest of the three types—indeed, the V5 prototype was 24kph (15mph) faster than British and French fighters then entering service. It had a smaller bomb load, with a maximum of two 250kg (550lb) or four 100kg (220lb) bombs compared with eight 100kg or three to four 250kg bombs of the He 111 and Ju 86, which both had longer ranges but were plagued by engine problems. The 1,000hp Daimler-Benz DB 600 provided the He 111B-1 with a performance worthy of its shark-like appearance, but there was to be no happy solution for the Ju 86. Its six-cylinder, liquid-cooled Jumo 205 diesel engines promised superior fuel consumption compared with petrol-driven engines of similar power, and these expectations were met when a constant level of revolutions was maintained, but the sudden bursts of power required in combat degraded performance, while at altitudes above 5,000m (16,500ft) the Jumo was prone to a sharp but unaccountable reduction in performance. The engines tended to overheat, leading to seized pistons and broken transmissions, and required very careful maintenance by specially trained engineers.

As the development of these aircraft began in 1933, an argument arose in the medium echelons of the *Luftwaffe* command over the most effective means of delivering bombs. The dispute continued until after the war broke out and had a severely debilitating effect upon the development of German bombers. From the beginning, the accuracy of horizontal bombing from high altitude was suspect, not only with the Goerz-Vizier 219 bomb sight used in first-generation bombers but even with the more accurate telescopic Lotfe (Lotfernrohr) 7/7D of the second-generation aircraft. There were good grounds for this concern, as the élite *III (K)/Lehr Geschwader* demonstrated in 1937 when its He 111Bs and Do 17Es could place only 2 per cent of their bombs within a 200m (660ft) diameter target from an altitude of 4,000m (13,000ft). Greater inherent accuracy was obtained when the bomber dived on the target; this attracted considerable naval interest, especially in the Pacific where the US Navy introduced the Curtiss F8C Helldiver and F11C Goshawk and the Japanese developed the Aichi D1A from the He 50. Junkers conducted trials with a K 47 in 1933–34 and convinced the *Luftwaffe* of the concept's viability, but Richthofen had misgivings over the vulnerability of aircraft to AA fire and the strain of dive bombing upon their airframes. These fears seemed to be confirmed when Udet was almost killed during an air show in July 1934 when the tail came off his Hawk II. Although a civilian, Udet had become a dive bomber enthusiast and used *RLM* funds to buy two Hawks (export versions of the Goshawk), which were officially demonstrated at Rechlin in late October 1933.[25]

In 1933 the *RLM* ordered 111 He 50s, which entered service in January 1935 as stop-gaps pending a definition of the *Luftwaffe*'s dive-bomber requirement. Exercises clarified this matter during the winter of 1933–34: every close air support option was examined, including the use of medium bombers and a revival of the wartime light attack (CL class) aircraft. The initial requirement of 11 May 1934 was for a twin-engine, multi-role aircraft with ominous similarities to the French *BCR*, but wiser counsels quickly prevailed. On 30 June the reconnaissance element was assigned to the Do 17, while close air support was split between light and heavy dive bombers.

The former was an updated CL class aircraft, a single-seater relying upon small bomb loads and machine guns and later redesignated *Schlachtflugzeug*. This role was close to that of the fighter-bomber, and Ar 65 or He 51 fighters were used to form many dive bomber *Staffeln* pending re-equipment of the light dive bomber units with the Hs 123 from 1936. While fighters were used in the ground attack role in Spain, this was an act of desperation (see Chapter 5) and the *Luftwaffe* did not deploy dedicated fighter-bombers until the Battle of Britain.

The heavy dive bomber (*Stürzbomber*, later *Stürzkampfflugzeug* or '*Stuka*') requirement was refined during 1934 and issued in January 1935. It was for a two-man aircraft capable of carrying a more powerful bomb load and was based upon the Ju 87 ordered by the *RLM* in September 1933. Richthofen remained dubious about the heavy dive bomber concept and on 9 June 1936 reportedly recommended to Wimmer that the project be cancelled, but the same day Wimmer was replaced by Udet as head of the *Technisches Amt*. Udet, who had joined the *Luftwaffe* as Inspector of Fighters and Dive Bombers and was an enthusiastic supporter of the Ju 87, reportedly reinstated the project. Richthofen remained as Head of Development until November 1936, and when he was dispatched to Spain it was to supervise the operational evaluation of both fighters and dive bombers. By then the hawk-like *Stuka* was in production and offered an extremely cost-effective means of precision bombing.[26] The Ju 87As of *IV (St)/Lehr Geschwader* could place 25 per cent of their bombs in a 200m diameter circle from 2,000m (6,600ft), yet the *Stuka*s were themselves originally regarded as stop-gaps until the third-generation bomber became available.[27] Under Udet, however, a concept once seen as a means to an end became an end in itself—and ultimately a deadly obsession.[28]

The *Luftwaffe*'s long-term bombing needs were also addressed, and on the day Hitler came to power the *Heereswaffenamt* issued an outline requirement for a long-range heavy bomber (*Langstrecken-Grossbomber*) capable of reaching targets in northern Scotland or in the Urals from German bases and thus nicknamed the 'Ural Bomber'. Although the competing designs were never capable of reaching the Urals, which were some 2,500–3,000km (1,500–1,800 miles) from air bases in eastern Germany, they could have struck any British industrial target. Limited industrial resources and the priority allocated to the medium bomber made progress slow, and many staff officers felt that the heavy bomber was an expensive luxury. At first Wever agreed, although as his aviation education progressed he changed his mind. On 11 May 1934 a seven-year development and production programme (a similar time scale applied to British heavy bomber programmes) for the four-engine aircraft was authorized but in mid-1935 this was accelerated so that prototypes of the competing Do 19 and Ju 89 would fly by the end of 1936.

Surprisingly, Blomberg gave greater support than Göring, as Wimmer discovered when he accompanied them to separate inspections of Junkers' Dessau works. When Göring saw the Ju 89 prototype he asked, 'What on earth is that?' Wimmer, who knew that Göring had been briefed on the project during the winter of 1934–35, explained, but the 'Iron Man' had forgotten and blustered, 'Any major project such as that can be decided only by me,' and stormed off. When Blomberg saw the prototype he asked when it would be ready and was told 1939 or 1940. He simply commented, 'That's about right.'

Both designs proved to be underpowered and by the spring of 1936 Wever himself was lost faith in them as day bombers. Although lighter than their potential contemporaries such as the Handley Page Halifax, Short Stirling, Boeing B-17 and French Centre (formerly Farman) NC 223, they would have had inferior speed, range and bomb loads. Even the Stirling, which carried a similar 1.5-ton maximum bomb load to that of the German aircraft and was notoriously underpowered, had 37 per cent more range than the Do 19, 2,000 miles (3,200km) compared with 1,250 miles (2,000km). Their nearest contemporary was the NC 223 night bomber, which could carry a payload of 4 tonnes and had a maximum range of 1,450 miles (2,300km).

From November 1935 the project was gradually reduced to demonstrator status, with a few prototypes in *Lieferplan Nr 2* (one of the Junkers aircraft was to be the Ju 90 airliner). On 17 April 1936, six months before the first 'Ural Bomber' prototype flew, Wever issued a new heavy bomber requirement, for 'Bomber A'. Germany had just reoccupied the Rhineland and, despite a muted British reaction, Wever had the foresight to consider striking English industrial targets, most of which could not be reached from bases in western Germany. He wanted 'Bomber A' to be capable of carrying a 900kg (2,000lb) bomb load a distance of 6,700km (4,160 miles) at a speed of 540kph (335mph)—a performance making it a true 'Ural Bomber'. His death occurred when the 'Ural Bomber' programme was downgraded to one of evaluation while that of 'Bomber A' had barely begun. His successors, having naturally re-evaluated the programme, continued to pursue his dream with 'Bomber A'.[29]

Bomber escorts had always been anticipated, and on 10 August 1932 a requirement for 'heavy', two-seat, long-range fighters was issued. Göring supported the concept in 1933 but Wimmer's definition rapidly turned it into a German *BCR*. The *Kampfzerstörer* (combat destroyers) were to smash enemy fighters by flying offensive patrols in relays ahead of the bombers. They were also to provide close escort for the bombers, conduct long-range interceptions, carry out high-speed bombing attacks, fly reconnaissance sorties and even (during the winter of 1933–34) perform ground attack missions. Wever recognized the dangers in the project, referring to the *Kampfzerstörer* as a 'forlorn hope', but he was unable to persuade Göring to abandon the project. In May 1934 an outline requirement was issued for a twin-engine, cannon-armed, multi-role aircraft, five months before the *Armée de l'Air* issued a similar document for a three-seat long-range fighter. The French selected the Potez 63, which was then adopted for a variety of roles, including that of night fighter (Potez 631 C3), light bomber (Potez 633 B2), corps aircraft (Potez 637 R3) and ground attack aircraft (Potez 639 A2).

Richthofen quickly realized that the German requirements were incompatible and on 22 January 1935 the specification was split into three—a heavy fighter, a high-speed, long-range reconnaissance bomber and a *Schnellbomber* ('fast bomber'). The first two competitors, the FW 57 and Hs 124, were multi-role aircraft with all the weaknesses of the *BCR*. The third, the Bf 110, won a development contract apparently through the influence of Udet, although he was still a civilian at the time. It flew on 12 May 1936 (eighteen days after the Potez 63), but it took the introduction of the 1,050hp DB 601A-1 to make it battleworthy and it entered service only in February 1939.[30] By the end of 1936 there were three contestants for the *Schnellbomber* requirement (Bf 162, Hs 127 and Ju 88) and the programme had made a confident start.

The issuing of the light (or single-seat) fighter requirement was delayed by Hammerstein-Equord. Upon learning of Milch's *51-Staffeln* programme in July 1933 he demanded that all the *Jagdstaffeln* be assigned to support the Army, leaving home defence to anti-aircraft guns and what he referred to as 'police fighter squadrons'. This led to a specification in the autumn of 1933 for a *Heimatschutzjäger*, a requirement for a combined lightweight fighter/trainer similar to the *Heitag* of the 1920s, which was met by the FW 56 *Stösser* (Bird of Prey), although in the event this aircraft was used only for training. The true light fighter requirement was issued in December 1933 and simply specified an all-metal monoplane with retractable undercarriage. The story of the Bf 109 has been told many times, but its overwhelming victory represented both a technical and a human triumph. That fighter traditionalists such as Udet and Greim should be won over despite the aircraft's technical sophistication is especially worthy of note. Indeed, it was Udet's superb flying skills which demonstrated conclusively the aircraft's superiority. During a mock battle on 3 July 1936 in front of Göring and Blomberg at Rechlin, he 'shot down' not only the bombers but also their He 51 escorts, yet he failed to dent the *Luftwaffe* leadership's confidence in the bomber's superiority. For Messerschmitt the victory was also a personal triumph over Milch, who bided his time and had his revenge six years later over the Me 210 issue.

Machines need men to operate and maintain them, and while the *Luftwaffe* was well served by *Reichswehr* preparations the semi-voluntary nature of its organization until March 1935 restricted expansion. (A similar problem was suffered by all the British services.) Some 2,000 men, including 500 aircrew, were transferred to the *Luftwaffe* by the *RWM* and supplemented by personnel from civilian organizations such as Lufthansa and DVS as well as by 'Old Eagles' (such as Udet) and reservists who were persuaded to enlist. It remained an air force in 'mufti', everyone having nominal civilian titles and Lufthansa or DLV uniforms (the latter forming the basis of the true *Luftwaffe* uniform from 1935). DLV provided much young blood, although the *Luftsportlandesgruppe* were not officially assigned to *Luftkreise* until March 1935. Increasingly Loerzer's organization acted as the primary flying training establishment and by October 1935 it had eight *Fliegerausbildungsstellen* and twenty *Fliegerübungsstellen* processing 30 men and youths at a time. DLV was also responsible for reservist training.

From 1 November 1933 formal flying training was conducted under naval air ace *Oberst* Friedrich Christiansen as *Inspektor der Schulen der Luftwaffe* with six former DVS schools, which retained their titles and their old functions. The unveiling of the *Luftwaffe* in March 1935 saw a substantial expansion of the training organization, each *Luftkreis* receiving a *Flieger Ersatz Abteilung* (*FEA*) by October 1935. While DLV provided *ab initio* flying training, the *Fliegerschulen* attached to the *FEA*s provided A to B (or sometimes C2) standard training. By mid-1936 there were a dozen *Fliegerschulen*, supplemented by a similar number of specialist schools, *Waffenschulen* and a *Blindflugschule*. Courses in the last did not begin until April 1935, but from November 1933 *'Reichsbahnstrecken'* (*RB-strecken*) were established to provide long-range, blind-flying and navigation training under the auspices of Lufthansa and based upon *Behelfs KG 172*. The *RB-strecken* were later absorbed into *Streckenschule Berlin*, which had some fifteen Ju 52s and was dissolved only in 1939.[31]

By March 1935 the *Luftwaffe* had some 18,000 men, of whom 7,000 were in the *Flakwaffe* (AA Arm). Although Hitler's reintroduction of conscription on 14 March 1935 promised sufficient manpower for the *Luftwaffe*'s ambitions, an immediate draft was needed to prepare for further expansion. Göring had to go, cap in hand, to the Army for technically trained men. Between May and September 1935 he received 6,000, including 4,300 volunteers for aircrew duties, especially flight engineers, who accounted for 44 per cent of the total. The remainder were signallers, administrators and construction engineers, but in October 1936 the drafts ended as the Army prepared for its own expansion.

The new generation of officers came from the *Luftkriegschulen* (*LKS*), three of which were established on 1 April 1936. Their graduates were supplemented from 1936 by promoting qualified NCOs and reservists, so that by 1937 the *Luftwaffe* was receiving 2,500 officers a year. The most serious problem was the lack of qualified staff officers, the Army having no more to spare. To solve the problem the *Luftwaffe* created the *Luftkriegsakademie* (Air War Academy) in November 1935, but the courses involved no more than 25 men at a time and the shortage retarded the creation of *Geschwader* staffs. By September 1939 the *Luftwaffe* could meet only half its needs and, like the Army, had to fill the gap with enthusiastic but unqualified officers.

The biggest disappointment for the *Luftwaffe* was its failure to secure Italian aid, both Fiat CR 30 fighters and fighter training facilities to replace Lipetsk, despite promises made by Mussolini to Göring on 11 April 1933 during the latter's visit to Rome. No fighters were forthcoming but in July 1933 Pflugbeil, who had been involved with bomber development in Lipetsk in 1928, led a party of German pilots to an Italian fighter training course. The men, ostensibly from South Tyrol (acquired from Austria by Italy after the war), included future *Inspekteur der Jagdflieger* Adolf Galland and wore *Regia Aeronautica* uniforms while on the course. They qualified in the autumn, but German–Italian tension and the ability to conduct such courses at Schleissheim meant that the experiment was not renewed.

The Italians, while sympathetic to their fellow fascists, feared that a resurgent Germany would absorb Austria. The Italian Chief of the Air Staff, Marshal Italo Balbo, pointedly told Milch that a force of fighters and reconnaissance aircraft would be adequate for Germany's needs. At that time the *Regia Aeronautica* had 90 squadrons evenly divided into bombers, fighters and reconnaissance aircraft.[32] Italian fears grew when the Austrian Chancellor, Engelbert Dollfus, was murdered by Austrian Nazis with German support in July 1934. His successor, Kurt von Schuschnigg, understandably took a more nationalistic stance and under his leadership the expansion of Austrian air power continued steadily, aided by Italian patronage.

A fighter squadron was activated at Graz-Thalerhof as *Gruppe C* after the arrival of six Fiat CR 20s in July 1933, while the following month the reconnaissance *Gruppe A* was raised with the Fiat A 120. The new air force edged crab-like into the open from March 1934 with the establishment of two units, *Lehrabteilung II* (*LA II*) at Graz and *LA III* at Vienna. The former controlled the operational units (*Gruppe A* became *Gruppe A1)* and the primary training squadron (*Gruppe P*) while *LA III* was largely responsible for administration and training but controlled *Gruppe A2* for advanced flying training. At the beginning of April 1934 the Austrians had 440 men and some 30 aircraft. A month later the *Luftschutzkommando* was created to supervise all Austrian air activities, including reconnaissance sorties against Austrian Nazis in July 1934 after Dollfus was shot. Similar flights were flown five months earlier by *Polizei* aircraft following a Socialist uprising in Vienna.

On 1 May 1935 the *Luftschutzkommando* became the *Kommando der Luftstreitkräfte* (*KoLu*) under *Generalleutnant* Löhr, who had *Major* Schöbel as his Chief of Staff. They planned to create ten *Staffeln* (six fighter, one bomber, three reconnaissance) and unveiled the *Österreichische Luftstreitkräfte* (*LStrKr*) on 19 June 1935 to token protests by the Allies. By the end of the year it had 686 men and 78 aircraft and was costing öS3.6 million ($670,000) a year to run. British D.H.60 Moth Major trainers were being imported, while a licence to produce the FW 44 *Stieglitz* (Goldfinch) trainer was obtained by Hirtenberger Patronen, which took over the Hopfner Flugzeugbau when it went into liquidation. The weakness of the Austrian economy restricted expansion in 1936, but by the summer the *LStrKr* had 1,000 men and Vienna ordered 45 Fiat CR 32*bis* fighters and five Caproni Ca 133 bombers. These developments were not regarded with the concern shown about the *Luftwaffe* which began flexing its muscles in 1936.

The expansion of the *Luftwaffe* began cautiously but became ever bolder. The *Reklamestaffeln* were the cadre for the first *Jagd-* and *Aufklärungsstaffeln*, while in October 1933 Lufthansa aircraft were earmarked for the *Behelfsbombergeschwader* (later *Behelfskampfgeschwader 172*). This unit had secondary training (and possibly transport) roles and from November 1933 it was under Christiansen's command before being transferred to *Flugkommando Berlin*. By the end of the year the *Luftwaffe* had a dozen *Staffeln*, and while their aircraft carried civil designations none appeared on the official register of civil aircraft.[33] The expansion of the *Luftwaffe* saw the creation in 1934 of the *Luftkreise*, *Fliegerdivision 1* and *FdL*. The *Luftkreise* were responsible for the infrastructure such as bases, supplies, signalling and navigation facilities, motor transport and air defence as well as air traffic control and meteorological services. Air defence was provided by the *Flakwaffe*, an Army organization under *Luftwaffe* operational administration until February 1935, when Göring took total control. During 1933 sixteen batteries were raised, and this figure was doubled the following year.

By 1 March 1934 the *Luftwaffe* had only 77 aircraft (including 27 bombers and twelve fighters) but a year later, as the work of Milch and Christiansen bore fruit, this grew to 800, whose squadrons had 'civil' designations of legitimate aviation organizations such as DVS, RDL and Lufthansa.[34] The relentless expansion continued at a faster rate when the mask dropped (see Appendix 6). There were a nominal 1,833 first-line aircraft (with 822 medium bombers and 251 fighters) by 1 August 1935, although many *Staffeln* were non-operational because of shortages of personnel and equipment, while by 1 April 1936 there were 2,680 aircraft (with 1,000 bombers and 700 fighters). This expansion made devolution essential, and on 1 April 1935 operational command within the *Luftkreise* was transferred to *Höhere Flieger-kommandeure*, complemented, six months later, by specialist *Höheren Kommandeure der Flakartillerie*. Exactly a year later administrative devolution took place with the establishment of *Luftgaue*, which were also responsible for air defence.

It was now impossible to conceal Germany's air rearmament even from casual visitors, and especially from diplomats. The British Air Attaché, Group Captain J. H. Herring, reported that Milch's assistant Bolle admitted on 10 June 1933 that Germany had begun air rearmament, which was confirmed by 'secret information'. Eight days later Herring visited an air display at Tempelhof and, while sitting with an important aviation official and his bored wife, observed two 'express postal aircraft'. They were, in fact, an He 51 and a Junkers A 48, which he had never seen before. He asked the official about them and the wife replied, 'Oh, those will be two of the new single-seat fighters I suppose.'[35] Intelligence also provided reports. In April 1933 the Foreign Office received details of orders for 190 military aircraft placed with Arado, Dornier and Heinkel,[36] while during 1934 the RAF's Director of Plans obtained documents on Milch's *51-Staffeln* programme and *Fliegerdivision 1*. However, the flow of information then became a trickle, and the RAF consistently underestimated the potential strength of the *Luftwaffe*: in 1935 it believed that by October 1936 the Germans would have 114 squadrons with some 1,300 first-line aircraft.[37] As a result, in July 1934 approval was given by the British Cabinet for Expansion Scheme 'A' and by the French Parliament for Plan I, each designed to provide first-line strengths of more than 1,000 aircraft by mid-1939.

The actions of the *Gestapo*, later under former *Luftstreitkräfte Feldwebel* pilot Heinrich Müller, and tighter security at the *RLM* helped to plug the leaks, yet there remained security problems. The *RLM*'s need for bureaucrats and officials provided an opportunity for many ministries to remove dead wood—the incompetent, the incorrigible and the corrupt—while Göring's friends and acquaintances also joined. Although the regime appealed to the nationalism of many Germans, others had moral, political or personal qualms. The most famous, and most effective, was Harro Schulze-Boysen, a nephew of *Admiral* Tirpitz. Although never a Communist, he was a charismatic figure with naïve views of the Soviet Union which, on the death of a friend, led him into espionage for ideological reasons (Soviet intelligence called such people 'crap eaters'). Family connections ensured his release from *Gestapo* custody and enabled him to become a naval air observer. He was transferred to the *RLM* where his diligence, intelligence and linguistic abilities (he spoke six languages) made him popular with his superiors. The popularity increased when he married the daughter of one of Göring's friends in June 1936 and the *Luftwaffe* commander was a witness. Through his influence the 'rehabilitated' Schulze-Boysen was commissioned and began creating a spy network that later became part of what German counter-intelligence called *'Rote Kapelle'* ('Red Orchestra').[38]

Schulze-Boysen worked only for the Russians, and a less radical dissenter left a typed report at the British Consulate in Oslo on 5 November 1939. The Oslo Report revealed details about the Ju 88, radar developments, the *Y-Geräte* blind-bombing aid and surface-to-surface missile development and provided valuable pointers for British Intelligence. There were also examples of unreliable aircrew. A Communist served as a pilot in *I/StG 165 'Immelmann'* then fled in 1936 and by the end of the year was leading an International Brigade company. A captured member of *KG 26* with no sympathy for the war, identified by the British as 'A231', volunteered details of another blind-bombing aid, *Knickebein,* in June 1940.[39] Later in the war the British acquired one of the latest Ju 88 night fighters through a pre-arranged defection. Yet these were rare incidents and *Luftwaffe* aircrew willingly risked their lives in their nation's cause. Professionally they lived for flying and, like young men the world over, spent their recreational time laughing, drinking and chasing girls. They were proud to be shielding their people from the horrors of air attack, reasserting German rights and righting the injustices of the Versailles Treaty.

Germany sought Allied agreement for a limited rearmament, including a small air force, when the World Disarmament Conference opened in February 1932. The delegation wrung a reluctant concession from the French in December 1932 concerning a general 'equality of rights', but the British strongly objected to a German air force which would ultimately pose the greatest threat to them. Because it was impossible to define bombers, proposals to abolish them were doomed and the Conference reached deadlock. A clumsy attempt to break through the impasse and legitimize the *Luftwaffe* was made in June 1933 when it was claimed that unidentified aircraft 'from the east' had dropped leaflets on Berlin. Göring talked of building 'police aircraft' and Milch blustered about 'equality of rights in the air', but the British Foreign Secretary, Sir John Simon, reflected international opinion when he described the incident as 'fictitious'. A few days later the *RLM* enquired about purchasing a pair of demilitarized Hawker Harts, but the Foreign Office firmly rejected the move.[40] Eventually, on 14 October 1933, Germany walked out of the Conference, which dissolved in May 1934, lifting the ethical restraints upon British and French air rearmament.

German air rearmament continued apace but by late 1934 the pretence of its being a purely civil activity was straining credulity and becoming counter-productive. If the objective was to deter the Allies and dominate Germany's neighbours then the deterrent had clearly to be revealed. This decision was made probably after 12 February 1935 when the Army formally conceded the *Luftwaffe*'s demand for the *Flakwaffe*. On 26 February the decree formally establishing the *'Reichsluftwaffe'* was signed by Hitler, Blomberg and Göring, the new service officially coming into existence on 1 March as the third element of the *Reichswehr* (renamed the *Wehrmacht* on 1 June). On 10 March Göring revealed the *Luftwaffe*'s existence when interviewed by Mr Ward Price, his favourite British correspondent. He claimed that it existed purely for defensive purposes, and this reassuring message was repeated to Sir John Simon when he visited Berlin on 25–26 March.

However, when Sir John met Hitler he was informed that the *Luftwaffe* had parity with the RAF, whose strength he put at more than 800. In fact the RAF had only 453 aircraft (690 if the FAA and the Auxiliary Air Force was included), while the *Armée de l'Air* had 1,042 first-line aircraft in Metropolitan France.[41] The Allies were quick to react and by the summer the French had accelerated Plan I for completion by the end of 1937 while the British replaced Scheme 'A' with Scheme 'C'. The British especially regarded the *Luftwaffe* as a direct threat, yet, with the exception of Wever, its leadership was so blinkered that it gave no serious thought of an air war against Great Britain until the summer of 1938.

Meanwhile the unveiling of the *Luftwaffe* was followed by dozens of ceremonies and parades in which units shed their disguises and the names of famed airmen were incorporated into the titles of *JG 132* (Richthofen), *KG 154* (Boelcke) and *StG 162* (Immelmann). *JG 134* was given the name 'Horst Wessel' after a Nazi hero: a pimp who had written the Party anthem, the *Horst Wessel Lied*, he was killed by a Communist pimp seeking to evict him and his girlfriend, who had worked for both men. The ceremonies were barely over when the *Staffeln* began large-scale exercises which exploited their mobility and the ability to move rapidly to dispersal fields (*E-hafen*). Such exercises, which used air transport and had aircraft acting as mobile signal stations, emulated those of the USAAC, which had conducted them since 1927.[42]

While expanding their own air forces, both the British and French took diplomatic action. The British were particularly enthusiastic about a non-aggression air pact involving the Locarno Treaty states, while the French signed a mutual assistance pact with the Soviet Union on 2 May 1935. The military implications of the latter for Germany were serious, the *VVS-RKKA* alone having a first-line strength in 1935 of 6,672 aircraft.[43] The pact proved extremely controversial in France, where it was opposed by the Right. Ratification by the French Chamber of Deputies began on 11 February 1936 and the following day Hitler casually ordered the Army Commander-in-Chief, *General der Artillerie* Werner von Fritsch, to prepare plans for reoccupying the Rhineland. Fritsch warned that the *Wehrmacht* was in no position to fight over the issue, but Hitler replied that he was really testing the signatories of the Locarno Treaty. The directives for *'Winterübung'* ('Winter Exercise') were drawn up by Blomberg on 28 February, the day after the French deputies ratified the pact and sent it to the Senate, and the scene was set for the *Luftwaffe*'s first trial of strength and the first of the bloodless 'Flower Wars'.

On 5 March 1936 the *'Winterübung'* directives were issued on the grounds that the Franco-Soviet Pact contravened the Locarno Treaty. Wever ordered two newly raised *Gruppen*, *III/JG 134* (Ar 65) and *I/StG 165* (He 51), to support the operation and stationed *Flak* units at the Rhine bridges. On 7 March three German battalions marched into the Rhineland while the excited (but inexperienced) pilots of *Hauptmann* Oscar Dinort's *III/JG 134* flew to Cologne, circling the famous cathedral at noon before flying to Köln-Butzweilerhof. Only there did each aircraft receive 1,000 rounds of ammunition, although neither guns nor sights were harmonized. That night *7/JG 134* flew to Düsseldorf while during the day *I/StG 165* moved two *Staffeln* from its home base at Kitzingen to Frankfurt/Main and a third to Mannheim. These were times of great tension for the *Luftkommandoamt* in particular and the *RLM* in general. There was alarm when reports came of French bombers in Alsace being prepared, and *KG 253* was reportedly alerted to attack Paris, which was the limit of the range of its Ju 52 bombers. Although the French mobilized thirteen divisions and occupied the Maginot Line, the Allies merely protested.

French defence policy, with its emphasis upon long-term rather than short-term action, forced Paris to stay its hand as the manpower shortages began to bite. The *Armée de l'Air* had a 23 per cent shortfall in its establishment of 39,000 (the RAF had 45,000 men), with only 6 per cent of the *capitaines* aged under 30 while 39 per cent were over 41 and of 336 recommended for promotion in July 1936 less than 9 per cent were under 36![44] By 1939 the *Armée de l'Air* lacked 31 per cent of its requirement for specialist NCOs, while the average number of active officer pilots in *groupes de chasse* and *bombardement* were six and ten respectively.[45] Although French aircraft were superior to those of the *Luftwaffe*, the supply of spares had not kept pace with production and, for this reason, on 10 March 1936 only 60 per cent of the 158 modern bombers were serviceable.[46]

It was equipment rather than manpower which restricted the RAF, whose bomber 'deterrent' would have had difficulty reaching Paris. Even if the aircraft could have reached a German target, it remained British policy through most of the 1930s to produce no bomb larger than 500lb (225kg) at a time when the USAAC was using 600lb (272kg) and 1,000lb (454kg) bombs. Had heavier bombs been available, there would have been no improvement, for by 1935 the average bombing error had grown from 150 to 500 yards and even a trained crew could be 50 yards out when attacking from altitudes of 200–300ft (60–90m).[47] The problems of locating and destroying the target continued to be neglected, and, as late as May 1940, of the 2,742 officers in the General Duty Branch only 109 (3.9 per cent) had passed a navigation course.

Nevertheless, British planning continued to emphasize the bomber, and after the appointment of Group Captain Arthur Harris as Deputy Director Plans late in 1935 the RAF vigorously campaigned for ever larger heavy bomber forces. However, an awareness of the British Isles' vulnerability to bombing led to a scientific investigation into home defence from the end of 1934. The first radar experiments were conducted in February 1935 and by the following September the construction of a radar network was being recommended to the British Government. A similar concept was examined by the French as the bi-static (separate transmitters and receivers) DEM for territorial defence, but this proved less effective than radar.

Of greater concern was the growing obsolescence of the *Armée de l'Air*, 42 per cent of whose 50 *groupes* in 1936 had *BCR* aircraft. Despite improvements in the development and procurement organization, new designs were slow to appear, while industry could not meet the demands of Plan I, for which only 607 aircraft had been received by June 1936. As unit strength declined, capability came to be measured in terms of *groupes* rather than *escadrilles*. The solution of the Popular Front Government, elected in May 1936, was a fundamental reorganization of the industry from July 1936, but this disrupted production completely and only 187 aircraft were ordered during the year.

The problem of next-generation aircraft was to plague the *VVS-RKKA* in the mid-1930s as it accepted modern aircraft such as the I-16 and SB. In 1937 the *AON* received a new medium bomber, the Ilyushin DB-3, but Soviet industry was unable to design powerful aero-engines and relied upon licence-built versions of American and French designs, while the production of aluminium-type alloys for airframes was grossly inadequate. As Spanish experience was to demonstrate, these failures made the *Luftwaffe*'s second-generation aircraft superior to their Russian opponents. In contrast, the US aircraft industry was getting its second wind and developing the P-36, B-17 and B-18 for a USAAC which from 1933 onwards concentrated its squadrons, including all its home-based strike units, under the GHQ Air Force. However, financial restrictions restrained re-equipment and expansion until the late 1930s.

Meanwhile Germany's situation was enhanced by the occupation of the Rhineland: the industrial Ruhr had better air defences, while the *Kampfgruppen* were nearer to their potential targets. As early as November 1935 the British War Office and Air Ministry conducted a wargame which was hastily abandoned after a German 'advance' from the Rhineland into France and the Low Countries led to the 'collapse' of Allied forces. This prospect grew greater following the supine Allied reaction to the occupation, which led Belgium to terminate its military arrangements with France and adopt a policy of strict neutrality.

The Rhineland occupation meant a further expansion of the *Luftwaffe*, with *Luftkreis IV*'s *Luftgaue VI* (Münster) and *XII* (Giessen) and *Luftkreis V*'s *Luftgau V* (Stuttgart) extending their infrastructure west of the Rhine. From 1 April 1936 they and the other *Luftkreise* expanded the *Luftwaffe*'s strength by 18 per cent (500 aircraft). 'Mother' *Gruppen* and *Geschwader* gave birth to 54 'daughter' *Staffeln* based on cadres of men and equipment which were then fleshed out during the next three months.

For officers, the administrative burden was increased by the need to absorb the *Luftwaffe*'s doctrinal documents. *Die Luftkriegführung* ('Conduct of the Air War') was published in 1935 and *LDv 16* in May 1936. The latter made it clear that, while the *Luftwaffe* was autonomous and the bomber was the key air weapon, the Service was expected to operate in conjunction with the Army and Navy after destroying the enemy's air power. It was also clear that the *Luftwaffe* was to function at an operational level, but this was not defined in *LDv 16* and when Deichmann sought the Air Staff's views on what this meant he received a multitude of answers. Although there was growing *Luftwaffe* interest in the value of strategic bombing, this was not covered by *LDv 16*, although Paragraph 22 stated that 'The fight against sources of power has a decisive influence on the course of the war. It strikes the root of the enemy's will to fight and to resist.' Paragraph 150 noted that 'The battle against the production of military resources promises decisive success if the enemy is unable to replace his losses through imports or supplies from production facilities situated beyond our reach.' However, Paragraph 186 emphatically stated that 'Attacks upon towns for the purpose of terrorizing the population are fundamentally rejected,' although retaliation for similar attacks was permitted.[48]

To evaluate and to develop the new generation of aircraft and their tactics, the *Luftwaffe* followed the Army's example to create *Lehr* units. The *Lehrgruppe* was raised at Greifswald in April 1936 with twelve Ju 52s and twelve Do 23s but on 1 October it was expanded into the *Lehr Geschwader*. The *Lehrgruppe* and *II/KG 152* were merged into *II(K)/Lehr Geschwader* and *I/StG 162* became *IV(St)/LG*, while *1.(F)/122* became *9.(A)/LG*, although it returned to *Aufklärungsgruppe 122* in September 1937. The new generation of aircraft began to arrive early in 1937 and the *Geschwader* would be further expanded into an élite formation over the next two years. Soon it would be receiving vital data from Spain where, two months after the publication of *LDv 16*, the *Luftwaffe* made its combat début and the *RLM* discovered how effective had been the nation's multi-million investment in air power.

NOTES TO CHAPTER 4

1. Irving, *Milch*.
2. *Ibid.*, p.357/n.16. See also Irving, *Göring*, p.113.
3. PRO Air 118/33: Report of British Air Attaché.

4. Irving, *Milch*, p.32.
5. See Homze, p.61; Irving, *Milch*, pp.327–9; and Mason, pp.164, 173–4.
6. Jones, pp.112–13, quoting PRO Air 2/2961.
7. Völkers, *Luftwaffe*, p.52.
8. Quoted by Irving, *Milch*, p.359/n.46.
9. There appears to be no biography of Stumpff, even in Mitchell, whose work is the basis of most biographies in this book. His work is supplemented by articles in *Landser* magazine, usually by *Dr* Gerd E. Heuer, brought to my attention by courtesy of *Herren* Hans-Eberhard Krüger and Hildebrand.
10. For Kesselring see Macksay's biography; Homze, p.60; and Mason, p.216.
11. Deichmann, p.69.
12. *Ibid*.
13. For Wever also see Homze, pp.60, 99–100, 133/n.4; and Mason pp.183–4.
14. Völker, *Luftwaffe*, pp.31–2.
15. See Deichmann, p. 53; Mason, p.212; and Völker, *Luftwaffe*, pp.32–3.
16. See Völker, *Luftwaffe*, pp.74, 86–7.
17. See Homze, p.56; and Völker, *Luftwaffe*, pp.28–31.
18. For German production plans and the aircraft industry to 1939 see Homze and also Green, *Warplanes*.
19. Green, *Warplanes*, p.111.
20. Homze, p.93/t.4, 110/t.7.
21. Yet Lufthansa purchased thirteen M20s and the eight survivors were requisitioned by the *Luftwaffe* in September 1939 (Davies, p.42).
22. For the Do 11/13/23 saga see Green, *Warplanes*, pp.110–12, 130–2.
23. Jones, pp.112–13.
24. See Karl Kössler's article on the Do 17; and Howson, pp.117–18.
25. See Ishoven, pp.289–90, 295–8, 306; and Smith, pp.8–10.
26. Future references to the Ju 87 in the text will be as '*Stuka*', although, strictly speaking, this term should also be applied to the He 50 and the Hs 123.
27. Homze, p.127.
28. For the dive bomber see Smith, pp.14–16; and Homze, pp.125–7.
29. For the German heavy bomber issue see Green, *Warplanes*, pp.127–9, 483–4; Homze, pp.121–3; and Mason, pp.191–2.
30. For the heavy fighter see Dressel and Griehl, pp.30–2; Green, *Warplanes*, pp.176–8, 383–4, 573–9; and Homze, pp.127–8, 130–1.
31. For *Luftwaffe* training see Ries, *Luftwaffen-Story*, pp.46–53, 64–71, and *Deutsche Flugzeugführerschulen*, pp.53–4, 86–7.
32. The Italian Air Force states that it has no documents on these events but has provided order-of-battle information. I am indebted to the British Air Attaché in Rome in 1992, Group Captain S. A. Wrigley, for his assistance with my enquiries.
33. Andersson, 'Secret Luftwaffe'.
34. For the order of battle and camouflage titles on 12 April 1934 see Schliephake, Appendix C.
35. PRO Air 2/1353.
36. Schliephake, Appendix F.
37. PRO Air 9/24.
38. The most detailed account of the 'Red Orchestra' is to be found in Höhne, pp.101–11. See also Irving, *Milch*, p.175.
39. Johnson, p.19.
40. PRO Air 2/1353.
41. PRO Air 9/24.
42. Maurer, pp.235, 245.

43. Tyushkevich, p.191.
44. Christienne and Lissarague, pp.271–3.
45. *Ibid*, pp.298, 304–6.
46. Gunsburg, p.29.
47. Jones, pp.134, 136.
48. *LDv 16* is reproduced in Völkers, *Dokumente*, pp.466–86.

CHAPTER FIVE

THE SPANISH TRIAL

July 1936 to March 1939

I n the summer of 1936 Spain was a rumbling volcano on the verge of eruption. Decades of unresolved social, economic and political problems had created internal pressures which burst forth with violent frequency. The political pendulum swung wildly, and in early 1936 the election of a Socialist Government stimulated plans for a *coup d'état* within the Spanish Army, a bastion of conservatism. The *coup* leaders, *General* José Sanjurjo and *General* Emilio Mola, realized that the weakness of the Army in metropolitan Spain made it essential that it should be quickly reinforced by the Army of Africa. This was the only body of professional troops under Spanish command and it was under the unchallenged leadership of *General* Francisco Franco, who had been transferred to the Canary Islands after the elections.

The rebellion began on 17 July and Franco was flown to Morocco—which was quickly seized—but metropolitan Spain was divided. The rebels, who soon called themselves Nationalists, had most of the north as well as Andalusian bridgeheads in Seville and Cadiz. The Government, or Republican, side held Madrid, the south and the Catalonian east, together with a northern coastal enclave consisting of the Basque provinces in the east, Santander and Oviedo. All the leaders recognized that they were on the verge of full civil war and sought foreign military aid in order to tip the scales. Mussolini had a long-standing agreement with Spanish monarchists to supply aid, but it was inevitable that the Nationalists would turn to Europe's other major Fascist power. On 20 July a Lufthansa Ju 52/3m, returning from West Africa and flown by *Flugkapitän* Alfred Henke, was requisitioned by the rebels to transport *General* Luis Orgaz from the Canaries to Tetuan, where it arrived during the night of 22/23 July.

The Nationalists were now in a desperate position: the Straits of Gibraltar were patrolled by a fleet which was loyal to the Republic and was blockading Morocco, Sanjurjo had been killed in an air crash and Mola lacked the manpower to storm Madrid. Franco decided to fly troops into the Seville bridgehead but he had only six aircraft, including two Dornier *Wal* flying boats. The Moroccan rebels knew that the German Foreign Ministry still supported the Madrid government but decided to appeal to Hitler through the Nazi Party's foreign organization. A rebel emissary, accompanied by two German Nazi Party members, was flown by a reluctant Henke to Berlin, where they arrived late on 24 July. An audience with Hitler the following day at Bayreuth was arranged through Hess, but the *Führer* proved surprisingly cautious despite an exhilarating performance of Wagner. Although he refused to send an expeditionary force for fear of confrontation with France, the knowledge that

Communists were gaining real control of the Republic led to the promise of some military aid.

Blomberg proved to be more enthusiastic than Göring, who was worried about overstretching the *Luftwaffe*. His objections ceased when he learned of Hitler's support for the rebels, and his enthusiasm grew when he learned that the *RLM* would be responsible for providing this support, although all activity was to be nominally under a Spanish organization, Hispano-Marroqui de Transportes SL (HISMA). Significantly, the meeting decided that aid would be supplied only to Franco, who was thereby set on the path towards supreme power in Spain. In the early hours of 26 July the meeting concluded by agreeing to restrict military support to aircraft (mostly Ju 52 bomber-transports and a few fighters). Strict security was to be maintained in order to prevent an international incident, but Hitler rejected Göring's suggestion that Lufthansa aircraft be used instead of *Luftwaffe* assets. The task of providing them was handed to Milch, who was briefed in Bayreuth during the afternoon and quickly organized maritime transport and a headquarters, *Sonderstab W*, under Wilberg. The aid programme was given the code-name *Unternehmen 'Feuerzauber'* (Operation 'Magic Fire'), and although it was a Sunday by the time Milch flew into Berlin, the *RLM* was buzzing with activity.

Henke, a very reluctant participant in the Spanish adventure, inaugurated the German airlift two days later on 28 July after returning the Moroccan delegation in the same *'Tante Ju'* he had used before. The interior of the airliner was stripped so that 35 men, sitting on the floor with their knees drawn up to their chins, could be crammed in. A second Lufthansa airliner was prepared and the two moved 1,207 men in the first week and 1,282 in the second.[1]

In Germany HISMA prepared to dispatch the first consignment of aid—nine Ju 52/3mg3es, six He 51B-1s, twenty *Flak 30* 2cm AA guns, and men. The Junkers were drawn from *I/KG 153* (Merseburg), *II/KG 155* (Neukirchen-bei-Ansbach) and *I/KG 253 'General Wever'* (Gotha), while the fighters came from *I/JG 132 'Richthofen'* (Berlin-Döberitz) and *I/JG 134 'Horst Wessel'* (Dortmund). The *'Feuerzauber'* party consisted of 25 officers (including *Leutnants* Hans Trautloft and Hans-Joachim 'Hajo' Herrmann) with 66 NCOs, other ranks and civilians. In many cases commanding officers volunteered men and informed them later, sometimes after making them sign declarations of secrecy,[2] while some officers were ordered to resign their commissions and join the reserve list before departure. All changed into civilian clothing and then went to Hamburg, where they met their commanding officer, *Major* Alexander von Scheele, a wartime *Schlasta* veteran who had emigrated to Latin America after the war but had returned to help the Fatherland recapture its old glory. Scheele's orders indicated that *'Feuerzauber'* was purely a training and transport mission. Combat missions were forbidden, although he could use the fighters to escort the transports until Spanish pilots were qualified to fly them. Ostensibly tourists, the party boarded the SS *Usaramo* following inspection by Milch and Wilberg and by the morning of 1 August they were at sea.

Meanwhile eleven demilitarized *Luftwaffe* Ju 52 bombers in Lufthansa colours flew to Morocco and Seville via San Remo in Italy. One accidentally landed at the Republican airfield of Badajoz, where the crew were arrested and taken to Madrid. Intervention from Berlin earned their freedom (Madrid was seeking German fighters at this time), but the aircraft was destroyed in a Nationalist air attack. During this period genuine Lufthansa Ju 52s were using Barcelona on commercial flights as late as 8 August, while on 15 August they evacuated German residents from Madrid.

On 6 August the *Usaramo* sailed into Seville and the following morning discharged her cargo during an air raid. Scheele assumed overall command and divided his tiny force into five groups. *Hauptmann* Rudolf 'Bubb' *Freiherr* von Moreau commanded the Ju 52 force at Tetuan while the grumbling Henke was to train the Spanish to use ten Ju 52s as bombers. Fighter training was conducted at Tablada airfield, Seville, by *Oberleutnant* Kraft Eberhardt, who was also responsible for communications and supply. A temporary maintenance depot and communications centre was established at Salamanca, while *Leutnant* Herrmann found himself responsible for training with the 2cm guns. The Lufthansa depot in Seville was initially responsible for mainte-nance, but a separate company was later set up for this work. The Germans wore the white company uniform of HISMA, and this organization provided them with food and accommodation.

The transports flew from Tetuan to Seville, a journey of 110 miles (180km), for the first five days until a temporary airstrip at Jerez de la Frontera became the terminus, cutting the journey by 30 miles (50km) and the flight times by one-third, to 40 minutes. A shortage of fuel and facilities bedevilled the operation, and despite scrounging fuel from the French in neighbouring Tangiers, Moreau's transports were grounded for two consecutive days. The fuel shortage ended with deliveries by sea from Germany and Italy in mid-August, but refuelling remained a problem for there were no bowsers and each Ju 52, with a capacity of 1,370 litres (300 gallons), had to be filled by hand. Later sherry pumps were adapted, but even they could transfer only three litres of petrol a minute.

Flights were made in the mornings and evenings because the overloaded aircraft could not operate in the midday heat when high winds were an extra hazard. The maximum number of sorties were crammed into these periods, the aircrews' only breaks occurring when the aircraft were refuelled or loaded. As night fell maintenance was conducted using either lanterns or vehicle headlights. The threat of AA fire from Republican warships forced the transports to fly at 8,000–11,500ft (2,500–3,500m), but only one aircraft was lost on the airlift: this crashed at Jerez on 15 August, killing *Unteroffiziere* Helmut Schulze and Herbert Zech.[3]

During the week 10–16 August a peak of 2,853 men (about 400 a day) were brought in, then the weekly figure dropped to 1,200–1,600. From September the emphasis was on bringing over material, and this effort peaked during the week 14–20 September, at 69.5 tonnes. When the airlift ended on 11 October the Germans had flown 13,900 men and 270 tonnes of equipment (including 36 guns) across the Straits of Gibraltar—an operation involving 868 sorties. As Hitler rightly commented later, 'Franco ought to erect a monument to the glory of the Junkers 52. It is this aircraft which the Spanish Revolution has to thank for its victory.'[4]

Despite Berlin's injunctions, Scheele used his initiative to deal with the threat from enemy warships. After an Italian bomber had damaged the destroyer *Lepanto* he converted two Ju 52s with crude bomb racks after the two sets of bombing equipment sent in the *Usaramo* were lost. Piloted by Moreau and Henke, the two bombers flew to Malaga on 13 August to make a dawn attack upon the anchored battleship *Jaime*

I. Moreau got lost but Henke hit the warship with two 250kg bombs which blew huge holes in the upper deck and killed 47 sailors. The survivors shot their remaining officer prisoners that night before withdrawing to Cartagena.[5]

Scheele's men now wished to emulate the Italians and fight, ostensibly as members of the Spanish Foreign Legion, and although Berlin refused permission it could not prevent greater German involvement in the war. Eberhardt's Spanish pupils, who included future leading ace *Capitán* Joaquin Garcia Morato, found the He 51 too demanding compared with the Italian Fiat CR 32 and four were written off in accidents (although Morato claimed a victory in one He 51). Rather than see the remaining fighters stand idle, Eberhardt, Trautloft and *Leutnant* Herwig Knüppel formed an operational Flight, and on 24 August this escorted Spanish-manned Ju 52s attacking the Getafe air base. The following day both Eberhardt and Trautloft claimed their first victories, with another four added over the next two days. Moreau had been campaigning for a fortnight to attack Madrid and finally, on the night of 27/ 28 August, he bombed the War Ministry. The following day Hitler formally authorized the use of German personnel on combat missions.

With *Luftwaffe* and Italian air support, the Army of Africa swept from Seville and secured western and southern Andalusia before advancing north towards Madrid. On 3 September it took Talavera de la Reina, the last major town before Madrid, but then Franco diverted his forces to Toledo where they relieved the beleaguered garrison in the Alcàzar on 27 September. A month earlier, on 23 August, Henke and Moreau had dropped two tonnes of food to sustain this garrison.

The growing German military involvement convinced Blomberg that Scheele was losing control of events and late in August he dispatched *Oberstleutnant* Walter Warlimont to control all *Wehrmacht* forces in Spain. Warlimont took up his appointment on 5 September as head of *Unternehmen 'Guido'*, which had diplomatic as well as military responsibilities.[6] Blomberg also knew that the Republic was receiving French military aid, including aircraft, some sent by Paris, some sent by sympathetic French aircraft workers and others smuggled in despite a formal agreement by European governments not to intervene in the civil war. Unofficial military aid also reached the Nationalists from Germany, with Mola receiving small arms and ammunition in a commercial transaction with Josef Veltjens, the wartime ace and holder of the *Pour le Mérite*. These were shipped into La Corunna on 26 August, the vessel in question returning with a consignment of iron ore and the bodies of Schulze and Zech. This freelance transaction worried Wilberg, and during a visit to Franco he ensured that future German aid would be Government-sponsored.

His visit led to more aircraft being dispatched to Spain, bringing the total by the end of October 1936 to 146 (see Appendix 7). Bomb racks and 158 tonnes of bombs were unloaded at Lisbon in mid-August through the intervention of the Portuguese dictator Antonio de Oliveira Salazar, while the following month the Germans delivered a *Flak 36* 8.8cm battery under *Hauptmann* Aldinger, some searchlights and 86 tonnes of bombs. Many aircraft went to the Nationalist Air Force, but in September the Germans created their first bomber *Kette*, the 'Pablos' (Pauls) under Moreau, with three Ju 52s, while in October a floatplane unit was formed with He 60s to protect German shipping. A *Versuchskommando*, also known as the *Mechelis Kette*, was raised with He 50 and Hs 123 dive bombers yet saw little service until the end of the year, a 1938 report blaming unspecified 'material difficulties'. Although grateful for the aircraft, the Spanish, and soon the Germans themselves, rapidly became disillusioned with their quality. The He 46s, dubbed *'Pava'* ('Hen Turkey'), proved little better than the Bre 19A-2s they replaced, while the He 51s seemed inferior to every other contemporary fighter. However, there was nothing else available and Eberhardt's flight was re-equipped and expanded with new He 51s. The older aircraft were handed to the Spanish.

Eberhardt was based in Caceres (although the fighters operated from a forward airstrip at Talavera), as were Moreau's *'Pablos'*, which flew their first raid against a fuel depot on 15 September then supported the advance to Toledo. After Toledo was captured Warlimont rested his airmen for nearly a month and also arranged for a small tank force to be dispatched to Spain. With considerable tact, he prevented this becoming a bone of contention between Blomberg and Göring by placing the tanks under nominal *Luftwaffe* command while subordinating the *'Feuerzauber'* mission to *'Guido'*. Shortly afterwards Wilberg expanded Scheele's forces, the fighters becoming *Gruppe 'Eberhardt'* (with fourteen He 51Bs by the end of September), while the creation of a *'Pedros'* ('Peters') unit and an expansion of the *'Pablos'* facilitated the raising of *Gruppe 'Moreau'* with twenty Ju 52/3mg3e bombers and two He 70F-2 reconnaissance aircraft.

During the second half of October Scheele's squadrons became an aerial fire brigade operating at both ends of the front. They helped to relieve Oviedo in the north on 17 October and simultaneously to attack enemy aircraft at Malaga using Ju 52s, He 60s and Hs 123s. During the night of 26/27 October five *'Pedros'* and *'Pablos'*, in company with three Italian Savoia S.81s, bombed Cartagena in a vain attempt to stem the flow of Soviet war material just as the Nationalists were poised to take Madrid. The threat of such intervention was one of the arguments Warlimont used when he flew to Berlin in late September to press for an expansion of the *Wehrmacht* in Spain. Aware of Germany's industrial weakness, Göring opposed him, but Hitler took a contrary view because Franco had formed a Nationalist Government on 30 September which Germany had recognized. Spain was a useful source of iron ore for Germany (11 per cent of the latter's imported iron ore came from Spain in 1929 and 9 per cent in 1938, compared with 14 per cent from France) and occupied a strategic position both at the mouth of the Mediterranean and on France's southern border.

On these grounds Hitler decided to create a full expeditionary force in Spain. The decision was made after Warlimont returned to Spain, and he learned of it from Franco and the visiting head of German Military Intelligence, *Admiral* Wilhelm Canaris. *'Feuerzauber'* and *'Guido'* were replaced by *'Winterübung Rügen'* ('Rügen Winter Exercise') and Scheele's command was to be absorbed by a new force originally called *'Eiserne Rationen'* ('Iron Rations') and then *'Eiserne Legion'*. However, Göring's preference for bird names led to its being renamed the Condor Legion.[7]

Generalmajor Hugo Sperrle was selected to command the Condor Legion, with *Oberstleutnant* Alexander Holle as his Chief of Staff. The Legion had a *Kampfgruppe* (*K/88*), a *Jagdgruppe* (*J/88*) and an *Aufklärungsstaffel* (*A/88*), with corps and reconnaissance *Ketten* and a *Küstfliegerstaffel* (*AS/88*). They were supported by a *Flak Abteilung* (*F/88*) with three heavy and two light batteries, a communications battalion

(*Ln/88*) and maintenance, transport and guard units. However, with its resources already stretched to the limit, the *Luftwaffe* could not afford the luxury of dispatching a fully equipped force to Spain, so in-country facilities and equipment, including vehicles, were used whenever possible. Consequently the 1,500 vehicles with which the Legion began its campaign were of 100 different types, which created a maintenance nightmare. Also as a result of this policy, Eberhardt's and Moreau's units were absorbed into the Condor Legion, becoming the *4. Staffel* of, respectively, *J/88* and *K/88*. Henke finally left Spain at this time to re-join Lufthansa, for whom he was to fly an FW 200 across the Atlantic to New York in 1938. Scheele became the Air Attaché in Salamanca, where he remained until January 1939.

Men and equipment were shipped from Stettin and Swinemünde for Cadiz and Seville, the men wearing mufti while sailing through the Channel and the Bay of Biscay. Most were volunteers, units being asked to send bachelors, who learned their destination en route. They arrived in early November and were given RAD-like uniforms with Spanish insignia, then spent the first fortnight preparing equipment. Meanwhile 33 Ju 52/3mg3es for *K/88* flew from Lechfeld via mainland Italy, Sardinia and the Balearics to Serrania de Ronda, with Italian warships stationed along the route to direct them like traffic policemen. For safety, and to avoid swamping the Italians, the Germans never dispatched more than ten aircraft a day across the sea. Sperrle himself left Germany by air on 31 October and arrived in Seville via Rome on 5 November.

The Russians were informed about the *Luftwaffe*'s involvement through Schulze-Boysen's growing circle of spies. Initial contact was by post, which alerted the *Gestapo*, but eventually Soviet Intelligence established more professional means of communication for a flow of information whose quantity exceeded its quality because Schulze-Boysen was incapable of assessing it. Stalin's reaction varied: in 1939 he was delighted to receive technical data on the Ju 88, but two years later he scrawled obscenities across a message warning of the imminent invasion of the Soviet Union.[8] Sperrle had 120 aircraft but only 5,000 of the 5,500 officers, men and officials on his establishment tables. Yet the Legion established its own depots and airfields, some of the latter in the south in case the main bases became unusable, as well as radio and land-line communications throughout Nationalist Spain. There were even specialist brothel arrangements, although on 1 December Richthofen noted that there had been 29 cases of sexually transmitted disease since August.[9] Sperrle had more pressing problems to consider, for the Legion's first four months must be reckoned a total failure.

As Franco's forces tried to take Madrid, the Germans could not provide adequate air support, despite the loss of 20 per cent of their first-line strength, leaving *K/88* with only 26 Ju 52s and two He 70s by the end of January 1937. The crisis was due to the enemy's winning total air superiority just as the Legion entered the fray and to the failure of the German aircrew to change the situation. This left them demoralized, and it is a tribute both to Sperrle's leadership and to their own dedication that morale did not collapse completely. The enemy's success could be ascribed to the arrival of modern Russian aircraft, including the nimble I-15, dubbed '*Chato*' ('Snub Nose') in Spain, and the SB, nicknamed '*Katiuska*' (a Russian diminutive of Katherine). There was also the advanced I-16, whose distinctive whining note earned it the nickname *'Mosca'* ('Fly') while its agility earned it the alternative sobriquet *'Rata'* ('Rat'). Concentrated in a powerful strike force on the Madrid front with experienced Russian pilots, these aircraft replaced the heterogeneous collection of warplanes and 'militarized' civil aircraft which had been distributed in penny packets along the Republican front. The Russians inflicted a steady stream of casualties upon the Germans, beginning on 4 November when *Leutnant* Oskar Kolbitz's bomber was shot down and the pilot killed. Within a week Eberhardt, too, was dead, while on 8 November two bombers were lost. The *Kommandeur* of *K/88*, *Major* Robert Fuchs, had a narrow escape but managed to get his damaged bomber home.[10]

K/88 flew its first mission against the ports of Cartagena and Alicante during the night of 15–16 November, but despite the use of a navigation beacon established at Melilla this attack was a failure, as was the first daylight raid on the main Russian air base at Alcala de Henares two days later. From 19 November *K/88* joined other Nationalist bombers attacking enemy defences to support the assault upon Madrid. Inevitably civilians died in the bombing, Republican propaganda painting horrifying descriptions of the 'slaughter' and claiming that the first raid (in which 40 tonnes of bombs had been dropped) had killed 1,000 people. But municipal records show that between 14 and 23 November only 303 buildings were hit (some by shell fire), with 1,119 casualties, including 244 dead. The bombing failed to break the defences and became ever more difficult in the face of aggressive enemy fighters, especially the *Mosca*s, who constantly penetrated *J/88*'s shield. The bombers' defensive armament was increased to five machine guns, but on 26 November, three days after the Nationalists had abandoned their attempt to storm Madrid, *K/88* temporarily gave up daylight bombing raids.

It switched to night attacks on chokepoints around Madrid, on bridges and in the narrow streets of towns. One or two *Staffeln* would bomb each night, supported by harassing attacks with single aircraft, but they had little effect. The crews faced extra dangers and discomfort because their bases were separated from the Madrid front by the 8,000ft (2,500m) high Sierra de Guadaramma and Sierra de Gredos mountains. In the depths of winter the pilots crossed them with one hand on the controls and the other beating their thighs to maintain circulation. Low clouds made flying even more hazardous as the heavily loaded bombers often flew close to the peaks, but there was virtually no Spanish meteorological service and several months elapsed before the *Luftwaffe*'s own system became effective. By good fortune only one transport was lost, on 17 January in the Sierra de Gredos, killing five men. In these circumstances, even restricting *K/88*'s crews to one sortie a day failed to boost morale—which was further undermined by outbreaks of influenza and pneumonia caused by poor living conditions and the fact that the men had only one uniform. Three men died of illness during January.

The fighters of *Hauptmann* Hubertus von Merhart's *J/88* were increasingly used for ground attack missions, which exposed the pilots to intense ground fire. On 6 January, while supporting an attack trying to cut the Corunna–Madrid road, *J/88* lost *Leutnant* Hans-Peter von Gallera and his wing man *Unteroffizier* Kurt Kneiding, who both fell in flames. Gallera's body was found by a company commander in the International Brigade's Thaelman Battalion who had been a former comrade in *StG 165 'Immelmann'*. Frequently German bravery was wasted, for Nationalist troops emerged from their trenches not to exploit the success but to applaud the attack! A frustrated Merhart refused to send his men into battle in inferior aircraft and stood his ground when Sperrle sought to browbeat him.[11]

The He 51 rarely reached the *Katiuska*s, which were 110kph (70mph) faster than the German aircraft, although on 12 December five pilots of *4.J/88* (the original German fighter unit in Spain) each claimed one on the northern front. The first was by *Leutnant* Kurt von Gilsa and the last by Knüppel, his eighth victory. No more were claimed until July 1937, and the woefully inadequate He 51 was relegated to the bomber escort role. With frequent moves anticipated, the *Gruppe*'s personnel were installed in a nine-carriage train with living accommodation, mess hall, kitchen, headquarters, maintenance and spares. Both *J/88* and *K/88* were used in vain attempts to trap the Russian airmen on the ground but all efforts were frustrated by careful dispersal and cunning tactics. In desperation the Legion spent much of January operating in the north and south where the Russian presence was weaker.

The importance of the new Nationalist offensive in the Jarama valley, south of Madrid, from 6 February saw the Legion gamble again on daylight bombing to isolate the battlefield by interdicting the roads. Opposition from the *Chato*s and *Mosca*s was fierce—they twice broke up *K/88* formations—and after just four days the bombers returned to the safety of the night. Only two He 51s were lost and their pilots baled out, but others on escort duty had the humiliation of being forced to shelter behind the bombers' guns! Throughout February *J/88* recorded only one victory, by *Unteroffizier* (later *Leutnant*) Hans-Jürgen Hepe over a *Chato* on 12 February. Following this débâcle, Berlin ended the Heinkels' fighter role and *J/88* was confined to low-level close air support missions, initially on the quieter sectors of the Madrid front. The only success recorded by the Legion in the Jarama battle was the support *F/88*'s guns provided for a Nationalist brigade, and direct support for the infantry became the unit's secondary role throughout the war.

Modern aircraft were vital for the Legion's future, and in December Moreau, who had nursed a badly damaged bomber back to base the previous month, was dispatched to Berlin by Sperrle with a plea for some of the new bombers. New-generation fighters were already in Spain with the assembly early in December of three Bf 109 and one He 112 prototypes, together with one of the Ju 87s, the He 112 V4's twin 20mm cannon armament earning it the nickname '*Kanonenvogel*' ('Gun Bird'). All were assigned to the new *VJ/88* but saw limited operational use and the Bf 109s returned to Germany in February 1937, followed by the *Stuka* five months later. The 'Gun Bird' stayed attached to *2.J/88* until it was written off in an accident in July 1937.

Accompanying these aircraft was Richthofen, the former Head of Development, who was to supervise their operational evaluation. Now an *Oberstleutnant*, he apparently pulled strings to obtain his posting, partly for professional reasons and partly because of a desire to fight Communism. After three weeks in Spain he returned to Berlin just before Christmas and on 6 January he had a conference with Milch, Kesselring, Udet and Wilberg, who gave him special orders for Sperrle. These probably confirmed that new aircraft would be sent to Spain, but a comment that he would be 'Chief' in Richthofen's diary on 9 January suggests that they included his appointment as Sperrle's Chief of Staff in place of Holle. Holle was a very able staff officer who had enlisted in the infantry during the Great War, joined the *Reichsheer* as an officer and transferred to the *Luftwaffe* in 1934. The reasons for his departure from Spain are obscure, although Richthofen's diary on 15 January suggests that Holle disliked delegating authority and was too pessimistic. He left on 22 January, but after acting as *Kommandeur* of a *Stukagruppe* he was appointed Chief of Staff of *Luftwaffenkommando Ostpreussen* in the summer of 1938. In 1940 he became the Chief of Staff of *Fliegerkorps IV*, later commander of *Fliegerkorps X* in the Mediterranean, and in 1944 he replaced Sperrle as commander of *Luftflotte 3* (renamed *Luftwaffenkommando West*).

Richthofen's appointment may also have been related to the arrival of the new aircraft which he had helped to develop. Sperrle is reported to have regarded Richthofen as a ruthless snob, but this might have been a hangover from their disputes about aircraft in 1934. Professionally they appear to have had few disagreements, but their personal relationship did not run smoothly. Richthofen disliked his superior's coarse wit and table manners, yet this did not stop them frequently ending the day playing cards nor meeting socially after their service in Spain. Because Richthofen and Franco got on well together, Sperrle increasingly left day-to-day affairs in the former's capable hands.[12]

The new aircraft began to arrive at the end of January 1937 and Moreau's *4.K/88* became *VB/88* with twelve Do 17E-1s, He 111B-1s and Ju 86D-1s. The new *Staffel* went into action on 9 March when the He 111s attacked enemy airfields to support the Italian offensive north-east of Madrid towards Guadalajara. Later the same month sixteen Bf 109B-1s, minus their *FuG 7* radios, arrived to re-equip *2.J/88* under *Oberleutnant* Günther 'Franzl' Lützow, who used personnel from *4.J/88* (disbanded in April). Lützow was the scion of a famous military family and would stand up to anyone, including Göring. He was one of the leading fighter staff officers and was described by Galland as 'the best fighter leader of the *Luftwaffe*'. He perished when his Me 262 was shot down in April 1945. Although aware of the urgency of exploiting the new fighters, Lützow demonstrated his authority by ensuring that his pilots were completely familiar with the new aircraft before they went into action and organized a conversion course which concluded with two one-hour flights.

Changes also occurred in *A/88*, where physical and psychological exhaustion among the He 45 crews of the corps *Kette* had led to the latter's disbandment and the return home of its men (the aircraft were transferred to the Spaniards). The *Staffel* retained the He 70s, although they were increasingly diverted to bomber missions, but the *Blitz* proved to have a disappointing performance and poor cockpit visibility and it was difficult to maintain. Only in the spring was the importance of reconnaissance recognized, and from April the Heinkels were supplemented by a *Kette* of four Do 17F-1s while the corps *Kette* was re-formed with four He 45s.

While new aircraft would lead to the Legion's renaissance, the support organization continued to cause concern. A train was converted into a mobile maintenance depot, but the failure to standardize on equipment remained a major problem for the Legion—and the *Luftwaffe* in general well into the Second World War. This was especially so with motor transport, where the number of vehicles was inadequate and where the profusion of types created a spares crisis, with up to 50 per cent of vehicles unserviceable. The situation was aggravated by poor roads, a lack of driving experience and the inevitable recklessness of youth, spurred on by the consumption of alcohol. There were frequent accidents, which brought a steady stream of casualties with broken bones, lacerations and concussion, not to speak of twenty fatalities.

Replacements and equipment were usually shipped to Spain from Hamburg, with an average of five ships sailing each week. The men were dispatched periodically in drafts of 300–500, increasingly volunteers, who assembled at Gatow, embarked at night and were usually confined below decks if any vessel was remotely near their ship. The ships' appearance was often altered after they had passed through the Channel, and once in Spanish waters the Nationalist Navy would escort them into port. Mail (addressed to Max Winkler, the Junkers representative in Spain) and key personnel flew there in one of four Ju 52s which ran a weekly service from Berlin. One of these aircraft disappeared over the Mediterranean on 30 April 1937 while flying from Seville to Rome, with the loss of eight men including Gilsa, the first German pilot to shoot down an SB, and his colleague *Unteroffizier* Ernst Mratzek. Another nine men were lost on 24 February 1939 when their transport crashed in Roubilon, France, the dead including Scheele, who was on his way home after three years in Spain. Once personnel reached Spain they were often promoted one grade upon disembarkation, and with good pay rates many a Spanish veteran returned to Germany tanned and with enough money to purchase an automobile.

The arrival of new aircraft breathed life into the Condor Legion and enabled it to re-join the Nationalist spearhead as Franco adopted a new strategy. With the Italians' defeat at Guadalajara, the Nationalists abandoned attempts to take Madrid and in late March they decided to destroy the Republican enclave along the Bay of Biscay while remaining on the defensive along the eastern front. The enclave's industry and minerals would strengthen the Nationalists' war effort and help to pay for imported military equipment, while the strategy had the advantages of allowing them to conquer an area which Madrid would find difficult to reinforce.

Richthofen was briefed by the Spanish Air Force Commander, *General* Alfredo Kindélan, and learned that the first phase would be in the east and involved taking the Basque capital, Bilbao. The shock troops would be royalist Navarese, who would be supported by 150 aircraft, including the Condor Legion and the Italian *Aviación Legionaria* (*Generale* Vicenzo Velardi). As the Nationalists had only 200 guns, few of them heavy, air power would have to make up the difference, and for the campaign the Condor Legion assembled 1,000 tonnes of bombs. Richthofen's price for his support was a demand, accepted by the Spanish, for close co-ordination between air and artillery bombardments and for prompt infantry assaults when they had been concluded.

Sperrle moved his headquarters to Vitoria while 62 aircraft (including six He 111/ Do 17Es, seven Bf 109s and three Hs 123s) were assembled at Burgos by 21 March. The groundcrews, usually called '*Schwarze Männer*' ('Black Men') because of their black overalls, worked hard to ensure a high serviceability rate, which sometimes reached 90 per cent during the campaign. The Nationalists faced a mere 30 fighters and corps aircraft, including a handful of *Chato*s, but to ensure that no reinforcements were sent from Santander in the west the Legion struck airfields around that port before the main assault began on 31 March.

The main blow fell upon the centre of the Basque line around Orchandiano and was supported by all the Condor Legion. Later in the day Sperrle's bombers struck Durango (some ten miles north of Orchandiano) after Mola had reported enemy reserves assembling in the town, but by the time the bombers arrived the troops had withdrawn and the bombs killed some 250 civilians and injured 500. Sperrle protested at this waste of resources, but his bombers inflicted further loss of life upon the luckless population through inaccurate attacks upon the town's small-arms factory, these raids merely strengthening the Legion's ogreish image.

As the battle for Orchandiano raged until 4 April there was strong tactical support from the He 51s of *J/88*, which bombed and strafed positions only 50 yards from the Nationalist lines for fifteen minutes before the infantry attack, although the sorties were sometimes hampered by rain showers. These close air support missions cost *J/88* one He 51 shot down and six badly damaged, so Richthofen understandably found it galling when the Spanish troops failed to exploit them. On 2 April there was a bad-tempered exchange with Mola, who spitefully demanded the abandonment of close air support in favour of attacks upon Bilbao's factories. Richthofen regarded the demand as 'nonsense', for he knew the value placed on these factories by both Spanish and German leaders. Although he could have ignored Mola, for appearances' sake he 'compromised' by stating that he would attack an explosives factory when there was a lull at the front.

Air power helped in the final battle for Orchandiano by disrupting the enemy's attempts to rally, but the fluidity of the situation after the town had fallen led first to an accidental attack by the Germans upon their allies and then to a brief moment of farce when Sperrle was arrested as a spy while wearing civilian clothes and visiting the front. Afterwards rain, low cloud and fog restricted the Legion's operations and during the lull Sperrle visited Franco to press for an envelopment rather than the traditional step-by-step campaign currently being pursued by Mola. It was a concept which also found favour with Richthofen because it would provide air power with the key role of isolating the battlefield and destroying enemy reserves. By late April the stage seemed set for such an operation as the Basque line lay only nineteen miles (30km) east of Bilbao and the defenders thought only of shelter in the 'Ring of Iron', a line of fortifications around the city. But what began as a unique opportunity for air power ended in a tragedy which muddied the *Luftwaffe*'s name for decades.

The Basques' resolution buckled under the weight of an attack launched on 20 April and within five days Richthofen, from his new headquarters at Durango, could watch them retreat towards Guernica, a rail terminus which lay at the lowest crossing point on the River Oca. Surprisingly, there was no target file on the town, although it held the world-renowned Astra-Unceta small-arms factory, and when Richthofen asked his staff 'Do any of you know anything about Guernica?', all shook their heads.[13] Yet during the morning of 25 April an He 45 of *A/88* flew over the town's Rentaria Bridge, which Richthofen provisionally earmarked for attack. However, he was more interested in the village of Guerricaiz (to which he refers in his diary as Guernicaiz), some six miles (10km) south-east of Guernica and six miles north-east of Durango, through which the Basque left flank would have to retreat.

In the evening of 25 April he ordered a morning attack upon the village by *K/88*, but even as this began, air reconnaissance reported heavy traffic in and around

Guernica, where, it appeared, enemy troops were massing. Although there were some 2,000 soldiers in the town, which the Basque Government had decided to fortify, most of the 'troops' were actually refugees, but Richthofen saw a fantastic opportunity to block the retreat of the entire Basque left which could then be enveloped. He rushed to give the news to the Navarese Chief of Staff, *Coronel* Juan Vigón, who gladly gave him permission to take the bombers off other missions to meet the new situation. As Richthofen's diary for 26 April makes clear, the objective was to block the roads south and east of Guernica. *A/88* and *J/88* were to comb the roads between Marquina and Guerricaiz while the bombers (and Italians) were to concentrate 'on to roads and the [Rentaria] bridge (including the suburbs) immediately east of Guernica.'[14]

But instead the Condor Legion, now with 80 aircraft, struck Guernica itself in an action widely portrayed as a deliberate attempt to cow the Basques by obliterating a historic town. This credits the Germans with an understanding of Basque culture which they did not begin to possess, while the subsequent destruction of the town astonished Richthofen, as his diary entries for 27 and 30 April clearly show. The mission was assigned to *VB/88* and *K/88* rather than the Hs 123s and Ju 87s because the dive bombers could not carry an adequate bomb load over the mountains from Burgos, yet the divergence between Richthofen's intentions and the way they were executed strongly indicates that the orders were garbled in transmission from Durango or misunderstood upon arrival.

The attack was made in the late afternoon by 26 bombers, which dropped 45 tonnes of bombs (including 2,500 incendiaries), and sixteen fighters (*1.* and *2.J/88*), which afterwards strafed people fleeing along the roads from the town. Because there were no AA guns, *VB/88*'s target-marking attack was made from 4,000ft (1,200m), yet although there was little wind Moreau's bombs fell into the town on each side of the river, hitting the railway station west of the river and an olive oil plant in the east. The latter burned so fiercely that the town was quickly enveloped in a cloud of smoke and dust. Most of *K/88*'s bombs hit the eastern part of the town, although some fell west of the river. No one knows how many died—estimates range from 250 to 1,500 people, of whom the majority were civilians—but 80 per cent of the houses were either destroyed or seriously damaged.

The following day bad weather grounded the Legion as an international furore developed and Blomberg repeatedly demanded to know who had bombed Guernica. Perhaps out of security considerations, Sperrle's headquarters replied 'Not Germans' and the men were now told to say nothing about the mission.[15] Only when his troops crossed the Rentaria Bridge to take Guernica on 29 April did Mola learn the truth, and in a fury he banned attacks upon towns or villages and demanded there be no repeat of the incident. Curiously, an official *Luftwaffe* evaluation repeated Nationalist propaganda claims that the Basques had destroyed the town, and this became the official line even in secret documents. Although bewildered, Richthofen later ignored the evidence of his own diary to state in his report that the 'concentrated attack' on Guernica 'was the greatest success'.[16] He does not appear to have blamed anyone, and he concluded on 30 April that smoke had prevented the destruction of the bridge, adding unfairly that the Junkers had 'just dropped bombs anywhere'.[17]

It is worthy of note that on the first day of the northern offensive the Operations Room bulletin board included a notice stating that the Legion would attack only military targets but would do so 'without regard for the civilian population'.[18] Hundreds of civilians certainly died at Guernica, which appalled world opinion and inspired Picasso to produce his world-famous mural, yet the presence of soldiers and the small-arms factory would have made the town a genuine military target even if it had not also been a communications centre. However, in the popular mind it was to head a long list which included Warsaw, Rotterdam, Coventry and Belgrade where civilian loss of life was heavy. All the latter were alleged victims of *Luftwaffe* spite, yet in most cases they too were legitimate targets.[19]

In fact, the destruction and abandonment of Guernica stiffened Basque resolve as they conducted a slow, fighting retreat towards the 'Ring of Iron', and a frustrated Richthofen noticed that his bombs no longer drove them from their positions. From early May *K/88* was extremely active, flying three sorties a day against the 'Ring of Iron' as well as enemy positions, and by 10 May it had dropped 681 tonnes of bombs. The He 51s were now copying the Spanish *'cadenas'* (chain) tactics developed in Aragon a month earlier, in which formations of aircraft attacked in succession at low level until their ammunition was expended. The 'Gun Bird' of *2.J/88* was also used for ground attack and claimed two tanks. Towards the end of May the front reached the 'Ring of Iron', but an assault was delayed by bad weather and this allowed the Republic to dispatch some 45 aircraft including *Moscas*, *Chatos* and *Katiuskas* to the enclave, of which Bilbao received eighteen fighters. A battle for air superiority developed and the Nationalists lost eleven aircraft, some of them from the Legion, which secured no victories. Mola refused to allow air or artillery bombardment closer than 3km (3,300yds) from Bilbao and this order stood even after he was killed in an air crash on 3 June.

The assault began on 11 June, with the main attack coming in from the south the following day, supported by 150 guns including three batteries of *F/88*. In perfect flying weather, waves of aircraft struck the Basque positions, dropping 100 tonnes of bombs while *5. (Leicht) F/88* broke up a counter-attack with its cannon. By the end of the day the Ring had been broken, although the success was marred by the accidental bombing of an *F/88* battery, which lost four dead as a result. The collapse of the Basque forces meant an intensification of Nationalist air attacks, and some Legion pilots flew seven sorties on 14 June. Three days later the Basques began to abandon Bilbao and their Justice Minister, Jesús Maria Leizaola, made a show of chivalry uncharacteristic of the war by handing over 1,000 political prisoners. These included *Leutnant* Joachim Wandel of *J/88*, who had been shot down by ground fire and taken prisoner while attacking enemy columns on 13 May. This may have paved the way for the city's surrender, on 19 June, to the Legion's liaison officer, *Leutnant* Gockel.

The campaign had cost about eight aircraft and nine aircrew, including two of Wandel's comrades executed at Orchandiano on 6 April. Another sixteen men had been killed in accidents or during the ground battle, the most disturbing losses being the deaths of three *K/88* groundcrew in two accidental bomb explosions. The Legion was to lose a bomber and an He 59 to the same cause in 1938 and the problem was not confined to Spain. In the first nine months of the Second World War the *Luftwaffe* would be plagued by similar incidents which cost at least a dozen aircraft and two dozen lives, the cause probably being the electric fuzes.

For Fuchs the problem was academic, for during the summer he was replaced as *Kommandeur K/88* by *Major* Karl Mehnert, while in *J/88*, at the beginning of July, Merhart was replaced by *Major* Gotthard Handrick, who joined just as the Madrid front burst into life. The Republican Government had made little attempt to aid the Basques because it was preoccupied with internal disputes and the need to rebuild its armed forces. Now it sought to use those forces both to regain the initiative and to re-establish international credibility through an offensive. At the Communists' insistence this involved a double envelopment of the First Nationalist Corps holding a bridgehead in Madrid city, and it began early on 6 July supported by 200–300 aircraft. The southern thrust was quickly contained but the northern one had spectacular success and took the town of Brunete to expose the main Nationalist supply route. Nervous at their success, the Republicans stopped their advance while they consolidated their rear, and this gave the Nationalists under *General* José Enrique Varela time to build a new line which effectively contained them.

At Franco's request Sperrle immediately dispatched *2.J/88* and *VB/88*, which went into action on 7 July, by which time the scale of the offensive was recognized. Sperrle now not only committed all 80 Legion aircraft but was also given command of the Nationalist air forces operating on the front, a total of 161 aircraft (33 Spanish), and established his command post alongside Varela's headquarters. Sperrle decided to keep the enemy bridgehead under constant pressure using small bomber formations, and under blazing skies the Nationalist air offensive began from 8 July. The vigorous enemy air defence actually made the offensive a twenty-four-hour operation because the *Mosca*s continued to force the older Nationalist bombers (Ju 52s and S.81s) to operate at night. In the day the modern bombers (including Do 17s and He 111s) hunted enemy troop columns, which were usually discovered by corps aircraft whose pilots radioed back their co-ordinates so that within 30 minutes of discovery the Republican troops would find themselves under attack. They quickly learned to march in well-dispersed formations, constantly watching the sky and noting ravines or ditches in which they might take shelter. The scale of the Nationalist bombing effort may be gauged from the fact that the Italians alone dropped 106 tonnes of bombs, an average of 5.5 tonnes per day, during the battle. Fighter-bombers also joined these attacks, although *J/88*'s He 51s also sought out, and carpet-bombed, enemy AA batteries.

In an effort to secure air superiority the Nationalists briefly struck enemy airfields on 10 July before returning to the battlefield, where the first Bf 109 was lost in air combat on 12 July. The pilot, *Unteroffizier* Guido Höness, was shot down and killed by a *Mosca* after dispatching two Czech Aero A-100 light bombers, but this was an increasingly rare success for the Republicans, who were unable to break through Varela's defences and were under constant air and artillery bombardment. On 13 July they went on to the defensive, but Varela and Franco were determined to regain Brunete and prepared a counter-offensive which began on 18 July. Sperrle's airmen succeeded in gaining air superiority and claimed 21 Republican aircraft on the first day, but the breakthrough and victory came only on 24 July after a week of intense pressure. Once again the He 51s provided close air support for the infantry while bombers struck enemy reserves sheltering in the numerous ravines which criss-crossed the battlefield, destroying morale as well as equipment.

The Republic had suffered a terrible defeat which cost it 20,000 casualties and much equipment, including 100 aircraft, but the Nationalists had paid a high price, suffering 17,000 casualties and losing 23 aircraft, a third of them (eight aircraft) from the Legion—which lost four on 25 July alone. Nine men were killed, but their deaths were avenged by *2.J/88*, which claimed sixteen aircraft (the majority I-16s but including three SBs), four at the hands of *Feldwebel* (later *Leutnant*) Peter Boddem, who was to leave Spain with a tally of ten. The air victory had owed much to the *Luftwaffe*'s new generation of combat aircraft, whose success strengthened Sperrle's case for the total re-equipment of his *Staffeln*.

But he was under a cloud because of a feud with the German Ambassador, *General* Wilhelm von Faupel, who had replaced Warlimont as the senior German representative in November 1936 and proved to be as opinionated as Sperrle. The vendetta was undermining German interests in Spain and Hitler decided to comply with Franco's demanded for Faupel's recall. Sperrle was scheduled to follow shortly afterwards, but the military situation was judged too delicate to remove him and he did not depart from Spain until 31 October. However, Blomberg was already considering a replacement in mid-July and proposed Kesselring, who had just taken command of *Luftkreiskommando 3* (see Chapter 6). As events in Italy five years later were to demonstrate, this would have been a fine choice, but Göring was unwilling to see any upheaval in the main command facing the French and he selected instead his chief military administrator *Generalmajor* Volkmann, who had been at the *RLM* for four years and was scheduled for front-line command.[20]

Whether or not he knew he was on borrowed time, Sperrle was preparing to complete the elimination of the northern enclave, the next phase involving the capture of Santander. None of the aircraft lost at Brunete had been replaced (*J/88* had only 27 fighters) and many of the 68 survivors required repair or maintenance, which they received over the next fortnight. The offensive began on 14 August and the Legion was not stretched, for the Nationalists assembled 220 aircraft, mostly Italian, and faced only 45 enemy machines, including eighteen *Mosca*s and *Chato*s. The Legion lost no aircraft, although its He 111s and Bf 109s were used for strikes deep in the enemy rear, the fighters proving superior to the Russian defenders and claiming fourteen victories. Santander fell within a fortnight, and after a brief pause the Nationalists began to advance into the Asturias on 1 September. The mountainous terrain proved difficult to overcome and progress from east to west proved painfully slow until 15 October, when the defence suddenly collapsed, allowing the Nationalists to occupy Gijon on the 21st of the month.

Before the last campaign the Legion spent several days softening enemy resistance by striking communications and troop concentrations, while on 1 September *K/88*, supported by the Bf 109s, attacked mountain passes, ports and airfields. The close air support proved especially devastating and the Republican commander, *Coronel* Adolfo Prada, later said that bombing was known to bury complete units, while before even meeting the enemy, air attacks cost others up to 10 per cent of their men. The He 51s often used low-level carpet bombing against strong enemy positions which they would attack from the rear by flying along valleys and then over crests. They flew up to seven sorties a day, refuelling on every other landing, and it was so hot that the pilots wore shorts or swimming trunks as well as their boots and goggles.

However, ground fire was deadly, and half the twelve aircraft destroyed or badly damaged were lost to this cause. The losses represented 17.5 per cent of the first-line strength, while 22 men, including a *Flak* gunner, were killed, of whom the most prominent was *Oberleutnant* Hans-Detlef von Kessel, *Staffelkapitän* of *A/88*, shot down over Llanes on 4 September. In return *J/88* claimed nineteen victories (eight to *Oberleutnant* Harro Harder) and *K/88* sank a 735grt coaster and the destroyer *Ciscar*.

With the end of the Asturias campaign, Sperrle returned to Germany, where he was promoted to *General der Flieger* and in February 1938 assigned command of the new *Luftwaffengruppenkommando 3*. A final dispute with Richthofen marred his departure when he refused to allow the Chief of Staff to make a farewell address. Volkmann and Richthofen appear to have worked well together, but the latter increasingly felt that the *Luftwaffe* had learned everything it could and wished to see the Condor Legion withdrawn. He knew that his successor had been selected, but he remained Chief of Staff for three months so as to ensure continuity and then returned to Germany, where he was promoted to *Oberst* and later became *Kommodore* of *KG 257*. He and Sperrle had proved to be an extremely successful team which carried the Legion through a period of crisis and then provided convincing demonstrations of the potency both of *Luftwaffe* equipment and of doctrines.

However, Richthofen's doubts were shared in Berlin from the summer onwards, and as late as November 1937 Blomberg opposed the dispatch of further military aid.[21] Hitler disagreed and Volkmann's departure provided Berlin with an opportunity to renew its commitment to the Nationalist cause by dispatching much needed men and equipment, bringing total deliveries of aircraft in 1937 to some 280. Both personnel and *matériel* had been at a premium during the summer and in the early autumn the shortage of spares had been so acute that the 'Black Men' had had to scavenge parts from wrecked aircraft and vehicles. As *Oberleutnant* Douglas Pitcairn, *Staffelkapitän* of *3.J/88* and a founder member of the Legion, later observed, 'We had the feeling we had been sent to Spain and then deserted.'[22] Mehnert's *K/88* absorbed *VB/88* and received sufficient He 111s to re-equip all three *Staffeln*, and Handrick's *J/88* received more Bf 109Bs to re-equip *1. 'Marabu' Staffel*, whose He 51s re-formed *4. 'Pik As'* ('Ace of Spades') *Staffel*, the *3. 'Mickymaus' Staffel* coming under *Oberleutnant* Adolf Galland. *A/88* converted all the reconnaissance *Ketten* to Do 17s, but the corps *Kette* retained the He 45. The Legion now had 100 aircraft, while *F/88* received an extra two heavy batteries and two 3.7cm *Züge* (platoons). Yet there was a shortage of high-octane fuel, forcing the Condor Legion to pay particular attention to fuel economy when planning its operations.

Volkmann arrived as the Spanish used imported and captured equipment to expand their armed forces, which were then concentrated along the eastern front in anticipation of a war-winning campaign. The Germans and Italians proposed striking into Aragon to split industrial Catalonia, with its capital Barcelona, from the rest of the Republic, but Franco believed that Catalonian resistance would be too strong and opted for a knock-out blow at Madrid. While the strategic options were discussed, the Legion was assembled around Soria, ready to support either one. On 4 December it joined a campaign to erode enemy air power through attacks upon airfields and for the first time *J/88* clashed with the I-152, known in Spain as the *'Super Chato'*, which proved inferior to the Bf 109B. Fog and snow not only disrupted the campaign but led to the enemy's capturing a Bf 109B on 10 December and an He 111 a week later. Both aircraft were thoroughly evaluated by the Spanish, the Russians and the French, the last producing two detailed dossiers with photographs,[23] but fortunately for the Germans the reports were considered so secret that the French Air Ministry failed to distribute them among designers.

On 15 December, a week before Franco's attack upon Madrid was scheduled, the Republic launched a spoiling attack at the isolated town of Teruel in southern Aragon, 110 miles (180km) south-east of Soria. The garrison was quickly surrounded, but only when blizzards had smothered all hope of relief did they surrender on 8 January 1938. By then Franco had taken up the gauntlet, although it was only on 22 December that he reluctantly abandoned hopes of taking Madrid. Political machismo made him contest any loss of territory, and immediately after the Republican offensive had commenced he began to transfer forces from Madrid to Teruel despite opposition from his allies, including Volkmann. Soon afterwards Franco informed his allies at a planning conference attended by Richthofen that he intended to launch a counter-offensive at Teruel on 29 December.

The proximity of the Legion to Teruel meant that it was committed from 17 December, two heavy *Flak* batteries being the first units dispatched, while the headquarters was transferred to Bronchales in the Sierra de Albarracin, 22 miles (35km) north-west of Teruel. Plummeting temperatures and snowstorms confined air operations to close air support and the bombing of enemy reserves. The Nationalist counter-offensive began on schedule but was stopped within two days by blizzards. With the temperature down to −18°C—a portent of the Russian winter in 1941–42—German aircraft engines froze and snow blocked the runways. The 'Black Men' had not only to clear the snow but also to devise temporary shelters in which they could warm up the engines before take-off.

Richthofen did not see the renewed counter-offensive, for on 11 January he was relieved by *Major* Hermann Plocher, who had celebrated his 37th birthday six days earlier. The broad-shouldered Swabian had joined the Army as an officer candidate in October 1918 and been commissioned three years later in *13.Infanterie Regiment*, where his colleagues included Erwin Rommel and Hans Speidel. He had been one of the first Lipetsk 'Young Eagles' and, in 1935, one of the last staff officers to transfer to the *Luftwaffe*, where he had worked in the organization branch. He had already been in Spain a month, but he had to await a lull in the Teruel campaign before taking up his new appointment. The blizzards were now easing, and despite the intense cold the Nationalists resumed their attack on 17 January. Progress was slow, but the advance sucked in Republican reserves, especially from the Alfambra valley north of Teruel, where the defenders were stretched to the limit. On 7 February the Nationalists exploited the situation and struck into the valley, where resistance quickly collapsed, allowing the attackers to gain much territory and take many prisoners in a day-long campaign.

The entire Legion supported the Alfambra offensive, and in the four days before the assault *K/88* helped to soften up the defences, the proximity to the target permitting more than 100 sorties a day, with 120 tonnes of bombs dropped on 6 February. However, on the previous day a *Staffel* accidentally hit a Nationalist division, which suffered 400 casualties (it would suffer 30 in the attack itself). The Republican Air Force, some 300 strong, responded vigorously and there were fierce air battles, while the Legion's base at Almazan was attacked and many aircraft were damaged. Although *J/88* had only limited success (34 victories) during the Teruel campaign as a whole, 7 February saw a magnificent victory when it jumped a formation of twelve *Katiuska*s and shot down ten together with two *Mosca*s, four of the bombers falling to *Oberleutnant* Wilhelm Balthasar and two to *Oberfeldwebel* Reinhard Seiler.

Bad weather delayed the Nationalists' bid to recapture Teruel until 17 February but within six days the town was back in their hands. Initially the Legion supported this attack only with the He 45 corps *Kette* and Bf 109s, but then came the début of the Ju 87 *Stuka*. A trio of Ju 87A-1s from *11. (St)/LG 1* under *Leutnant* Hermann Haas had arrived in Spain in mid-January and been designated *5.J/88* or, more commonly, the *Stuka Kette*. They were held back until the final assault upon Teruel, when they swooped on enemy strongpoints and displayed an average bombing error of only five metres. The *Stuka*s' success marked the demise of the two surviving Hs 123s, which were transferred to the Spanish. The battle concluded with spectacular air combats in which *J/88* claimed ten fighters as the Republican Air Force struggled to shield its retreating troops.

The Condor Legion suffered trifling losses at Teruel—five aircraft and ten dead (including accidents)—so it was ready to participate in the new offensive with which Franco planned to exploit his latest victory. On 24 February he announced that the Nationalists would strike into Aragon rather than return to Madrid and he assigned 26 divisions to *General* Fidel Davila's Northern Army, which held the line between the Pyrenees and Teruel. Volkmann was astonished but delighted, for this was the strategy he had advocated since arriving in Spain and he had believed that Franco would renew the Madrid offensive.

The campaign began on 9 March as a single thrust on a relatively narrow axis but rapidly extended into multiple advances along the whole of Davila's front. These thrusts occurred both simultaneously and sequentially, usually being broken off when the enemy defences became too strong. The Moroccan and Italian Corps struck the first blow south of the Ebro, with the Navarese and Aragon Corps attacking north of the river on 22 March, the second forcing the defenders to draw troops away from the Moroccans, who reached the sea near Vinaroz on Good Friday, 14 April. However, the success rebounded upon the Nationalists for the reinforced Republican Army was now able to hold Catalonia, despite the loss of Lerida on 3 April. The thwarted Nationalists now concentrated upon taking Valencia and from 25 April began driving southwards, but a combination of bad weather, difficult terrain and fierce resistance led to frequent delays and slow progress, and Castellon de la Plana fell only in late June. A converging attack from Castellon and Teruel began on 1 July but, despite overwhelming air and artillery support, this, too, made little headway and on 25 July the offensive was ended because of the Republican attack across the Ebro.

Spectacular progress was made in the early weeks because the Republicans' weakness was exploited by small, mechanized task forces and overwhelming Nationalist air superiority. The latter was based upon quality rather than quantity, for the Nationalists had 460 aircraft (including 220 Italian and 140 Spanish) and the Republicans an estimated 550, of which 300 were new models. The Nationalists received tactical and operational-level air support, the latter involving attacks upon communications and reserves deep in the rear which destroyed not only *matériel* but also morale. As one Republican officer commented, 'Terror from enemy attacks from the air was greater than that inspired by the pistols of our own officers.'[24]

In the first advance the Legion supported the Moroccan Corps, with *K/88* used against enemy fortifications up to A-Day, when it dropped 88 tonnes of bombs in two missions, leaving the surviving defenders sobbing and shaking when the Nationalist infantry reached them. The aircraft assumed the role of cavalry, exploiting the infantry's success and disrupting every enemy attempt to rally: the *Stuka*s flew four sorties each, *K/88* dropped another 45.5 tonnes of bombs and even the Bf 109s flew strafing runs. The Germans were so successful that *A/88* was unable to find worthwhile targets, and on 16 March Plocher agreed that the Legion would operate north of the Ebro, leaving *AS/88*'s seaplanes to support the Moroccan Corps as best they could. The poor weather dashed German hopes of repeating north of the Ebro their success to the south, and during the assault on Lerida the new *Staffelkapitän* of *3.J/88*, *Hauptmann* Hubertus Hering, and his wing man were killed when their aircraft collided. *Hauptmann* Adolf Galland, who was about to return to Germany, resumed command of the *Staffel*. The strain upon Volkmann was exacerbated by the *Anschluss* Crisis, which increased his fears of the Legion's being isolated in the event of a European conflict. He pressed Berlin to withdraw the Legion, and although the crisis quickly blew over, the *RLM* began to doubt Volkmann's suitability for senior command. The doubts grew following reports of friction between Franco and Volkmann, who was constantly proffering gratuitous advice which was largely ignored.[25]

When the Nationalists reached the sea the Legion's headquarters were transferred to the coastal town of Benicarlo, near the mouth of the Ebro. For the advance upon Valencia the twin-engine aircraft used Saragossa bases while single-engine ones moved to bases around Vinaroz, although the main fighter base at La Cenia could be observed from enemy-held hills north of the Ebro. Heavy rain, high winds and low cloud played havoc with air operations and once grounded the He 51s for nine days, but the Republican Air Force exploited its all-weather peacetime bases to bomb the Legion's headquarters. In the newly arrived I-16 Type 10 the Republic had a fighter whose speed matched that of the Bf 109B but could operate at a higher altitude. Even at night there was no peace, for older aircraft harassed the Nationalist rear, a tactic the Russians were to repeat in 1941 and which inspired the *Luftwaffe*'s *Störkampf/ Nachtschlachtstaffeln*.

The battle raged not only on land but also at sea and along the coast, with *AS/88* heavily engaged. Originally forming only a reconnaissance unit under *Major* Karl-Heinz Wolff supporting the Nationalist Navy, *AS/88*'s He 59s began to carry torpedoes ('Eels') from January 1937 and on 30 January *Oberleutnant* Klümper used one to damage the Spanish merchantman *Delfin* (1,253grt). In the second half of 1937 *AS/88* began to bomb warships and merchantmen, and these operations increased from January 1938 when Wolff's successor, *Major* Hefele, was replaced by *Major* Martin 'Iron Gustav' Harlinghausen. Harlinghausen, aged 36, was a short, slim man with nerves of steel and a sense of humour who had joined the Navy as a seaman in 1923, had become a pilot in 1931 and an observer in 1934 and then worked in training schools.

Under his vigorous direction *AS/88* also attacked coastal communications as well as ports with bombs and 20mm cannon. A favourite tactic was to approach land targets from the sea at high altitude with the engines switched off, drop bombs at 1,000ft (300m) then start the engines to make a hasty departure. By early August up to sixteen sorties a week were being flown, sometimes in *Kette* formations, with each aircraft carrying a tonne of bombs, but twelve He 59s were lost in 1938 and 1939, three to night fighters. Although the He 59 remained the backbone of *AS/88* when Harlinghausen was replaced by *Hauptmann* Smidt in March 1939, two He 115A-0s also arrived for evaluation.

To support land operations the Nationalist interdiction of maritime traffic intensified from March 1938, largely at Mussolini's direction, and post-war records show attacks upon 166 ships (51 foreign) between March 1938 and January 1939.[26] Contemporary records indicate that between January 1937 and early August 1938 in the Mediterranean and the Bay of Biscay 33 vessels (82,580grt) were sunk and 73 damaged.[27] The Italians had the greatest success against shipping (see Table 7) because Barcelona, the Republic's largest port, was the prime target of their strategic bombing operations and it was defended by only 50 AA (30mm–105mm) guns. The ports of Cartagena and Valencia, against which the Legion tended to operate, had only 30 and 25 AA guns respectively, allowing the bombers to wreak havoc among ships tied up at wharves and slowly to throttle the Republic.[28]

Inevitably, *AS/88* did not escape unscathed, but between March and May 1938 the Legion lost only 26 dead, including six in an air crash. However, the lack of replacements renewed Volkmann's fear of isolation. The fears increased with the growing Czechoslovakian Crisis, and he expressed his concerns to Berlin, which recalled him late in May. Plocher was in charge during the advance upon Castellon, when the Legion encountered enemy airmen in aggressive mood. On 2 June six *Katiuskas* attacked La Cenia; *J/88* shot down five, although the following day 40 aircraft attacked the Legion's command post, which was forced to move to more salubrious premises. On 14 June *J/88* shielded the Nationalists around Castellon from formations of 40–60 aircraft but lost two Bf 109s, whose pilots were captured (one later escaped).[29]

The resurgence of the enemy air force provided further ammunition for Volkmann when he confronted the *Luftwaffe* High Command in Berlin on 10 June. He pointed out that only sixteen of the 30 Bf 109s were serviceable, the He 51s were almost unflyable and many of *F/88*'s guns had been worn smooth. Since 9 March he had lost 20 per cent of his strength, with twenty aircraft destroyed (including five He 111s and four Bf 109s), seven badly damaged and 38 aircrew dead or missing (nine were prisoners of war). He demanded either reinforcement or withdrawal, but the growing Czechoslovakian Crisis delayed a decision. He returned to Spain and on 16 June warned his officers that they might have to restrict operations since the units could not be brought up to strength.

In fact, the following day Berlin decided to provide some replacements and reinforcements, including enough Bf 109Bs and some Bf109C-1s to re-equip *Oberleutnant* Werner Mölders' *3.J/88* (Mölders had replaced Galland in May).[30] With their arrival in July, *4.J/88* was disbanded for the last time and the remaining He 51s were transferred to the Spanish. *J/88* was revitalized and on 15 July the re-equipped

TABLE 7: MERCHANT SHIPS OVER 500grt SUNK BY AIR ATTACK IN THE MEDITERRANEAN, 1937–1939

Type	No	Aggregate tonnage (grt)
Italian bombers	28	80,961
Italian seaplanes	1	2,413
German bombers	8	20,030
German seaplanes	23	55,161
Total	60	158,565

Note: Italian bombers also sank a 3,254grt auxiliary, while German bombers sank the gunboat *Laya* on 15 June 1938 at Valencia and the naval fresh water carrier *Aljibe 3* (680grt) at Cartagena on 17 April 1938. Both air forces also sank numerous fishing boats and sailing vessels, while the Germans sank three ships (7,305grt) in the northern campaign.

3.J/88 went into action, with *'Vati'* ('Daddy') Mölders and *Leutnant* Walter Oesau (*Stab J/88*) opening scorebooks which would ultimately show 115 victories (fourteen in Spain) and 125 victories (nine in Spain) respectively. Mölders proved to be an excellent fighter pilot and an able tactician, and he introduced the universally adopted 'finger four' flexible fighting formation based upon pairs of wing men. Increasingly *J/88* was able to conduct sweeps (*Freie Jagd*) of up to 32 aircraft and usually triumphed in the air battles they provoked. The Nationalist Army received the usual support as it tried to overcome the fortifications shielding Valencia. *K/88* dropped 56 tonnes of bombs on 5 July alone, despite having to fly 125 miles (200km) to the battlefield, while the *Stuka*s flew a couple of missions a day, but the Republican line held.

Meanwhile the attention of Volkmann and Plocher was increasingly drawn northwards as photographs from *A/88*'s Do 17s showed enemy activity on the northern bank of the Ebro. The Operations Officer, *Hauptmann* Torsten Christ, concluded on 23 July that the enemy intended to attack across the Ebro bend towards Gandesa but the Spanish ignored him. During the night of 24/25 July Christ's prediction came true and the Republican Army of the Ebro overran Nationalist outposts, although Gandesa kept the attackers at bay for a week until the Republicans lost the initiative and went on to the defensive. Franco decided to reduce their bridgehead and transferred 400 guns together with 434 aircraft to the new front, where the Nationalists launched the first of a series of limited offensives on 6 August in stifling heat. Other attacks followed throughout the summer and autumn, slowly regaining territory. The last took place on 30 October, and as snow dusted the battlefield on 18 November the Republican rearguard crossed the river. The initial victory proved a Pyrrhic one for the Republic, whose army and air force in Catalonia were wrecked, the material losses including 150 aircraft.

On Volkmann's initiative the Legion (with 70 aircraft) provided the initial air support at an unprecedented level, with 718 sorties (excluding fighters) during the first week of the battle compared with 299 the previous week (as Table 8 shows). The

TABLE 8: CONDOR LEGION SORTIES, 18 JULY–18 SEPTEMBER 1938

W/b	Bomber			Stuka	Recon	Corps
	K/88		A/88			
18 July	147	(183.75)	101 (50.5)	10 (5.5)	31	10
25 July	422	(486.00)	165 (80.5)	77 (37.5)	52	2
1 Aug	339	(414.00)	69 (34.5)	39 (19.5)	38	2
8 Aug	237	(284.75)	101 (48.8)	33 (16.0)	30	8
15 Aug	263	(341.25)	–	39 (19.5)	31	11
22 Aug	175	(243.75)	–	23 (11.5)	30	27
5 Sept	126	(148.75)	–	11 (5.5)	10	9
12 Sep	83	(104.85)	–	9 (4.5)	5	4

Note: Figures in parentheses are bombs dropped in tonnes.
Source: *Lagebericht* 6 July onwards in IWM Microfilm Ger/Misc/MCR/19 (1A).

bombers, including *A/88*, flew 587 sorties to drop 566 tonnes of bombs, while the *Stuka Kette* flew another 77 sorties (37.5 tonnes). Much of this effort, including 328 bomber sorties (56 per cent), was directed against six bridges. Intense anti-aircraft fire was encountered (the enemy had 27 guns of various calibres) but no fighters, because the Republican Air Force was inactive until the beginning of August. Even then it was largely confined to occasional fighter sweeps, corps flights and ground attacks, and only on 23 August did the *Katiuskas* appear. Nationalist fighter sweeps took a steady toll of enemy aircraft and established air superiority well behind the Ebro. The battle against the bridges consumed 1,713 tonnes of German bombs (representing some 1,600 sorties), but the skill of Republican engineers ensured that none was out of action for long.

The Czechoslovakian Crisis in September had a direct effect upon Volkmann, who was ordered that month to send 246 men home by sea to meet the *Luftwaffe*'s pressing need for experienced aircrews. The order cost him half his bomber crews and fighter pilots and a quarter of his reconnaissance and seaplane crews as well as both *Kommandeure*, with Handrick replaced in *J/88* by *Hauptmann* Walter Grabmann on 10 September and Mehnert in *K/88* by *Major* Fritz Härle. Spanish replacements were used to plug the gaps: *K/88* received nine crews and *3.J/88* had three pilots, while others joined *A/88*—all proving extremely capable, to the Germans' relief. Yet in mid-September Volkmann wrote to Berlin a very critical and pessimistic letter about the Spanish conduct of the war, and this was the last straw. As soon as the Munich Agreement had been signed, Volkmann was recalled and he left Spain on 13 November. He joined the *Luftkriegsakademie* but he clearly felt that his career would make no further progress in the *Luftwaffe* and with the outbreak of war he re-joined the Army. Volkmann was promoted to *General der Infanterie* and given command of a division but he was killed in a car accident on 21 August 1940. In the *Luftwaffe* he had helped to develop the first generation of combat aircraft, but he was an administrative officer rather than a front-line leader and even in 1940 his unit did not distinguish itself.[31]

A shortage of crews and aircraft as well as heavy rain meant that the Legion's bombing effort on the Ebro slowly declined after the first week, and the decline accelerated in mid-August following a decision to end the bombing role of *A/88*'s Do 17s so that they might concentrate on reconnaissance. Meanwhile the corps *Kette* became increasingly active ranging Nationalist guns.

There was an upsurge in air activity in the latter half of September when *J/88* claimed twenty victories, including three by Grabmann, but on 4 October *Katiuskas* destroyed one Bf 109 and damaged four in an attack upon the *Gruppe*'s base at La Cenia. Yet the Nationalist fighter shield was undented, and during October *J/88* alone claimed another 23 victories. By 1 November the Republican fighters had been ordered to avoid combat. By then, too, the *Stuka Kette* had received five Ju 87B-1s, which went into action on 30 October in support of the final push, these *Stukas* being exchanged for the Ju 87As which returned to Germany.

The Ebro battle had cost the Legion ten aircraft, most destroyed in accidents, with five aircrew dead and six missing (all prisoners), while another fourteen aircraft had been badly damaged and seven aircrew had been wounded. To provide much needed rest and reinforcement, most *Staffeln* were stood down immediately after the battle pending the arrival of new aircraft, including the Bf 109E-1/3, He 111E-1/J-1 and Hs 126A-1, which raised the Legion's strength to 96 (including 37 Bf 109s, 30 He 111s and nineteen Do 17s), or 21 per cent of the Nationalists' total air strength. Replacement German aircrew also arrived, and *3.J/88* bid '*adios*' to its Spanish pilots.

In the lull after the Ebro battle Plocher was briefly in command but on 1 December *Generalmajor* Richthofen returned to Spain accompanied by a new Chief of Staff, the 37-year-old *Oberstleutnant* Hans Seidemann. An officer cadet at the Armistice, Seidemann had joined a *Freikorps* unit which was later inducted *en masse* into the *Reichsheer*. Commissioned in 1922, Seidemann later became a staff officer and transferred to the *Luftwaffe* in 1935. He was to work many times with Richthofen in the coming years and he ended the war commanding *Fliegerkorps VIII* as a *General der Flieger*.

The new staff team had nearly a month to prepare for Franco's next operation, the capture of Catalonia. Franco had hoped to exploit the Ebro victory quickly, but fog and heavy rain forced frequent postponements until 23 December. The offensive was conducted with the usual powerful artillery and air support, although there was fierce resistance for nearly a fortnight before the Republican line was broken and the rout began. Tarragona fell on 14 January 1939 and Barcelona a fortnight later, with the Nationalists encountering the greatest resistance from heavy rain and snow, although they were able to reach the French frontier on 9 February.

The Legion made its usual contribution to the offensive, flying an average of four sorties a day, although the enemy air force, with only 106 aircraft (including 26 *Katiuskas*), remained as aggressive as ever. Because enemy aircraft were constantly moving, *A/88* had great difficulty discovering their bases, but the *Staffel*'s patience was rewarded. On 12 January *K/88* destroyed ten aircraft at Monjos and Pate by attacking from the sea, while on 6 February *J/88* struck Vilajuiga airfield, destroying eleven aircraft and damaging a further fifteen beyond repair for the loss of one Bf 109. This allowed *K/88* and the *Stukas* to attack enemy communications around Figueras at hourly intervals in what proved to be the last missions of the campaign, directed by Richthofen from his new headquarters at the former Republican air base at

Sabadell in the northern suburbs of Barcelona. The campaign had cost the Legion eleven aircraft and 21 German aircrew, while two Spanish crews were lost on 27 December when a Do 17F and an He 111B collided in mid-air after the Dornier had been hit by ground fire. Another ten Germans from *F/88* and *Ln/88* were also killed, including three members of a cable-laying party who were captured and then summarily shot.

The Legion moved to central Spain and established a new headquarters at Toledo to prepare for the final assault upon Madrid. It did not take place, for the war which began with revolt also ended with one when anti-Communist forces rose against the Republican Government and, after a miniature civil war, accepted Franco's demand for unconditional surrender on 26 March. The Legion's 314th and last victory came on 5 March when *Oberleutnant* Hubertus von Bonin (who had replaced Mölders as *Staffelkapitän* of *3.J/88* after the Ebro battle) shot down a *Chato* over Alicante. To demonstrate the Nationalists' superiority, *J/88* flew four sweeps on 13 March but the previous day a He 111 had exploded in mid-air over Madrid, killing all six men on board, including Härle, the *Kommandeur* of *K/88*. There was also a series of fatal accidents after the war had ended, the last of these occurring on 11 May when *Unteroffizier* Hans Nirminger of *J/88* crashed while stunting over Leon. Eleven days later the Legion held its last parade in Spain and on 26 May the 5,136 officers, men and civilians sailed from Vigo, taking with them 700 tonnes of equipment and most of their aircraft, including the Bf 109Es, He 111Es and Ju 87s. On 6 June there was a massive parade in Berlin, *Unternehmen 'Döberitz'*, attended by 14,000 of the 19,000 men who went to Spain.

The final report on the Legion, written on 20 May 1939, claimed 386 victories (313 in the air, 59 by *Flak* and fourteen on the ground) for the loss of 72 aircraft (including 42 fighters and fifteen bombers) to enemy action and 160 (including 78 fighters and 39 bombers) in accidents. *K/88* had dropped 21,045 tonnes of bombs, and the Legion claimed to have sunk 60 ships. The small German training mission had trained 500 Spanish aircrew, while another 40 pilots had been trained in Germany compared with 80 in Italy.[32] The Legion had lost 226 dead, some through accidents or illness (including 41 aircrew), of whom 167 had fallen to enemy action (including 131 aircrew and 21 *Flak* gunners), which had also wounded 139 men. Another 449 had been accidentally injured, mostly on the roads.[33]

Experience in Spain largely confirmed the *Luftwaffe*'s thinking, although Lützow warned that 'The fact that we suffered only slight losses despite being outnumbered is due to the inadequate training and erratic leadership of enemy airmen, and to the greater speed of our own fighters . . .' The supremacy of the bomber seemed to have been confirmed and the development of the *Schnellbomber* (later Ju 88) was thereby encouraged, but there was a clear need for improved defensive armament—which was reflected in the evolution of the Ju 88 design. While the Bf 109 proved capable of meeting all the bomber's escort needs in Spain, this did not obscure the *Luftwaffe*'s requirement for a long-range fighter such as the Bf 110, although none had been dispatched to the Iberian Peninsula. The experiences over Madrid in the winter of 1936–37 also led Udet to reduce the bomber/fighter production ratio from 3:1 to 2:1 in January 1937 and also to strengthen the Bf 109's armament, beginning with the introduction of the Oerlikon MG FF/M 20mm cannon in the Bf 109D.

However, the reports were not all optimistic: one bluntly warned that 'It proved impossible to inflict lasting damage on, or to put completely out of action, any enemy air force ground installation. It also proved impossible to knock out enemy air forces on the ground because of the high degree of flexibility of the enemy formations . . .' The relatively short advances made in Spain also failed to test the *Luftwaffe*'s ability to support deep thrusts by mechanized forces or even arrangements for tactical air support. On his own initiative Richthofen attempted to tackle both problems. He had limited success and little support from Berlin,[34] although the *RLM* did recognize the need for an armoured *Schlacht* aircraft and issued a requirement for one in December 1937 which led to the FW 189C and Hs 129. It was a need the Russians also anticipated, and a few months later the *VVS-RKKA* issued a similar specification which ultimately led to the Ilyushin Il-2.

The *Luftwaffe* was growing increasingly interested in strategic bombing, destroying the enemy's means of prosecuting the war, but Spain provided few clear lessons, although one report noted that attacks on industrial targets caused a 60–70 per cent drop in production and that heavier attacks would be even more effective. However, German bombing accuracy remained poor, and in March 1938 twenty He 111s flying at only 50m (160ft) failed to bring down a bridge over the Ebro. The only prolonged strategic bombing in Spain was carried out by Douhet's heirs in the *Regia Aeronautica*, which was already in terminal decline. Although the number of *squadriglie* (squadrons) grew from 95 in August 1930 to 205 in September 1938, the expansion was compromised by a thin industrial base which failed to make the complete transition from the technologies of the 1920s to those of the 1930s. The modern bombers were of composite construction, and the low power of the engines meant that three had to be installed instead of two. Even where modern construction techniques were employed, Italian fighters were based upon traditional philosophies and were underarmed. Procurement suffered the same problems as with the French, and the establishment of a *squadriglia bombardamente terreste* was only six aircraft, although in September 1938 these *squadriglie* made up 39 per cent of the *Regia Aeronautica*'s order of battle.[35]

The Italians committed 15 per cent of their total *squadriglie* (and 22 per cent of their bomber *squadriglie*) to Spain but the Majorca-based strategic bombing force rarely exceeded 40 aircraft. Attacks usually involved less than fifteen bombers, which struck harbours because the Republic was dependent upon imports of food, raw materials and military equipment. Although the Germans participated in half a dozen raids against Valencia and Cartagena in 1936 and late 1937, the Legion usually concentrated upon supporting land operations. The nearest thing to a strategic bombing offensive was *Unternehmen 'Neptun'* (Operation 'Neptune') against Cartagena on 16 April 1938 when *K/88*, operating from Seville but staging through Granada, flew 68 sorties and dropped 82 tonnes of bombs. Although the battleship *Jaime I* was hit (it accidentally blew up nine days later), little damage was caused and subsequent attacks upon ports must be regarded as operational-level rather than strategic missions.

The other legacy of the Spanish Civil War was to provide the *Luftwaffe* with a cadre of battle-hardened aircrew, although some may have learned the wrong lessons.[36] Thirty pilots with *J/88* were to win the *Ritterkreuz* in the Second World War, including

Werner Mölders, the top-scoring German ace in Spain (fourteen victories), who earned the *Ritterkreuz* with Diamonds, while scores of more than 100 earned Herbert Ihlefeld (nine), Lützow (five) and Walter Oesau (nine) the *Ritterkreuz* with Swords.

NOTES TO CHAPTER 5

1. Proctor, pp.20–2.
2. See Herrmann, pp.25–6.
3. Proctor, pp.25–6, 29–31.
4. For the airlift see Green, *Warplanes*, p.407; and Proctor, pp.26, 30–1.
5. For the first operational mission see Proctor, pp.28–9; Ries and Ring, p.17; and Dr Osborne's essay.
6. Proctor, pp.35–6.
7. *Ibid.*, pp 28–52, 56, 59.
8. Based upon Höhne, pp 112–13; and Zaloga, *Target America*, pp.43, 66.
9. Diary, BA MA N 671/1.
10. Richthofen's diary, 9 December 1936.
11. Proctor, pp.87–9.
12. For Sperrle and Richthofen see Thomas and Witts, pp.30, 42–3, 53; and Richthofen's diary.
13. Thomas and Witts, p.106.
14. Richthofen's diary.
15. Thomas and Witts, p.285.
16. *Ibid.*, p.282.
17. Proctor, p.130.
18. Thomas and Witts, p.63.
19. For Guernica see Thomas and Witts; Proctor, pp.127–31; Ries and Ring, pp.62–5; and Thomas, pp.624–9. There is useful information in the apologia by Kappe-Hardenberg.
20. Proctor, pp.157–8.
21. Salas, p.133.
22. Proctor, p.171.
23. They are in *SHAA* File 2B78, Dossier 2.
24. Thomas, p.801.
25. Proctor, p.182.
26. Osborne, p.64. Details of *AS/88* operations are to be found in César O'Donnell Torroba's article
27. Statistics by Mr Carl Christenson of the National Insurance Company of Copenhagen. Published in the *Journal of Commerce*, 10 August 1938.
28. See *SHAA* File 2B791.
29. During the summer of 1938 some twenty captured *Luftwaffe* aircrew were exchanged for Russians.
30. Proctor, pp.211–12.
31. *Ibid.*, pp.237–8.
32. Salas, p.206.
33. Losses based upon Proctor, p.253; Ries and Ring, pp.276–81; and Bartels and Proctor's article.
34. Murray, *Effectiveness*, pp.104–5.
35. Data on the *Regia Aeronautica* supplied through Group Captain S. A. Wrigley.
36. See Steinhilper, pp.126–7, 128, 158, 160.

CHAPTER SIX

THE STEEL FAÇADE

June 1936 to September 1939

In the 39 months preceding the outbreak of the Second World War the *Luftwaffe* projected a self-confident image of omnipotent power. The image was a powerful tool in German foreign policy, yet the steel façade concealed a flawed framework of command problems and economic chaos which slowly weakened the whole structure.

The command problems began on 3 June 1936 when Wever was killed in a tragic accident. For speed, and to increase his personal log of 200 hours, Wever had flown in his He 70D-1 D-UZON to lecture *LKS* Dresden cadets. However, he was so anxious to return to Berlin for a state funeral that he neglected a pre-flight check and failed to release the aileron lock. The aircraft was unable to take off and crashed.[1] The future head of *Luftwaffe* intelligence, Josef 'Beppo' Schmid, was with Göring when he received the news and reported that he 'broke down and wept like a child'. Göring's grief was genuine, for Wever was truly irreplaceable. He had been the cornerstone of the triumvirate providing a buffer between the fragile egos of Göring and Milch while bringing a shrewd and perceptive mind to the problems of air power. His death destabilized the *Luftwaffe*'s command structure at a time when it was facing domestic and international crises.

The problem of choosing a successor was made more difficult by the absence of a *Luftwaffe* General Staff, partly because of opposition from Wever himself. Two years earlier he had rejected proposals to form such a staff for fear of creating an intellectual élite in a small officer corps, which was itself immature—as graphically illustrated by both Jeschonnek and Schmid. He was supported by Göring and Milch, who both feared that the General Staff would undermine their authority. The German officer class traditionally had little time for civilian authority, an attitude aped by Göring, and within the *Luftwaffe* the officer corps wished to have a fellow officer, like Fritsch in the Army and Raeder in the Navy, rather than 'civilians' as leaders. Göring and Milch deeply resented such views, for they regarded themselves as military men, Milch especially perceiving himself as the natural choice as *Luftwaffe* commander following his promotion to *General der Flieger* on 20 April 1936 (Göring was promoted to *Generaloberst* on the same day).

Wever changed his own mind and Göring's within a year when the shortage of staff officers began to bite and it became clear that the Army would no longer plug the gap. The first *LKS* was founded under *Generalleutnant* Otto Stülpnagel to meet this need in the long term, while the *Lufttechnischen Akademie* was created under *Generalleutnant*

Karlewski to provide technically trained officers. Göring's change of heart came as much because the General Staff would answer to him as commander of the *Luftwaffe* as because Milch continued with his opposition. The ambitious *Staatssekretär*, who frequently deputized for Göring in public, was growing less restrained and no longer confined references to himself as *Reichsluftminister* to his cronies. Since Göring also ran the premier telephone interception service in the *Reich* (its headquarters were later moved to the *RLM* building), he could hardly have failed to be aware of this and sought to bring Milch to heel.

The *Luftwaffe* General Staff was scheduled for creation on 1 August 1936 under Wever and with his death Göring again turned to Blomberg for a replacement. The only candidate was *Generalmajor* Franz Halder, a gifted Bavarian, but he rejected the offer and in September 1938 succeeded Beck as Chief of the Army General Staff. The thwarted Göring decided to postpone the formal creation of the *Luftwaffe* General Staff and on August 1936 he appointed his administrative chief *Generalleutnant* Kesselring to lead the *Luftkommandoamt*. But the Directorate remained under Milch's authority, and out of pique at Kesselring's appointment he began asserting that authority. He sniped at Kesselring personally and interfered in the Directorate's activities on every possible occasion. One crucial incident occurred after several bombers of the *Lehr Geschwader* had crashed into the sea. Milch accused the *Kommodore, Major* Hans Jeschonnek (who had once been his adjutant), of forcing his pilots to fly too low and demanded a court martial. Kesselring told him to mind his own business and Milch could only administer a personal reprimand which earned him Jeschonnek's undying enmity.

Kesselring was also incensed by Milch's disclosures of the *Luftwaffe*'s strength to the British, on whom Milch regarded himself as an authority, having visited the country regularly since 1926. After visiting London in June 1936 he received Hitler's permission to begin official exchanges of information with the RAF, with the dual objects of monitoring and slowing RAF expansion. He actually provided false information, claiming, for example, that the *Luftwaffe* would have 1,500 aircraft by April 1937 and 2,340 by 1938, when in April 1936 it had nearly 2,700. Nevertheless, Kesselring denounced him to Göring for treason—to no avail—and perhaps this was the final straw, for, after nine months of unequal struggle, he offered his resignation.

On 1 June 1937 Kesselring was given the *Luftkreiskommando III* (*Luftkreise* were given arabic numerals in November 1937), but Milch's triumph was short-lived because the head of the *Luftkommandoamt* fired a Parthian shot. He was well aware that social and personal relations between Milch and Göring were collapsing; indeed, the former would slam down the telephone after heated arguments. In November 1936 there was a confrontation at Göring's home, Carinhall, when Milch demanded either full authority or permission to return to Lufthansa. The challenge was contemptuously dismissed by Göring, who informed the *Staatssekretär* that he would be informed when, and how, he would leave the *Luftwaffe*, although the option of suicide was left open. Milch's bluff was called and Göring pettily added to his humiliation by removing the *Staatssekretär*'s name from his Christmas list.

Kesselring exploited this rift and ensured that his successor would not be placed in the same invidious position by arranging for Göring to take personal charge of the *Luftkommandoamt*, which was renamed the *Luftwaffengeneralstab* (*Luftwaffe* General

Staff) on 2 June 1937.² As no replacement for Kesselring was forthcoming from Blomberg, Göring selected the congenial but reluctant *Generalmajor* Stumpff as the first *Chef des Generalstabs der Luftwaffe,* replacing him in the *Luftwaffenpersonalamt* with Greim. However, while Greim and Udet of the *Technisches Amt* were nominally under Milch, by mid-July 1937 they were being regularly consulted without his foreknowledge.³ Milch still had some influence over personnel and apparently engineered the replacement of Deichmann as head of Stumpff's *Führung Abteilung* and *de facto* Deputy Chief of the *Luftwaffe* General Staff,⁴ but Greim ensured that his replacement on 1 October 1937 was Jeschonnek, promoted to *Oberstleutnant*.

For Milch the only bright feature in a dark year which also cost him his appendix were official visits in his personal He 111 V16 accompanied by Stumpff and Udet to the *Armée de l'Air* (4–9 October) and the RAF (17–25 October). By chance the German delegation met Air Chief Marshals Sir Edgar Ludlow-Hewitt and Sir Hugh Dowding (the Commanders of , respectively, Bomber and Fighter Commands in September 1939) as well as Winston Churchill. The *Luftwaffe* delegation was unimpressed by either the aircraft of RAF squadrons they visited or the 'shadow' factories (Britain was producing only 145–200 aircraft a month in 1937), although the latter should have given them food for thought. Milch's foolish attempts to impress his hosts were to have disastrous consequences, for he 'revealed' a false figure for the *Luftwaffe*'s strength in 1938 which clearly demonstrated his own earlier deceit. The suicide of his credibility occurred as the British were considering Scheme 'L', the emphasis of which upon strategic defence was quickly accepted and therefore set the scene for the *Luftwaffe*'s defeat in the Battle of Britain.⁵ He compounded the error, according to his own account, by informing his British hosts that he was aware of their radar experiments—although there appears no confirmation of this claim in British records.⁶

Milch's arrogance and indiscretion were justification for Göring to reorganize the *Luftwaffe*'s command structure by New Year 1938, but there were also pressing administrative and military reasons. The *RLM* had become too cumbersome for one man's effective control, although Göring may have held his hand until the *Staffeln* received second-generation aircraft. By late 1937 Milch had wet-nursed them to maturity and it was time now for Udet to stamp his authority by guiding the third generation of aircraft to production. Udet and Greim were already reporting directly to Göring and, under the reorganization authorized on 18 January 1938 but executed on 1 February, this relationship was formalized with both directorates becoming autonomous. The remaining administrative directorates and departments were subordinated to *General der Flakartillerie* Otto Günther Rüdel, a 55-year-old staff officer, who became *Chef der Luftwehr*. Milch remained *Staatssekretär* and had the added honour of *Generalinspekteur der Luftwaffe*, but these were ceremonial roles for his status was equal to that of Greim, Rüdel, Stumpff and Udet, while Stumpff shared control of the ten inspectorates.

For the next year Milch, like an American Vice-President, was a leader in search of a role, but this exile demonstrated how badly the *RLM* missed his expertise. Göring quickly discovered that the new organization required too much of his time and, under the strain of imminent war, it demonstrated confusion between the command and support elements. These factors brought about another reorganization of the

RLM exactly one year later, and this was virtually unchanged until 1944. Effectively, Göring copied the *Luftstreitkräfte* organization, retaining the *Luftwaffe* General Staff in the new position of *Oberbefehlshaber der Luftwaffe* (*ObdL*), which re-created Hoeppner's *Kogenluft* while Milch adopted the *Idflieg* role. Milch regained much of his old power by becoming *de facto* Deputy Commander of the *Luftwaffe* (although he retained the titles of *Staatssekretär der Luftfahrt und Generalinspekteur der Luftwaffe*) and controlling all aspects of administration. This included Rüdel's organization and the inspectorates through *General der Flieger* Bernard Kühl as *Chef des Ausbildungswesens*. The subordination of Udet gave Milch administrative responsibility for aircraft development and procurement, although Udet remained the chief executive.

From 1938 onwards there was a steady rationalization of operational commands as power was concentrated into the hands of the most experienced *Luftwaffe* officers. Since June 1936 seven *Luftkreiskommandos* had existed, yet only four had airmen leaders (Felmy, Kesselring, Sperrle and Zander) while the remainder were under retired *Reichsheer* officers such as Kaupisch of the Berlin-based *Luftkreiskommando II*. On 5 November 1937 Göring attended a meeting at which Hitler revealed that the function of the *Wehrmacht* was to exploit any opportunity to strengthen Germany's position—specifically the south-eastern flank of Austria and Czechoslovakia—even if this meant war. Blomberg and Fritsch protested, but Göring remarked only that *Luftwaffe* troops in Spain should be withdrawn.

The Army's faintheartedness was noted, and on 4 February 1938 Hitler purged it of 'defeatists', exploiting Blomberg's unfortunate second marriage and a false accusation of homosexuality against Fritsch. Some 40 officers were retired, including most of the *Luftkreise* commanders and the 'academics' Stülpnagel and Karlewski as well as a disgruntled Wilberg, although all were compensated by their promotion to *General der Flieger*. At the outbreak of war only Wilberg re-joined the *Luftwaffe*; the others were recalled to the Army for administrative duties, although Kaupisch held a corps command until 1940.[7] However, the *Anschluss* Crisis delayed the reorganization of the command structure until 1 April 1938, when the *Luftkreise* were merged into three *Luftwaffengruppenkommandos* and two *Luftwaffenkommandos* (another *Luftwaffenkommando* being created for Austria), while the *Lehrverbände* became a separate command (see Table 9).

TABLE 9: *LUFTWAFFE* REORGANIZATION, 1 APRIL 1938

Luftkreis	Headquarters	New designation
1	Königsberg	*Luftwaffenkommando Ostpreussen*
2	Berlin	*Luftwaffengruppenkommando 1*
3	Dresden	*Luftwaffengruppenkommando 1*
4	Münster	*Luftwaffengruppenkommando 2*
5	Munich	*Luftwaffengruppenkommando 3*
6	Kiel	*Luftwaffenkommando See*
7	Brunswick	*Luftwaffengruppenkommando 2*
–	–	*Luftwaffenkommando Österreich*
–	–	*Höherer Kommandeur der Lehrtruppen*

Göring's hopes of replacing Blomberg were dashed when Hitler himself became head of the *Wehrmacht*, although he promoted the 'Iron Man' to *Generalfeldmarschall* as compensation. Simultaneously, a nominal tri-service command, *OKW*, was created under Keitel, now a *General der Artillerie* (*Generaloberst* from 1 November), with *Generaloberst* Alfred Jodl as his Operations Officer, Warlimont in charge of inter-service planning and Canaris responsible for Intelligence. However, Hitler made *OKH* the real centre of command.

As war clouds gathered during the summer of 1938 the *Luftwaffengruppenkommandos* replaced the *Höherer Fliegerkommandeure* on 17 July with *Fliegerdivisionen* which acted as mobile task forces of air, *Flak* and support units. Nine months later, on 1 April 1939, the *Luftwaffengruppenkommandos* were renamed *Luftflotte*, with *Luftflotte 1* absorbing *Luftwaffenkommando Ostpreussen*, while *Luftwaffenkommando Österreich* was upgraded to *Luftflotte 4*. Each *Luftflotte* had two or three *Fliegerdivisionen* and *Luftgaue*. There were by then eight *Fliegerdivisionen*, *Fliegerdivision 7* (attached to *Luftflotte 2*) having been activated in June 1938 to command airborne forces, while in April 1939 the *Höherer Kommandeur der Lehrtruppen* became the autonomous *Luftwaffen-Lehrdivision* with two *Lehrgeschwader* and plans for four.

The New Year also saw a new Chief of Staff with the appointment on 1 February of the 39-year-old Jeschonnek. Hans Jeschonnek was one of ten children born to an East Prussian headmaster and the youngest son born to the latter's first wife. At the age of 15 he had joined the Army and in 1917 had transferred to the *Luftstreitkräfte*, where he served with *Jasta 40*, with whom he had two victories. He flew for the Upper Silesian *Freikorps* before joining the *Reichsheer*, where, once, he was in the same cavalry squadron as Wolfram von Richthofen. His elder brother Paul (Wilberg's Operations Officer in 1917) was deeply involved in illegal air activity, but the pair worked together for only a short time before Paul was killed in a crash in 1929. Hans had come top of his General Staff training course before joining *T2 V (L)* in April 1928, and he made a name for himself with a series of important staff papers. He was Milch's first adjutant, but the two men fell out and Jeschonnek had to be transferred to *KG 152 'Hindenburg'* as *Kommandeur*, and there he attracted the eye of Wever, who saw in him a future successor.

When Jeschonnek became the General Staff's Head of Operations in October 1937 he exercised tremendous influence over aircraft development through drafting the operational requirements. As the first *Kommodore* of the *Lehr Geschwader*, he was aware of the German weakness in horizontal bombing, and in dive bombing he saw the means of obtaining greater accuracy. Jeschonnek was a dedicated, hard-working officer and his star went into the ascendant in November 1938 just after his promotion to *Oberst*. In the aftermath of the Munich Crisis, Hitler demanded a huge expansion of the *Luftwaffe* and on 7 November Jeschonnek published plans for an air force with 10,700 first-line aircraft by 1 January 1942, including 5,000 bombers (of which 500 would be He 177s). Although the industry was hard pressed through national shortages of *matériel*, this became the basis of *Lieferplan Nr 9*, despite universal opposition at an *RLM* conference on 28 November. The conference preferred the pragmatic proposal by *Oberst* Joseph Kammhuber, the organizational head at *RLM*, for a 4,000-aircraft force. Jeschonnek stubbornly stuck to his guns and observed, 'In my opinion it is our duty to support the *Führer* and not oppose his wishes.' Milch saw an opportunity to humiliate the young officer before Göring but was dumbfounded when the 'Iron Man' supported Jeschonnek, whose plan, he claimed, was feasible. Kammhuber promptly demanded a front-line assignment, although several months elapsed before he replaced *Oberst* Heinz-Helmutt von Wühlisch as Felmy's Chief of Staff in *Luftflotte 2*.

The antipathy between Milch and Jeschonnek may have favoured the latter's promotion, but Göring did prefer to work with younger, less experienced officers. However, Jeschonnek did not get on with the *Reichsmarshall*'s inner circle, which included Keller and Bodenschatz (who headed Göring's private office and was the *Luftwaffe*'s liaison officer to Hitler). His lack of seniority made it difficult to impose views upon more experienced officers such as Kesselring, Sperrle and Richthofen, to whom he would often defer. Nor did Jeschonnek inspire his subordinates, for he was a naturally reserved man who could be bitingly sarcastic with subordinates (a trait not unknown in German management), making spur-of-the-moment remarks which left deep wounds. He succeeded in making his close friend *Oberstleutnant* Otto Hoffmann von Waldau Head of Operations in the *Luftwaffe* General Staff, but the relationship, which was once on first-name terms, rapidly deteriorated since von Waldau was a perceptive staff officer who would not hold his tongue. He disagreed with his friend on the Ju 88 and, later, the invasion of the Soviet Union, and when he was killed in an air crash in 1943 Jeschonnek refused to attend the funeral. Yet the latter was not without his human side, for he had a wife and daughter as well as a younger brother, Gert, who elected to join the Navy.[8]

Unlike Wever, Jeschonnek underestimated the importance of technology upon air operations. His denigration of technology and engineers was shared by many *Luftwaffe* leaders from Göring downwards, and even in the RAF engineering officers were treated as second-class citizens. It reflected a significant educational weakness in the Imperial German Army, which drew most of its officers from classically biased schools (*Humanistische Gymnasien*) rather than technically orientated ones (*Real-gymnasien*). Only 5 per cent of the *Luftwaffe*'s generals and General Staff officers had technical degrees, although 17 per cent of the generals' fathers were from technical professions while 66 per cent of the generals had rural backgrounds. Consequently, the *Luftwaffe* reflected the *Wehrmacht*'s major weakness, which increasingly emphasized tactics and operations at the expense of vital features such as intelligence and supply.[9] In defending their status, the engineers were to have a terrible revenge by reorganizing the development and procurement organization.

Despite his faults, Wimmer was extremely effective in the *Technische Amt*, but he was unpopular with Göring, who replaced him with Udet on 10 June 1936. Wimmer was side-tracked into minor positions until April 1939, when he succeeded Keller as commander of *Luftwaffenkommando Ostpreussen*. He served under Kesselring and commanded *Luftflotte 1* for a few months in 1940 before taking over *Luftgau Belgien-Nord Frankreich* (whose first commander was his former production assistant *Major* Friedrich Loeb). He remained there until the general retreat of 1944, as a result of which he narrowly escaped an undeserved court martial instigated by Göring.

Oberst Ernst Udet, the former Inspector of Fighters and Dive Bombers, was recommended by Hitler, who felt that Germany's most famous pilot and a former manufacturer should be responsible for aircraft production. Udet's new position required dedication, organizational skills, attention to detail and technical knowledge at a time when there were major problems with the new generation of combat aircraft together with financial and material shortages. Instead, the *Luftwaffe*'s aircraft programmes came under the nominal control of an amiable dilettante and brilliant cartoonist who failed in almost everything he did.

A stocky, balding man with an expanding waistline, Udet had lived life to the full and looked ten years older than the 40 he celebrated in April 1936. Like Jeschonnek, he was a schoolboy wartime volunteer and later joined the *Fliegertruppe* as an observer. He became a fighter pilot and ended the war commanding Richthofen's old *Jasta 11* as the highest-scoring surviving ace (62 victories). The storms which buffeted the Weimar Republic failed to corrode this amusing, generous and bohemian character, who remained the ultimate barnstormer until he joined the *Luftwaffe* in June 1935. His skill and daring thrilled thousands at displays and in the cinema, while his experience led him to design acrobatic aircraft which trained hundreds of young Germans and Austrians. But he abandoned the company which bore his name when his partners sought to build airliners.

Udet's relations with the regime were ambivalent. The *Gestapo* regarded him with suspicion because he had emigré friends and would visit them, while another friend, the aviation writer Walter Kleffel, was twice jailed on trumped-up charges. His fatal depression in 1941 was aggravated by a *Gestapo* ban on flying for fear that he would follow Hess's example. Yet in May 1933 Udet became a Nazi Party member and soon was Loerzer's Vice-President in DLV. His old friend Greim persuaded him that he could help introduce dive bombers into the *Luftwaffe*, but ultimately he joined from a mixture of patriotism and comradeship and because it brought him a steady income.

His greatest achievement was ensuring the selection of the Bf 109 for his old friend Messerschmitt (whose BFW took over the Udet company's facilities). Udet loved to fly but he was no test pilot for he had little technical comprehension and his technique involved throwing aircraft around the sky. His ignorance almost killed him when he flew the sophisticated He 118 dive bomber prototype on 27 July 1936. He ignored, or forgot, a pre-flight briefing about changing the propeller pitch and in a near-vertical dive ripped out the propeller and was saved only by his parachute.[10]

Udet had no illusions about his capabilities and told Göring, 'This isn't for me. I don't know anything about production or big aircraft.' Unfortunately Göring was determined not to become too dependent upon Milch and believed that Wimmer's organization would carry Udet. The Directorate was efficiently organized into three departments, research, development and production, but the most experienced officer, *Major* Friedrich 'Fritz' Loeb, was soon to be requisitioned by Göring to organize industry for the Four-Year Plan. Loeb returned to the *Luftwaffe* as an *Oberst* in 1939 and the following year was given the newly formed *Luftgau Belgien-Nord Frankreich*. Tragically, he was killed on 21 June 1940 when his aircraft collided with another while landing at Bruxelles-Evère.

The other key officers, led by Lipetsk veteran *Hauptmann* August Ploch, were engineers who preferred technical discussion to the administrative chore of organizing production. Udet was left to develop overall planning, although he preferred to visit the factories and research establishments. He also indulged his interest in dive bombing, until it became an obsession—with disastrous effects upon third-generation aircraft such as the Ju 88, He 177 and Me 210. Udet had been a close friend of Milch, whom he had taught to fly, but the *Staatssekretär* was piqued at the appointment and began his usual intrigues upon learning that Udet was preparing *Lieferplan Nr 4*. Milch's unpopularity with Göring doomed his first schemes, but later intrigues contributed to Udet's suicide in 1941 when he blamed, among others, 'the Jew Milch'.

The efficiency of Wimmer's organization briefly maintained the momentum set by Milch, but within eighteen months of Udet's appointment progress was becoming more uncertain. In December 1937 Udet's engineers saw an opportunity to improve their lowly status and proposed a horizontal reorganization to create specialist departments focusing upon specific problems, a solution adopted by *Idflieg*'s *Flugzeug-meisterei* during July 1918.[11] Udet (a *Generalmajor* from April 1937) saw this as a means of easing his burden and reorganized the *Technische Amt* into thirteen departments between 28 March and 9 May 1938.[12] Yet, somewhat late in life, Udet had developed ambition, and this grew following his promotion to *Generalleutnant* in November 1938. Apparently at the suggestion of Greim (who was about to assume command of *Fliegerdivision 5*), he proposed reforming the *Flugzeugmeisterei* with himself as *Generalluftzeugmeister*, and on 29 January 1939 his wish came true and he was given total executive responsibility for aircraft development and production. Now he controlled the *Technisches Amt* (which remained under his personal control), the Supply and Economic Directorates, the Industrial Group and several development establishments, with nineteen departments and 150 engineers as well as hundreds of officers and officials. Udet was too weak a personality to co-ordinate this huge organization, and even Milch would have found it a challenge.[13]

The various blinkered attitudes each department adopted were well illustrated with the FW 190. The fighter's BMW 801 engine was plagued by overheating problems, which were unsolved when pre-production aircraft were delivered to *JG 26* in early 1941 and became so serious that an *RLM* investigation commission recommended the cancellation of the programme. It was saved by the *Geschwader*'s Technical Officer, Ernst Battmer, who demanded closer co-ordination between the airframe and engine manufacturers. This led to 50 modifications which finally cleared this excellent fighter for service.[14] The fragmented structure meant that manufacturers had to deal with half a dozen or more departments simultaneously. Winning approval from them all slowed development and inevitably created friction not only with industry but also within the Directorates.

Udet was also unable to eliminate the fundamental flaw in Wimmer's organization, which demanded frequent design changes even though this reduced production runs and increased costs. The Ar 68, for example, had three basic versions with different engines (Ar 68E, Ar 68F, Ar 68H) and twelve sub-versions with various equipment including three types of radio. Yet Udet's appointment was not the complete disaster it has been portrayed as and his first schedule, *Lieferplan Nr 4*, increased the ratio of single-seat fighters. He also tried, with some success, to tackle the extravagance of 'cost-plus' contracts and relaxed the military export restrictions imposed by Wimmer. Unfortunately German aircraft tended to cost some 30 per cent more than their competitors, forcing the *RLM* either to subsidize sales or to seek barter agreements for vital raw materials.

However, nothing could stop the malevolent effects of incompetent leadership and fragmented organization from spreading through the production process as stealthily as a cancer, although these were not felt in full until 1941 and 1942. Hitler formally approved production of the second-generation combat aircraft on 7 March 1936 and by the winter the Ju 86A-1 (in *KG 152*), He 111B-1 (*KG 154*), Do 17E-1 (*KG 153*, *KG 155*), Do 17F-1 (*Aufklärungsgruppe (F) 122*) and Bf 109B-1 (*JG 132*) were in service. They were sophisticated aircraft with powerful engines, increased instrumentation and complex electrical and hydraulic systems. The Bf 109B, for example, retained the traditional fixed-pitch, twin-blade, wooden propeller but also had slotted flaps, automatic wing leading-edge slots, a retractable undercarriage and a radio while the Bf 109D received a three-blade, metal, variable-pitch propeller. Engineering costs alone were 50 per cent greater than those of their predecessors, so series production, and the constant demand for improved versions, was difficult and costly. In August 1936 the *RLM* estimated that the introduction of the new aircraft would add another RM2,600 million ($1,000 million) to the approved expenditure.

Inflation fuelled costs, and while the *RLM* recouped excess profits from the major manufacturers, the component producers were left untouched until 1942. The inflation was driven by excessive Government expenditure as well as an adverse balance of payments which eroded gold and foreign exchange reserves, in turn reducing imports of iron ore (from Sweden, France and Spain) and copper. This had profound effects upon the aircraft industry, which employed 204,000 people in October 1938, for the shortage of steel prevented the building of factories as well as the production of machine tools, airframes, engines and anti-aircraft guns, while copper for electrical wiring was at a premium. Only bauxite for aluminium was available in abundance, thanks to barter deals with Yugoslavia.

The irresistible force of Nazi ambition was deflected by the immovable object of material and financial shortages. Long-term production was hamstrung, and industry had to place men on short time or even lay them off. Ambitious schedules continued to be produced—industry received eight between January 1937 and September 1939, and two were being drafted when war broke out—but all were obsolete before the ink was dry. The services competed for materials like dogs after a bone, and in an effort to bring order out of chaos Göring was made head of the Four-Year Plan in October 1936. The intention behind the Plan was to maintain the momentum of rearmament by organizing the economy, ensuring that Germany would be self-sufficient, especially in oil and rubber. I. G. Farben, for example, developed a process for converting Germany's abundant coal reserves into synthetic oil which could be turned into petrol. The first order was placed in December 1933, and by 1935 the company had developed 87–89 octane fuel at RM185.30 ($75.63) per tonne. However, natural fuel (of which 60 per cent had to be imported) cost only RM98.52 to 124.25 ($40.21–50.71) a tonne. Most of this came from the United States, but Romania contributed nearly 37 per cent while the Soviet Union provided 12 per cent. A 100-octane fuel was also developed but, surprisingly, I. G. Farben and the *RLM* seemed initially uninterested in developing this grade—which was to be the prime fighter fuel during the Second World War. Germany also had domestic sources of natural oil, production of which increased from 1.78 million tonnes in 1936 to 2.34 million in 1938 thanks to the acquisition of Austrian fields.[15]

While the *Luftwaffe* was a major beneficiary of the Four-Year Plan in the allocation of personnel, raw materials and finance, it could not escape the economic facts of life. Göring told a conference on 2 December 1936 that Hitler wished to maintain the pace of rearmament, and he added that, from New Year's Day 1937, the expansion of the *Luftwaffe* was to continue as if for mobilization. The emphasis was to be upon expanding first-line strength, even if the vital infrastructure of bases and factories was neglected. When the *RLM* budget was slashed in 1937 the axe cut most sharply on aircraft procurement and repair as well as spares, with a debilitating effect upon the *Luftwaffe*'s ability to fight a prolonged war. Several years elapsed before the damage was repaired. Penny-pinching failed to save successive *Lieferplänen*, where either production was reduced or the deadlines were stretched. In order to maintain the flow of aircraft, quality was diluted by retaining the older aircraft in the schedules.[16]

The problem was aggravated by the chronic unreliability of the Ju 86's Jumo 205 engines, which earned the aircraft the nickname 'The Flying Coffee Grinder' and made the redrafting of production plans essential. Between January and April 1937 *Luftwaffe* units reported 25 emergency landings caused by engine failure, while during the year the aircraft was involved in 56 accidents, but the *RLM* and Udet blamed inexperienced pilots and were slow to act, as Condor Legion veteran 'Hajo' Hermann discovered. When he complained to Wimmer and *Oberst* Helmuth Förster (the 'Old Eagle' *Kommodore* of *KG 253*) in April 1937 about two engine failures in 30 hours, he was reprimanded and accused of inexperience![17] The *Technisches Amt* could not afford to be so cavalier and *Lieferplan Nr 5* saw the Ju 86 phased out in favour of the Do 17 and He 111, but, to maintain levels of activity at the plants most severely affected, Udet accelerated the development of third-generation aircraft. These were placed on *Lieferplan Nr 7* production schedules, which also included improved second-generation aircraft such as the Bf 109E, Bf 110C, He 111H, Do 17M/P and Ju 87B.

For industrial as well as operational reasons, therefore, the Ju 88 was a key programme. The 'Wonder Bomber' requirement demanded a 3,000km (1,850-mile) range, a maximum speed of 500kph (310mph) and a two-tonne bomb load. When drafted, it was envisaged as a horizontal bomber, but as operational experience seemed to confirm that horizontal bombing was inherently inaccurate and a waste of resources, a 20–30-degree dive bombing angle was added to the requirement in December 1936. A year later a fourth crew member and a heavier defensive armament were added, and these forced 50,000 design changes while increasing weight from 5 tonnes in the first prototype (Ju 88 V1) to 10 tonnes in the dive bomber prototype (Ju 88 V4). The effect upon performance was dramatic, and at one time the maximum speed was reduced to 300kph (186mph), although it was later increased to a more respectable 460kph (286mph) in the production Ju 88A-1, whose range was reduced to about 1,750km (1,100 miles).

By his own account, Milch had doubts that the Ju 88 would be the 'Wonder Bomber' that Göring, Udet and Jeschonnek anticipated, but Junkers apparently understated the expected development problems and on 30 September 1938, as the Czechoslovakian Crisis ended, the company received the *RLM*'s biggest pre-war production contract. To accelerate development and production, Koppenberg was given plenipotentiary status and he created an empire of eight airframe plants and

77,700 workers (53 per cent of Germany's total of airframe workers). Other manufacturers suspected that he was exploiting his position to regain Junkers' dominance in the industry, and he certainly boasted of stopping work on the Do 217 and of targeting the He 177.[18] As the Ju 88 underwent operational evaluation, its combination of a highly stressed airframe and unreliable dive brakes led to a series of fatal accidents—the most prominent being that involving the Condor Legion bomber ace 'Bubb' Moreau—and it took the *Luftwaffe* some time to get used to what the Americans would have called 'a hot ship'.

There were also severe difficulties with the *Luftwaffe*'s heavy bomber programme, which would have encouraged Koppenberg. Testing of the 'Ural Bomber' prototypes proceeded through the winter of 1936–37 and confirmed that they were under-powered, although *Hauptmann* Deichmann, the *Luftwaffe* General Staff's operations chief, recognized the potential of the heavy bombers. However, on 29 April 1937 Göring and Kesselring took the much criticized decision to end the 'Ural Bomber' programme. After Germany's defeat the two men were blamed for having deprived the *Luftwaffe* of heavy bombers at the outbreak of war, but the decision was unavoidable. Even if the Do 19 and Ju 89 had proved successful, British experience indicates that there could not have been an effective force until 1942–43. Quantity rather than quality made the *Luftwaffe* important to Germany's diplomatic aim of dominating its immediate neighbours, and Göring was advised that 2.5 twin-engine bombers could be produced for the same amount of aluminium as a single four-engine one—hence his oft-quoted remark that 'The *Führer* does not ask me how big my bombers are but how many there are.' German industry was stretched to breaking point in meeting the demand for twin-engine aircraft, and the inclusion of heavy bombers would have seriously compromised the production schedules.

Nevertheless, the *Luftwaffe* remained interested in the strategic bomber, and *LKS* lecturers, like their counterparts in the USAAC Tactical School, increasingly emphasized their importance within an air force which would support land and sea operations. The *Luftwaffe* also recognized the need for a long-range bomber as the prospect of war with Britain or the Soviet Union loomed and the practical range of the Ju 88 slowly contracted. In June 1936, therefore, definition began of a 'Ural Bomber' replacement, 'Bomber A' (He 177 from November 1937) and a year later a mock-up was ordered of Heinkel's *Projekt 1041*. Heinkel was apparently selected because he had lost the light fighter contract and would be underemployed when the Ju 88 replaced the He 111, and the He 177 was written into the development schedules by the end of 1937.

Although the aircraft had four engines they were coupled together in two nacelles with each pair of DB 601 engines (designated DB 606) driving a single propeller. This was partly to improve the aerodynamics, but the arrangement also reflected Jeschonnek's dislike of four-engine bombers and preference for twin-engine ones which could be used as dive bombers for greater accuracy.[19] This, and the dive bomber requirement, was to plague the aircraft's development: Thirty (later forty) prototypes and pre-production aircraft were ordered. The specification called for an aircraft capable of diving at angles up to 60 degees, and this required extensive reinforcement of the airframe, with production aircraft having an anticipated loaded weight of nearly 27 tonnes. Historians have condemned this as madness, but it was a mania shared by the Royal Air Force, which included requirements for both dive bombing and catapult-launch in a twin-engine, multi-role (medium bomber/torpedo bomber/reconnaissance) aircraft to meet Specification P.13/36. While no dive bombing angle was specified, a minute of 20 April 1937 by Group Captain R. D. Oxland, Deputy Director of Operational Requirements, notes: 'The angle of dive requirement may be altered from 70° to 25° . . .'[20] The requirement led to the Avro Type 679 (later the Manchester, which was redesigned to become the Lancaster) and the Handley Page H.P.56 (later the Halifax), which had both flown by October 1939.[21] The RAF quickly abandoned the dive bomber and catapult-launch requirements and the *Luftwaffe* did the same in the He 177, apparently on Göring's express instructions, but the designs inherited great structural strength—which made the British bombers capable of the violent manoeuvres which were their prime means of survival. In the case of the He 177, weight increased from nearly 24 tonnes in the early prototypes to 28 tonnes in the He 177 V6, while the first production versions (He 177A-1) would have a loaded weight of 30 tonnes.

However, the Achilles' heel of the He 177 was the DB 606, the configuration of which made it prone to overheating, and engine fires became a common occurrence. Heinkel recognized that this would be a chronic problem and in mid-1939 began designing the He 179, a version with four separate engines, but Udet and Daimler-Benz were determined to solve the DB 606's problems for reasons of prestige. Although the development of four-engine aircraft was slashed by Udet on 12 September 1939, the following year Heinkel renewed development as the private venture 'He 177B' (He 277), but this did not receive official sanction until May 1943 and was too late to influence the outcome of the war.[22]

Meanwhile, despite shortages of raw materials, German industry slowly increased production, with a monthly average in 1936 of 426 (127 combat aircraft) rising to 467 (220) in 1937 then declining to 436 (279) in 1938. This was a remarkable technical and human achievement, but while it permitted modernization and expansion, airframe and engine reserves were inadequate and there was only a 30 per cent engine reserve. Yet the delivery of new aircraft to the *Staffeln* was merely part of a complex re-equipment programme lasting two or three years.

The complex new aircraft made some trades redundant and groundcrews had to be dispatched on training courses, or in some cases re-training courses. Base facilities needed upgrading, with increased accommodation, the provision of more electric power points, concrete aprons for routine maintenance and hangars for servicing. Before the aircraft arrived, spares, work benches, jigs and machine tools had to be delivered, unpacked, checked and returned if faulty. This activity and the paperwork blizzard added a colossal load to *Geschwader* and *Gruppe* staffs as well as to specialist officers and the *Staffelkapitäne*. The last had also to supervise flying and tactical training, a process made more complex when aircraft such as the Ju 86 posed maintenance problems.

Between 5 May 1937 and 5 May 1938 the *Luftwaffe*'s strength rose from 178,209 to 214,419, the majority conscripts but with many volunteers, most of whom entered at the age of 17 to serve for two to twelve years. Special labour certificates, subsidies and loans were available for those who completed their service, while *RLM* patronage helped former personnel to find civilian work associated with aviation. While the excitement of working with aircraft was undoubtedly the driving force, the *Luftwaffe* also represented the dissolving of class barriers, which was a feature of Nazi propaganda. Relationships within the officer corps tended to be less rigid than in either the Army or the Navy, although reservist officers tended to exploit their rank and senior officers seemed to resent junior officers' having automobiles. Yet senior officers still vetted, through the Officers' Committee, the *fiancées* of their juniors and a veto apparently spurred the career of *Major* Helmut Wick, the *Luftwaffe*'s top *Experte* (56 victories) at the time of his death in November 1940.[23]

From 1935 virtually all *Luftwaffe* personnel found, like the RAF motto, that hard work led to the stars. Ditch-digging with the paramilitary *RAD* usually preceded entry into an *FEA* (most were redesignated *FAR* in April 1939) for six to twelve months' basic training. Afterwards those with flying aptitude would attend a two-month *ab initio* course at a *Fluganwärterkompanie* before joining the *Fliegerführerschule A/B* at the *FEA/FAR* airfield for 100 to 150 hours' flying training. A chronic shortage of instructors meant that the instructor/pupil ratio appears to have been 1:6 when 1:4 was more efficient. With a greater burden on the pupil, the 'wastage' rate may have been 25 per cent, as in the RAF, where quality rather than quantity was emphasized, but no actual figures are available.

Fighter and *Stuka* candidates would then go to *Waffenschulen* for, respectively, three and four months' operational training. Bomber and reconnaissance candidates attended a *Fliegerschule C* (60 hours) and a *Blindflugschule* (50–60 hours) before undergoing three months at *Waffenschulen* or specialist schools. Observers received flying training to C standard, then a 9–12 month course at *Aufklärungsfliegerschulen* (including instrument training) for, like USAAC bomber commanders, they were expected to be capable of performing every crew member's task. In theory, pilots joined units with 250 hours in their log books, but the *Luftwaffe*'s continuing expansion meant cutting corners in training, especially instrument and operational training, which forced units to make up the lost time themselves.

At least the *Luftwaffe* addressed the problem of providing sufficient men to meet all aircrew trades, in contrast to the British and French, who neglected most aspects of aircrew training, except for pilots, until the late 1930s. Even pilot training was limited, and few were fully instrument-rated, while an overemphasis upon quality meant that the RAF faced a desperate shortage of fighter pilots during the Battle of Britain. In Germany the *Luftwaffe* took total responsibility for flight training from 17 April 1937, when the DLV was replaced by the *NSFK*, which became purely a civilian organization. Most DLV schools were transferred to Christiansen's *Kommando der Fliegerschulen*, which now had 21 *Flugzeugführerschulen* that became specialized during 1937, with six each providing A, B and C standard training while three multi-standard schools supported the *Küstfliegerstaffeln*. New aircraft received included the Bü 131, FW 44, He 72 and Kl 35 for A2 courses as well as the Ar 66, Ar 96 and Go 145 for B1 courses. The B2 and C1 courses retained old single-engine commercial aircraft such as the Junkers F 13, W 33 and W 34, although the FW 58 was increasingly used, while the Ju 52/3m became the backbone of the C2 syllabus although supplemented by obsolete bombers.

From November 1938 the A and B courses were conducted at *Flugzeugführerschulen A/B*, each with some 60 single- and 30 twin-engine trainers. By the outbreak of war there were sixteen of these schools, usually attached to the *FEA/FAR*, whose own numbers increased from sixteen in November 1938 to 27 by April 1939. The three naval *Flugzeugführerschulen See*, each with 50 aircraft, and the *LKS* (of which there were five by the outbreak of war) were directly under the *Luftwaffengruppenkommandos/ Luftflotten*, as were the eight *Flugzeugführerschulen C* (each with some 50 aircraft) and two *Blindflugschulen* supported by *Streckenschule Berlin*. Operational training was conducted at ten fighter, bomber, reconnaissance and naval *Waffenschule* with 30–40 aircraft, the bomber schools (*Kampffliegerschulen*) soon becoming merged into *Grossen Kampffliegerschulen*.

However, the status of training declined in proportion to the expansion of the establishment. Christiansen's organization was downgraded to *Inspektion der Flugzeug-führerschulen* in 1938 when the *Luftwaffengruppenkommandos* were created and accepted responsibility for the schools through six *Kommandos der Fliegerschulen und Ersatzabteilungen* (naval schools were under *Luftwaffenkommando See*). A year later they, too, disappeared and most flying training became a *Luftgaue* responsibility through the *FAR*, with quality monitored by *Oberstleutnant* Rudolf Meister. Although his department was under the *Chef des Ausbildungswesens der Luftwaffe*, *General der Flieger* Kühl, the latter was primarily concerned with operational training. By contrast, the RAF had the equivalent of the *Generalleutnant* (acting as the equivalent of a *General der Flieger*) responsible to the Air Council for training while the USAAC had an officer equivalent to a *Generalmajor*. It was not until 1942 that the *Luftwaffe* finally realized the importance of flying training through the appointment of *Generalleutnant* Werner Kreipe.

The problems with training were reflected in what Göring called 'a plague' of accidents—48 fatal in the second half of 1936; 147 men killed and 2,422 injured (108 aircraft destroyed and 1,290 damaged) in 1937; and eight fatal in March 1938 (involving the destruction of seven aircraft and damage to two more). These accidents partly reflected the *Luftwaffe*'s increased flying hours—603,000 in 1935 (40,000 by front-line units and 278,000 by schools) and 750,000 in 1936 (290,000 by units and 285,000 by schools)—and the number of hours per flying accident was probably comparable to the RAF's and the *Armée de l'Air*'s figures, which are shown in Table 10. Many accidents were unavoidable, while a host of minor mishaps could be ascribed to an unfamiliarity with new features such as retractable under-carriages—indeed, 75 Bf 109s, Do 17s and He 111s were damaged in 1937. Yet a depressing number of valuable men and machines were being lost to youthful high spirits, the most tragic losses occurring as pilots stunted over home towns to impress parents and girlfriends.[24] Flight discipline improved with the creation of an Inspector for Flight Security and Equipment, but throughout the war the *Luftwaffe* suffered a disproportionate number of landing accidents, suggesting serious faults in its training.

In contrast to the *Luftwaffe*, the *LStrKr*'s expansion was a more sedate affair and it had only 151 aircraft by New Year 1936, although by the following Christmas Löhr had ambitious plans for fifteen *Staffeln* (six *Bombenstaffeln*) by 1938 with 737 aircraft, 145 (including 45 fighters, twelve Ju 86s and twelve Ju 87s) from Germany. The Ju 86s were received in 1937 together with another 40 aircraft, mostly trainers, while the *LStrKr* evaluated the Bf 109, He 112, Ju 87 and Hs 126 before selecting the He 112

TABLE 10: BRITISH AND FRENCH AIR FORCE ACCIDENT STATISTICS, 1933–1938

Year	Service	Fatal accidents	Deaths	Hours flown	Hours per fatal accident
1933	Britain	40	54	380,600	9,515
	France	2	3	11,125	5,562
1934	Britain	20	31	390,500	19,525
	France	–	–	12,712	–
1935	Britain	26	43	457,400	17,592
	France	1	1	14,646	14,646
1936	Britain	58	98	598,300	10,315
	France	1	1	14,825	14,825
1937	Britain	93	156	812,800	8,739
	France	3	4	16,773	5,591
1938	Britain	140	218	1,057,400	7,552
	France	–	–	17,672	–

Note: RAF figures are based upon the RAF Accident Report Summary of 1938 (PRO Air 8/253); French figures are from *Inspection Général Technique de l'Air* statistics of 1938 (*SHAA* File 2B133).

to meet the fighter requirement. Italy was regarded as a prime source of aircraft, and Löhr's 1939 plan envisaged twenty *Staffeln* and a front-line strength of 200 aircraft, including three *Einheits (Kampf) Staffeln* with Breda Ba 65 dive bombers.[25]

But the beginning of 1938 saw the excesses of the Austrian Nazi Party and its strident cries for the long-mooted union (*Anschluss*) with Germany provoke a crisis. Chancellor Schuschnigg believed that the Germans were plotting a take-over and sought to defuse the situation by visiting Hitler at Berchtesgaden in February 1938. However, under threats of *Luftwaffe* intervention which were underlined by Sperrle's presence, he was forced to accept the Austrian Nazi leader, *Dr* Arthur Seyss-Inquart, as his Interior Minister, but even this did not end demands for an *Anschluss*.

Schuschnigg decided to resolve the situation with a plebiscite on Austrian independence, and when the campaign began on 9 March it aroused Hitler to new heights of fury. On 10 March he ordered the activation of contingency plan (*Sonderfall*) '*Otto*' for the invasion of Austria and as *8.Armee* was mobilized Schuschnigg received a telephoned ultimatum demanding his resignation. Sperrle's *Luftkreis 5* alerted *I/JG 135*, three Do 17E *Kampfgruppen* from *KG 155* and *KG 255*, two reconnaissance *Staffeln* and three corps *Staffeln*, while Kesselring's *Luftkreis 3* raised *Transportgruppen Chamier* and *Fleischhauer*, together with *Kampfgruppe zb V Ziervogel*, and fleshed them out with some Lufthansa Ju 52s to fly in troops and equipment. Most units came under *Generalmajor* Ludwig Wolff's *Höhere Fliegerkommandeur 5*, but the newly activated *Stofl 8.Armee* controlled the corps *Staffeln* and six *Flak* battalions. Simultaneously Sperrle prepared airfields and moved support units closer to the border.

Above: Contrary to modern mythology, the *Luftwaffe* ordered dive bombers before Ernst Udet demonstrated a pair of Curtiss Hawks. The initial order was for single-seat He 50As such as this one, but these were regarded merely as interim equipment pending the definition of the *Luftwaffe*'s dive bomber requirement. (Greenborough Associates)

Below: During the winter of 1933/34 the *Luftwaffe* considered the dive bomber requirement and concluded that there should be two types. The Light Dive Bomber was to be a single-seat close air support aircraft which, like the CL class aircraft of the First World War, would rely upon machine guns and light bombs. The He 123 was selected for this role. (Greenborough Associates)

Above: The Ju 87A was developed to meet the Heavy Dive Bomber requirement for missions beyond the Forward Edge of Battle. This was a two-seat aircraft capable of carrying a 250–500kg (550–1,100lb) bomb, the specification being written around the Ju 87 prototype.

Below: The German Army—and most air forces—regarded corps aircraft such as this He 46C of *Aufklärungsgruppe (H) 112* as the foundation of air support. The corps aircraft was a maid-of-all-work, conducting photographic and visual reconnaissance, artillery location and fire control, ground attack and even supply duties, but it was too slow to survive combat with high-speed fighters. (Greenborough Associates)

Above: Wever and Wimmer wished the *Luftwaffe* to have a heavy bomber force and in 1934 issued the so-called 'Ural Bomber' requirement. This is the Do 19 V1, which made its maiden flight on 28 October 1936. However, months earlier Wever had recognized that this aircraft and the competing Ju 89 would never meet the *Luftwaffe* requirement and the 'Ural Bomber' programme had been relegated to demonstrator status. (Greenborough Associates)

Below: The Ju 89 V1 was the competitor for the 'Ural Bomber' programme and flew for the first time in December 1936. Two prototypes were completed for evaluation, while a third became the Ju 90 airliner prototype. Lufthansa purchased eleven Ju 90s in 1938 and they rendered sterling service as transports during the Norwegian Campaign. (Greenborough Associates)

Left, top: With the official revelation of the *Luftwaffe* by Göring in March 1935 there were innumerable parades at which units were formally inducted. At this one both Hitler and Göring are in attendance. (Alex Vanags-Baginskis)

Left, centre: A Ju 53/3m g4e of the Spanish Nationalist Air Force's *Grupo 2-G-22*. The Condor Legion originally used thc Ju 52 but it proved incapable of operating during the day against the I-16 *Mosca* and was transferred to night operations except when aerial opposition was negligible. The Ju 52s were largely responsible for the destruction of Guernica. (*Curatel General del Ejercito del Air, Service Historico y Cultural*)

Left, bottom: After serving with *A/88*, the He 70F-2s were transferred to the Spanish Nationalist Air Force, which called them *Rayo*s. They served with *Grupo 7-G-14*, and this one is pictured at Saragossa. (*Curatel General del Ejercito del Air, Service Historico y Cultural*)

Above: An He 111B-1 of *K/88* comes into land at a Spanish airstrip. (*Curatel General del Ejercito del Air, Service Historico y Cultural*)

Below: A few Hs 123As were sent to Spain at the end of 1936, serving with the *Michelis Kette*. They were not successful, and with the arrival of the Ju 87As they were transferred to the Spanish, who dubbed then *Angelito*s. (*Curatel General del Ejercito del Air, Service Historico y Cultural*)

Above: A close-up view of the nose of a *K/88* He 111B-1. (*Curatel General del Ejercito del Air, Service Historico y Cultural*)
Below: A briefing for *K/88* crews, who display an idiosyncratic collection of outfits. (*Curatel General del Ejercito del Air, Service Historico y Cultural*)
Right, top: A Do 17E-1 reconnaissance bomber of *K/88* with a distinctive *Staffel* marking on the cowling. (*Curatel General del Ejercito del Air, Service Historico y Cultural*)
Right, centre: Not a place for a smoke! Ground crews of *2.K/88* take a rest using bombs and fuel drums for chairs. (*Curatel General del Ejercito del Air, Service Historico y Cultural*)
Right, bottom: A Do 17E-1 takes off from a dusty Spanish airfield. In August 1938 the surviving aircraft in *K/88* were transferred to the Spanish, who used them in *Grupo 8-G-27*. (*Curatel General del Ejercito del Air, Service Historico y Cultural*)

Left, top: The wreckage of an He 111 of *K/88* which accidentally crashed at Candasnos, near Saragossa, in March 1938. It would appear that none of the crew was killed, for no fatalities are noted at this time in German records. (*Curatel General del Ejercito del Air, Service Historico y Cultural*)

Left, centre: The introduction of the Bf 109B-1, such as '6-15' shown here, allowed *J/88* to begin the long process of regaining air superiority. (Greenborough Associates)

Left, bottom: An He 59B-2 of *AS/88*. Originally sent to Spain for reconnaissance duties, these aircraft were later used as torpedo bombers and then as general bombers not only against shipping but also against ports and coastal roads. (Greenborough Associates)

Right, top: The Ju 87B made its combat début at the end of the Battle of the Ebro on 30 October 1938. It could carry a 500kg (1,100lb) bomb with a full two-man crew. The five *Stuka*s were allotted to *J/88* as *5.Staffel*.

Right, centre: The sign of the *Stuka*! During the Nationalist offensives of 1938 the Ju 87s were used to attack road junctions such as this one.

Right, bottom: Helmuth Volkmann played an important part in developing the *Luftwaffe*'s first generation of combat aircraft and in the autumn of 1938 he became Commander of the Condor Legion. He was not a success, constantly pressing for the Legion's withdrawal, and upon his return he was given an educational position. He re-joined the Army at the outbreak of war and died in a car accident in August 1940.

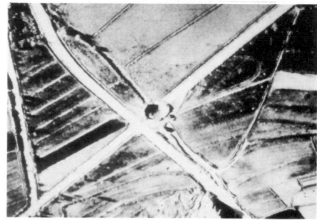

Above: The Ju 86D-1 was intended to be the third leg of the *Luftwaffe*'s bomber triad (with the Do 17 and He 111), but its unreliable Jumo 205 diesel engines earned it the nickname 'The Flying Coffee Grinder' and it was phased out of service during 1938. (Alex Vanags-Baginskis)

Below: Aircraft were built not only by their designers but also by other manufacturers under licence. These Bf 109C-2s are awaiting acceptance after having been produced at the Focke-Wulf factory at Bremen, which was already working on the FW 190. (Alex Vanags-Baginskis)

Right, upper: Following the *Anschluss*, members of the former *Österreichische Luftstreitkräfte* change allegiance and join the *Luftwaffe*. (Alex Vanags-Baginskis)

Right, lower: With the capitulation of the Czech president in March 1939, the Germans occupied Bohemia and seized airfields. Here a Czech air base commander reports to the new *Luftwaffe* base commander.

Above: A *Luftwaffe* sentry stands guard over former Czech Avia B.534 fighters. Many of these aircraft were sold to Bulgaria although a few were used briefly by *I/JG 71*.

Left: Ernst Udet became responsible for *Luftwaffe* equipment in 1936 and became *Generalluftzeugmeister* in 1939. A likeable ne'er-do-well, he took the job partly from vanity and greed, but he inherited a cumbersome organization which he could not control. Siegert, as *Idflieg*, was in exactly the same position in the First World War as *Flugzeugmeister*, and his failure to improve *Luftstreitkräfte* equipment earned him great criticism. The same failures were to drive Udet to suicide. (Alex Vanags-Baginskis)

Right: Alexander Löhr was the Commander of the *Österreichische Luftstreitkräfte* at the time of the *Anschluss*. He later became Commander of *Luftflotte 4*. (Alex Vanags-Baginskis)
Below left: Kurt Student helped to promote gliding in Germany and was later involved in air equipment development. Before he went on leave in 1938 he was told that he would command *Fliegerdivision 7*, but only when he was briefed by Stumpff did he learn that this was a paratroop command. Much of the paratroop development work was actually carried out by *Major* Heinz Trettner. (Alex Vanags-Baginskis)
Below right: Robert *Ritter* von Greim had the distinction of giving Hitler his first flight and was also the last Commander of the *Luftwaffe*. A Bavarian air ace of the First World War, he commanded *Fliegerdivision 5/Fliegerkorps V*. (Alex Vanags-Baginskis)

Above left: Ulrich Grauert commanded *Freikorps* air units and was intimately involved with clandestine air activity in the 1920s. As Commander of *Flieger-division 1* in September 1939 he played an important part in saving *8.Armee* when it was defeated in a surprise Polish attack. In 1940 his *Fliegerkorps I* was exchanged with *Fliegerkorps VIII* in *Luftflotte 2*. (Alex Vanags-Baginskis)

Above right: As a young officer Günther Korten participated in the Nazis' Beer Hall *Putsch* of 1923. Ten years later he joined the *Luftwaffe* and in 1938 he became Chief of Staff to *Luftwaffen-kommando Österreich* (later *Luftflotte 4*). He succeeded Jeschonnek as *Luftwaffe* Chief of Staff but died of wounds received during the Bomb Plot against Hitler. (Alex Vanags-Baginskis)

Left: Alfred Keller was an 'Old Eagle' who received the *Pour le Mérite* for bombing operations. He was involved with civil aviation but provided clandestine support for the *Reichsheer* and re-joined the *Luftwaffe* when he became Commander of *Fliegerkorps IV*. (Alex Vanags-Baginskis)

Right: Another 'Old Eagle', Otto Dessloch had a curious career in the *Luftwaffe*. In September 1939 he commanded *Fleigerdivision 6* but later in the year was given command of *Flakkorps II*. He continued to command *Flak* units for the next two years before being given *Luftflotte 4*. He commanded three *Luftflotten* in total, including *Luftflotte 4* twice. (Alex Vanags-Baginskis)

Below left: Bruno Loerzer was a close friend of Göring's and persuaded him to become an airman in the Great War. During the period of the Weimar regime his aviation activities were confined to training, but he joined the *Luftwaffe* in 1933 and eventually commanded *Fliegerdivision 2/Fliegerkorps II*, which distinguished itself despite him. (Alex Vanags-Baginskis)

Below right: Joachim Coeler came to the *Luftwaffe* via the Navy. He led *FdL Ost* and then *FdL West* but as an ardent supporter of aerial minelaying he was given command of the new *Fliegerdivision 9*. (Alex Vanags-Baginskis)

Above: Udet signs Heinkel's Visitors' Book. Although his position should have kept him desk-bound at Berlin, Udet preferred to visit the factories and fly new aircraft. The manufacturers encouraged this because they found him very easy to influence. (Greenborough Associates)

Below: The death of the 'Ural Bomber' programme did not mean the end of the *Luftwaffe*'s interest in heavy bombers, and it was soon followed by the 'Bomber A' requirement. The He 177 was ordered to meet this, and the He 177 V1 prototype first flew on 19 November 1939. (Greenborough Associates)

Milch, recalled to Berlin from a skiing holiday in Switzerland upon learning that his 'rich aunt' was dying, was briefed by Göring (a sign of improved relations) as the *Luftwaffe* mobilized. Operations were expected to be confined to supporting the Army's advance, seizing airfields and dropping leaflets, though some bombers would be held in reserve.[26] Milch was delegated to assist Sperrle, no doubt to the former's delight and the latter's irritation, as diplomatic pressure increased on the Austrians.

To avert an invasion, Schuschnigg resigned in favour of President Wilhelm Miklas, who briefly refused Hitler's new demand that Seyss-Inquart be appointed Chancellor. Miklas knew that the situation was serious, since border guards reported ominous activity, including aerial demonstrations. At 1000 hours on 11 March Löhr ordered reconnaissance flights at two-hour intervals along the German border and later extended this coverage into a night-time surveillance. During the afternoon he received reports of German aircraft crossing the border, but there was little he could do because a shortage of spares and aircrew had cut *LStrKr*'s first-line, serviceable strength to 33 out of a total of 210 aircraft (see Appendix 8).[27]

During the day pressure upon Vienna increased, directed by Göring over the telephone, and with no support even from Italy the isolated Miklas capitulated during the evening. During a diplomatic ball Göring authorized Milch to commence *Luftwaffe* operations the next day, and at 0820 on 12 March *KG 155* and the transport *Staffeln* took off, although it was the trainers of *Oberst* Vierling, Sperrle's *Kommandeur der Fliegerschulen und Fliegerersatzabteilungen*, who were first over Vienna dropping leaflets. Wolff flew in with the first wave and with two companies of the élite *RLM* guard unit *III/Regiment General Göring (RGG)* occupied Wien-Aspern airfield, where he set up his headquarters. The first aircraft to land was a Do 17E of *6./KG 155*, which touched down at 0940 after dropping leaflets.

The airfield was a hive of activity during the rest of the day, with bombers on leaflet missions and a stream of transport *Staffeln* disgorging 2,000 troops (mostly from *RGG*) to assist local *SA* members in the occupation. Despite the support of Austrian airmen, the facilities were saturated, but the problem eased since by the end of the day there were nine *Luftwaffe Staffeln* at Austrian air bases: *Stab, II* and *7./KG 155*, with a Bf 109 *Staffel* of *I/JG 135*, were at Wien-Aspern with two transport *Gruppen*; an He 51 *Staffel* with a Bf 109 *Kette* of *JG 135* was at Linz, with three Ju 52s carrying the groundcrews and supplies; two corps *Staffeln* of *AufklGr 15* and a reconnaissance *Staffel* of *AufklGr 25* were at Wels; and a *Kette* of *1. (F)/25* was at Innsbruck.

As his command grew, Wolff moved first to Löhr's headquarters, then to the Hotel Imperial and finally to the Hotel Astoria. On 13 March, as a transport *Gruppe* moved to Wiener-Neustadt, some 260 aircraft (including 80 transports) demonstrated over Vienna and two *Staffeln* of *KG 155* flew into Klagenfurt. Two days later there was a celebratory parade, with Hitler taking the salute at the Maria-Theresa Monument. This concluded with a fly-past organized by Milch and involving some 450 aircraft (including Austrian fighters) led by Wolff (with Löhr in the co-pilot's seat) and 270 aircraft led by Vierling.

Three aircraft were lost accidentally during '*Otto*', and while it appeared as an awe-inspiring display of air power both *Luftwaffe* and foreign observers noted serious problems. Staff work was often poor, while the assembly and transport of reservists was a shambles. The shortage of petrol bowsers forced the *Luftwaffe* to move fuel in large drums in impressed civilian vehicles and supplies were consumed faster than anticipated, while the poor discipline of reservists led to the loss of much equipment through carelessness. Whether or not the operation would have succeeded without the co-operation of the Austrian airmen is impossible to say, but active resistance would certainly have made it more difficult.[28]

The *LStrKr* was formally absorbed by the *Luftwaffe* at the beginning of April and *Generalleutnant* Löhr became *Kommandierender General und Befehlshaber des Luftwaffen-kommandos Ostmark*, which included *Luftgaukommando XVII*. He had an excellent German Chief of Staff in *Oberstleutnant* Günther Korten, who had participated in the Beer Hall *Putsch* and later learned to fly in the *Reichsheer* before graduating in aerial photography at Lipetsk. He had formerly been the Commander of *Aufklärungsgruppe (F) 22* and in August 1943 succeeded Jeschonnek as *Luftwaffe* Chief of Staff. Five *Luftwaffen Gruppen* were joined by *Stab Aufklärungsgruppe (H) 52*, which moved to Vöslau-Kottingbrunn as *Aufklärungsgruppe (H) 18*.

A mixed military commission under the German Military Attaché, *General* Wolfgang Muff, screened the Austrian forces for political reliability and half the senior officers (*Oberst* and above) were dismissed. They included Yllam, who proved too much an Austrian nationalist—although Löhr sought to persuade him to join the *Luftwaffe*—and died in retirement in January 1942 at Klagenfurt, where his squadrons had fought in 1919.[29] Most Austrian airmen switched allegiance, and by 12 April 3,100 had joined the aviation elements while 970 had joined *Flak*. *JaGschw II* became *I/JG 138*, the pilots retaining Fiat CR 32 fighters, although most of the groundcrews went to *1. (F)/28*, and while *JaGschw I* was scheduled to become a heavy fighter *Gruppe* its *Staffeln* were actually absorbed by *JG 132* and *JG 134*. The men of *Bomben Geschwader I* were absorbed by *KG 155* (redesignated *KG 158*), but other *LStrKr* personnel were dispersed throughout the *Luftwaffe*. The newly formed *I/StG 168* received a substantial influx of Austrian personnel and Fiat fighters (which also briefly equipped *I/JG 135*), but most of the latter were transferred to the Hungarian Air Force later in the year. About 125 Austrian aircraft were retained for *Luftwaffe* use, mostly German-made trainers, transports and liaison aircraft, while the *NSFK*'s *Gruppe 17* (Danube and Alps) received 61 trainers. The remainder were scrapped.

The *Anschluss* provided the *Luftwaffe* with recruits, a pool of skilled labour which was quickly exploited by BFW to produce the Bf 109 at Wiener-Neustadt and vital raw materials, including iron ore, oil and hydro-electric power. There were also gold and foreign exchange reserves worth RM305 million ($122.50 million) at a time when the Reich's foreign exchange reserves were down to the equivalent of RM90 million ($36.15 million), while Germany now threatened eastern Europe and in particular France's key ally Czechoslovakia. Military relations between these two nations were close, especially those between the air forces, who had held staff meetings twice a year from 1933 and arranged to base French heavy bombers in Bohemia in western Czechoslovakia together with 400 tonnes of bombs.[30] Bohemia was now in a German vice as the Nazi press raved about the oppression of the German minority in the Sudetenland, through which ran a fortified line which was Czechoslovakia's shield. On 28 May Hitler called a conference of his military leaders and informed them that he wanted Czechoslovakia 'wiped off the map'. Two days later the first operational directive was published calling for preparations to be complete by 1 October, and while the emphasis was upon Czechoslovakia there could be no ignoring the prospects of war in the west.

The *Luftwaffe* was to destroy the Czech Air Force, then support the Army at tactical and operational level by striking mobilization and administrative centres, the latter to paralyse the Czech Government. However, on 30 May Hitler specified that industrial centres were to be spared as far as possible, while reprisal attacks against the civilian population were to be subject to his approval. Contingency planning for an attack upon Czechoslovakia was completed in May 1935 following the Czech–Soviet mutual assistance pact as *Unternehmen 'Schulung'*, which was periodically updated. Following the *Anschluss* it was revised as *'Planstudie 1938'*, and in great secrecy a *Luftwaffe* operational plan in support of it was drafted by Kesselring and his Chief of Staff (*Oberst* Speidel) in June and published on 11 July as *'Planstudie Grün'*. Detailed planning was performed by Grauert's *Fliegerdivision 1* and Wimmer's *Fliegerdivision 2* as well as by *Luftgaue III, IV* and *VIII*, while later Sperrle and Löhr were drawn in together with Milch, the latter co-ordinating logistics support.

It was hoped to begin the *Luftwaffe* campaign with a surprise knock-out blow on the Czech Air Force followed by attacks upon Czech Army, but if surprise were lost the latter was the priority target. *Staffeln* were to remain at peacetime bases until the late afternoon of the day before the assault and would move into forward airfields upon receipt of code-word *'Geschwadertag Albert'* ('Geschwader Day Albert'). *Fliegerdivisionen 1* and *2* would provide cadres for a close support command (*Fliegerdivision zbV*), while the participation of *Luftwaffengruppenkommando 3* (Sperrle) and *Luftwaffenkommando Ostmark* (Löhr) would depend upon the international situation. Kesselring's forces were in reserve and he was assigned the newly raised *Fliegerdivision 7* under *Generalmajor* Kurt Student, the Inspector of Paratroops and Airborne Troops, to plan and to direct *Übung 'Freudenthal'* (Exercise 'Freudenthal'), an airborne operation 150 miles (240km) behind the Czech border and the fortifications. Planning here, as elsewhere, was greatly assisted by a clandestine photo-reconnaissance unit created in January 1935 under *Oberstleutnant* Theodor Rowehl and operating under Göring's command. Generally known as *Kommando Rowehl*, it was ostensibly part of Lufthansa, for whom it conducted 'route-proving trials'. Travelling in a variety of aircraft, Rowehl's men flew over much of Europe, including the British Isles, and their photographs were the basis of *Luftwaffe* target folders for several years.

With offensive power concentrated in the east, the western air defences were strengthened through the creation of *Luftverteidigungszonen* to provide area defence. The first was actually around Vienna, but the most important, *Lv-Zone West*, was activated on 1 June 1938 in Germany's western approaches under *Generalleutnant* Karl Kitzinger. It was some 375 miles (600km) long with a depth ranging from 12 to 60 miles (20–100km) and a nominal establishment of 1,300 guns, including 245 static weapons emplaced in the *West Wall* under *Höheren Kommandeur der Festungsflakartillerie III* (Wiesbaden), to provide a triple overlapping field of fire of up to 23,000ft (7,000m), exposing each aircraft to five minutes' concentrated fire by three batteries. In addition to guns there were barrage balloons, searchlights and fighters. Similar but smaller *Luftverteidigungskommandos* (*LvKdo*) shielded Berlin, Stettin, Hamburg, Düsseldorf and Leipzig.

These shields allowed the *Luftwaffe* to sharpen its swords, but the confidence of its officers in their offensive capabilities encountered considerable scepticism from Army generals, including Beck, and this caused friction. Beck especially angered *Luftwaffe* staff officers after his wargame *'Generalstabreise'* had concluded that operations against France and Czechoslovakia would involve a catastrophic defeat of the *Wehrmacht* because it would take so long to defeat the Czechs. At a post-exercise dinner on 10 May 1938 Jeschonnek denounced Beck. 'Thirty days for those ridiculous Hussites! As if there weren't any *Luftwaffe*,' he snorted. 'For Beck our squadrons are only a troublesome appendix. But all of you will see the most stupendous things.'[31]

Ultimately it was the *Luftwaffe*'s perceived ability to do 'stupendous things' which provided the most decisive victory of the 'Flower Wars', although the roots stretched to the aftermath of the *Anschluss* Crisis. In late March the British and French Governments independently asked their highly experienced Chiefs of Air Staff, Marshal of the Royal Air Force Sir Cyril Newall and *Général* Joseph Vuillemin, to assess their respective service's capabilities. The shocking response from both was that their bomber forces were incapable of effective offensive operations and would be rapidly destroyed by enemy defences; indeed, Vuillemin estimated that they would be wiped out within a fortnight. By mid-September 1938 the British and French had some 700 bombers but the only modern ones were 385 RAF Battles, Blenheims and Whitleys. Ludlow-Hewitt in October 1937 had recognized that his bombers, such as single-engine Wellesleys and twin-engine Harrows, were no match for the Bf 109 and his Annual Report of 10 March 1938 stated that 'our Bomber Force is, judged from a war standard, practically useless'.[32] Six months later he had only 200 fully operational pilots, a serviceability rate of about 50 per cent and a 33 per cent reserve.

If the swords were blunt, the shields were paper thin. The RAF calculated in September that it could mobilize only 288 fighters (48 Hurricanes) with a 33 per cent reserve, while a memo of 29 August estimated that by 1 October the squadrons would have only fourteen Spitfires and 70 Hurricanes because production was only 35 fighters a month.[33] The 400 French fighters were mostly braced monoplanes, supplemented in 1938 by 23 MS 405s. The British later estimated their first-line strength on 1 October to be 1,606 aircraft with a 25 per cent reserve, while the French had 1,454 with a 50 per cent reserve.[34]

Newall and Vuillemin assessed the *Luftwaffe*'s strength to be between 2,700 and 3,000 (on 1 August it was 2,928, including 81 transports), aided by accurate estimates of its order of battle,[35] but there were Jeremiahs aplenty to provide more alarming figures. The most influential was the world famous aviator Lindbergh, who lived in Europe after the kidnap and murder of his son. His estimate of 8,000 aircraft followed visits to *Luftwaffe* bases and factories but, as the British Air Attaché in Berlin, Group Captain J. L. Vachell, recognized, he was repeating deliberately the exaggerated reports of the US Military Attaché, Major Truman Smith, who hoped to stimulate American air rearmament.[36] Lindbergh's was not the only influential voice, and the Foreign Office received a report from Sir Frederick Sykes, the RAF's second Chief of the Air Staff, confidently asserting that the *Luftwaffe* had 12,000 aircraft!

Vuillemin's post-war reputation for 'defeatism', which was actually hard-headed realism, was underlined by his famed reciprocal visit to Germany between 16 and 21

August 1938 following one by Milch and Stumpff to France in 1937. This was an opportunity for Vuillemin to assess personally the reports from his excellent intelligence service, which 'enabled him to form a very accurate idea of the real strength of Nazi Germany'.[37] He visited Rechlin, air bases and factories, and at Augsburg and Oranienburg there were attempts to overawe him as the Bf 110B and the 'He 112U' (actually an experimental He 100) appeared to roll off the production lines. He was not deceived; indeed, his report makes no mention of the 'He 112U'.[38]

The Allies were conscious that the Czechs would be unable to pin down *Luftwaffe* strength for long: although Prague had an air force of 54 squadrons, virtually all its 800 first-line aircraft were obsolete. Modern designs, including the Avia B 135 fighter and Aero A 300/304 bomber, were being developed, but as a stop-gap 60 SBs were purchased from the Soviet Union and flown from the Ukraine pending delivery of licence-built aircraft during the autumn of 1938 as the Avia B 71.[39] Prague had a mutual assistance pact with Moscow, but even if London and Paris had desired Soviet support, the *VVS-RKKA* was becoming an increasingly slender reed. The Purges were wiping out its leaders and former leaders, including Alksnis and Rosengol'ts, while even designers such as Tupolev were not immune. The airmen were largely reduced to the tactical role after the disbandment of the *AON*.

These reasons helped to force defensive strategies upon the British and French airmen, but financial and industrial factors also contributed. The British did not wish rearmament to cripple their economy and decided that strategic defence was cheaper and more cost-effective than strengthening Bomber Command. The decision was made despite strenuous objections from the Air Staff in October 1937 by the Defence Co-ordinator, Sir Thomas Inskip, after exercises had shown the effectiveness of radar-based air defence. The French went on to the defensive after Vuillemin's pessimistic report to the *Comité Permanent de la Défense Nationale* on 15 March and after learning four months earlier that domestic monthly production was only 60 aircraft, less than 30 per cent of the British figure. Deliveries of combat aircraft to the *Armée de l'Air* per quarter in 1938 ranged from 61 to 126 (20 to 42 per month) due to the reorganization of industry, a lack of investment and a small work force (35,000 in 1938, rising to 171,000 in 1940). The March meeting agreed an ambitious new schedule emphasizing fighters, Plan V, using American production to make up the French shortfall, and orders were placed for Curtiss 75 fighters, Vought V-156 dive bombers and Douglas DB-7 and Martin 167 bombers while the British ordered the Lockheed Hudson maritime patrol aircraft, all of which were in service by 1940. However, further French contracts for the Hawk 81 (later P-40 Warhawk/Tomahawk) fighter and Consolidated Model 32 (later B-24 Liberator) heavy bomber were to benefit only the British.

It was not until the spring of 1939 that both air forces had adequate numbers of modern fighters and the British air defence system was complete. In September 1938 only five of the twenty radars in the Chain Home (CH) line were operational, supported by a rudimentary command and communications system. All nations were acutely sensitive to threats of all-out bomber attacks upon their capitals, which was a scenario in the 1937 *Wehrmacht* manoeuvres, but the British seem to have been especially pessimistic, with estimates of 400–600 tons of bombs being dropped and a casualty list of 1.5 million (for which the British stockpiled large numbers of papier-mâché coffins). Yet this did not prevent Britain and France publicly supporting the Czechs during the summer and mobilizing their forces. The French were more resolute in their commitment than the British, who, within a week of the invasion of Austria, had adopted appeasement based upon Czech territorial concessions. This, together with the knowledge that they would bear the burden of the land war, undermined the resolution of the French. German nerves were also shaky, especially in the Army, which constantly carped and criticized. *Luftwaffe* officers such as Kitzinger spitefully repeated the soldiers' views, which inevitably reached Hitler's ears. In an effort to calm their fears he called together all his senior commanders for a pep talk on 10 August, those present including the *Luftwaffengruppenkommando* leaders and Jeschonnek, but Beck was unconvinced and later resigned, to be replaced by Halder.

Two days later Hitler discovered that the *Luftwaffe* was more shadow than substance, following a survey of the *Wehrmacht* to assess its condition in early October, when *'Grün'* was scheduled. He discovered that production of anti-aircraft guns and ammunition of all types, as well as that of machine guns and cannon for the aircraft, was no more than 50 per cent of the immediate operational requirement. The target for monthly production was 1,179 aircraft, but, despite extending the working day from eight to ten hours in June, Udet warned that it would take between four and thirteen months to reach this figure and that only 493 aircraft would be produced in September. The introduction of modern aircraft such as the Bf 109D, He 111H and Do 17M/P without adequate spares meant that the *Luftwaffe*'s serviceability rate was low (57 per cent overall), and on 1 August only one of the 173 *Schlachtflugzeuge* (He 45/Hs 123) was operational. Up to 60 per cent of the aircraft required overhaul, while stocks of engines in maintenance and supply depots amounted to only 4–5 per cent of front-line strength. The supply organization later estimated that the *Luftwaffe* had only enough reserves for four weeks' operations, and while most *Geschwader* had only one technical equipment set (*T-Sätze*) the *Kampfgeschwader* had the 'luxury' of three. Through careful organization and by reducing flying hours, especially training, serviceability rates were massaged up to 90 per cent by 19 September and 94 per cent by 26 September, especially in the *Kampfgruppen*, where rates rose from 49 per cent on 1 August to 90 per cent on 26 September.

The biggest problem, shared with the RAF and the *Armée de l'Air*, was the shortage of experienced aircrews. As with the RAF, the constant expansion 'diluted' quality, and of 2,577 aircrews (69 per cent of establishment) only 1,432 (38 per cent of establishment) were fully qualified. In the core (fighter, bomber and reconnaissance) squadrons there were only 2,134 aircrews, of whom 1,168 (31.5 per cent of establishment) were fully qualified, including 27 per cent of bomber and dive bomber crews. Many bomber pilots were not instrument-rated, while the back-seat men in the *Stuka*s required instruction in using the radio or the machine gun. The effective bomber force was therefore reduced to 378 aircraft (32.5 per cent of strength), but the fighter force had 537 (83.5 per cent of strength), which ensured a shield for the *Reich*, although many fighters lacked radios.

Seven *Jagdgruppen* retained biplanes (mostly Ar 68s), while an eighth (*IV/JG 132*) was hastily raised with requisitioned Japanese He 112Bs on 1 July, although it was understrength like most of the Bf 109 *Gruppen*, which had an average of only 26 fighters. Deliveries of the Bf 110B-1 had hardly begun: only four were received by the end of August, and by December the production rate was barely ten a month, so many bomber missions would have been unescorted and suffered attrition even from the handful of modern fighters. It was little wonder, then, that the *Luftwaffe* sought to bluff Vuillemin and reinforced this by an ostentious display of confidence a few weeks later as the crisis deepened when both Milch and Stumpff were the official guests of the Swedish Air Force.

During August the *Luftwaffe*'s preparations continued with the readying of airfields and the creation from school instructors and aircraft of two *Staffel*-size *Reservejagdgruppen*, five *Schlachtgruppen* for close air support and four *Kampfgruppen* *zbV* to reinforce the three existing transport *Gruppen*. Preparations for *Fall 'Grün'* accelerated when Milch and Stumpff returned, and on 20 September the *Gruppen* moved towards their forward bases, although the *Luftwaffe* staff were nervous both about the prospects for the airborne operation and about the overall timing. Led by Jeschonnek, they twice met Kietel and Jodl of *OKW*, on 8 September and 14 September (the date of Chamberlain's first flight to Munich), and the generals had to fight hard to keep the airborne operation in the overall plan, which involved 49 divisions and 2,400 aircraft.

Kesselring was to support *2.*, *8.*, and *10.Armee* with 1,200 combat aircraft (*Fliegerdivisionen 1*, *2* and *zbV*) and 400 transports (*Fliegerdivision 7*), while to protect his bases in Silesia, Saxony and Thuringia from air attack he had borrowed two naval *Freya* radar sets. Sperrle's *Fliegerdivision 5* had 650 aircraft, including a close-support force under the newly created *Fliegergeschwader 100*, to support *12.Armee*, while Löhr had 180 aircraft to support *14.Armee* (see Table 11). The emphasis was upon bombers: even *Luftwaffengruppenkommandos 2* and *3* had been stripped and were left with some 175, and of the 400–500 fighters scheduled to participate in *'Grün'* a third were for home defence duties. Such was the *Luftwaffe*'s commitment to the offensive that only 400 aircraft (excluding naval assets) were left in the west.

Only now was serious consideration given to air operations against Great Britain, although Felmy, Commander of *Luftwaffengruppenkommando 2*, had made an outline study at Stumpff's request during the *Anschluss* Crisis. On 23 August Felmy was ordered to produce more detailed proposals, and during the next nine months these was followed by two more on 22 September 1938 and 13 May 1939, neither of which found favour with Göring. Felmy identified two serious problems, the bombers' lack of range with effective payloads and the absence of adequate bases. In his second report he noted that the He 111's range could be extended by 645km (400 miles), though only by reducing bomb loads to a puny 500kg (1,100lb). Effective attacks required twice that payload, and for this reason much of British industry was beyond the effective range of the *Kampfgruppen*. In the absence of heavy bombers the only solution, Felmy concluded, was to seize bases in the Low Countries before undertaking an air offensive against the British. Unknown to Felmy, Group Captain John Slessor reached the same conclusions in a report to his superiors.[40]

Felmy warned of the urgent need to improve navigation training for long-range missions, especially over water, and also that the bombers would have to fly unescorted since British targets would be beyond the range even of German-based Bf 110s. His last report, *'Planstudie 39'*, assumed that there would be war in 1942,

	FlDiv 1/2	FlDiv zbV	FlDiv 5	LwKdo Österreich
Gruppen				
Jagd	10 (4 HD)*	–	5 (2 HD)	1
Kampf	17	–	8	3
Stuka	4	–	2	1
Schlacht	–	3	2	–
Total	31 (4 HD)	3	17 (2 HD)	5
Staffeln				
Jagd	–	2	2 (HD)**	–
Reconnaissance	2	1	3	1
Total	2	3	5	1

Note: HD = Home Defence. * Assigned to *Luftgau*. ** *Reservejagdstaffeln*.

when Germany would have a heavy bomber force, and with this he advocated an immediate, intensive air offensive against the British. His most positive proposal, in September 1938, was for the conversion of FW 200 and Ju 90 airliners as a small, stop-gap heavy bomber force pending the He 177's entry into service.

Felmy's September report arrived the day after *OKW* was informed that Prague had accepted a peace plan proposed by Chamberlain. Orders were given to prepare for a peaceful occupation, but Hitler demanded more and on 26 September the German Army moved into its assault positions. In these circumstances Felmy's report incensed a nervous Göring, who reprimanded him for producing a discussion paper instead of detailed plans. The *RLM* had planned to make the Deputy Air Attaché in London, *Oberst* Kessler, Felmy's Chief of Staff but on 17 September Kessler visited his old friend Göring at Carinhall and begged him to avoid war with the British. This display of 'nerves' led to Kammhuber's being substituted for Kessler, who later became *Kommodore* of *KG 1* and was subsequently appointed Chief of Staff of *Luftflotte 1* before he succeeded Harlinghausen as *Fliegerführer Atlantik*.

In any event, Felmy's conclusions were academic. Chamberlain had always been determined to appease what he saw as the injustices done to Germany by the Versailles Treaty and now did everything in his power to avert war. Symbolically he flew for the first time, to open negotiations with Hitler. By 27 September he had abandoned Prague and forced the Czechs to make further concessions, which were incorporated in the agreement signed by Britain, France, Germany and Italy at Munich on 30 September permitting the German occupation of the Sudetenland. The *Luftwaffe*, which had received the *'Geschwadertag Albert'* code-word on 28 September, prepared to support this operation with reconnaissance squadrons, two *Jagdgruppen*, two *Stukagruppen* and a *Schlachtgruppe*. But as the Germans crossed the border in the afternoon of 1 October most *Staffeln* were grounded by twelve days of fog and rain. The corps units flew some sorties and there were a few *Staffel*-size fighter patrols.

On 3 October a communications aircraft supplied by Sperrle landed at the Czech airfield at Asch-Eger to provide a forward command post. Two days later fighters and bombers demonstrated over Czech cities and on 6 October Student's men displayed their skills in front of Hitler using 300 aircraft and gliders, German residents having prepared the 'landing zone' beforehand. By 10 October the Sudetenland was in German hands, and out of 500 aircraft only four were lost with eight men killed, mostly in accidents, although nervous *Flak* crews in Vienna shot down a courier aircraft. From 13 October the *Luftwaffe* units began to stand down and most of those mobilized in August were disbanded, although the *Schlachtgruppen* were reorganized into *Stuka-* or *Kampfgruppen* while *Fliegergruppe 20* became *II (Schl)/LG2*.

The Munich Agreement averted war and released a tidal wave of relief across Europe, nowhere more so than among the *Luftwaffe* leadership. The dreaded prospect of a two-front war before re-equipment was complete had been avoided, yet it was clear that the Allies would not allow themselves to face aerial inferiority in the future. On 15 October, two days after the mobilized units had stood down, Hitler demanded a fivefold increase in the *Luftwaffe* to 12,000 aircraft, as part of a massive new armament programme to be completed by early 1942. The emphasis was upon operations against Great Britain using both long-range bombers and fighters. The *RLM* held preliminary conferences on 24 and 26 October, the latter at Carinhall. At the former, for the first time, Göring mentioned the 'knock-out blow' which the British Air Staff feared above all else and said it would involve every aircraft, including those of training units. At the latter Jeschonnek, as Head of Air Staff Operations, persuaded Göring to authorize a force of 500 He 177s.

Udet warned that the grandiose plans were simply beyond Germany's resources but the Carinhall Conference led to the Jeschonnek Plan of 7 November. This envisaged that the *Luftwaffe*'s first-line strength in April 1942 would be 10,700 aircraft, including the heavy bombers and carrier-borne aircraft, and would cost RM60,000 million, ($24,000 million), equivalent to the total German rearmament expenditure between 1933 and 1939. Such a force would also require 200,000 tonnes of fuel per month and the import of 2,910,000 tonnes, equivalent to 85 per cent of world production.[41] The united opposition of Stumpff and Milch was unable to overcome Jeschonnek's confidence and, as mentioned above, to everyone's amazement Göring approved the plan.

The planning carousel turned again to produce *Lieferplänen Nr 9, 10, 11, 12, 13* and *14*. But each time material reality shattered the dream, forcing even Jeschonnek to cut back his bomber and heavy fighter requirements in February 1939. Once again production runs were reduced or their deadlines were stretched, and while new aircraft such as the Bf 109F and Me 210 heavy fighter were included, the older models often had to make up the numbers. Yet by the outbreak of war Germany was producing 690 aircraft a month and Hitler believed the Ju 88 programme to be so important that he gave it third priority behind munitions programmes and demanded 300 a month.

In January 1939 there were 269 first-line squadrons and by 1 July six more had been added, together with a *Seenotstaffel* (rescue squadron). Only 30 squadrons (11 per cent) in July were corps units, while 90 (33 per cent) were bomber units and 27 (10 per cent) were *Stuka Staffeln*. There were 40 (14.5 per cent) light and 27 (10 per cent) heavy fighter squadrons, although the shortage of Bf 110s meant that many of the latter operated Bf 109s. The *Flakwaffe* remained the heart of the air defence system, with 2,600 heavy and 6,700 light/medium guns and 3,000 60cm and 150cm searchlights, while accounting for some 20 per cent of the *Luftwaffe*'s 370,000 men (including 16,900 active and 4,900 reserve officers) in April 1939.

In March 1939 came the last of the 'Flower Wars' when Hitler, still furious at being 'thwarted' by Chamberlain at Munich, decided to finish off Czechoslovakia. The Czechs had already reluctantly accepted that they were now in the German sphere of influence and soon after the Munich Agreement prepared to supply the *Luftwaffe* with second-line aircraft, principally trainers. Meanwhile Berlin encouraged the Slovaks in the east of the country to press for independence protected by Hitler. The 67-year-old Czech president, Emil Hácha, went to Berlin on 14 March to defuse the situation but Hitler informed him that Germany intended to occupy the remainder of Bohemia. Although in tears, the old president stubbornly refused to capitulate and this worried Göring, Milch, Udet and Bodenschatz for as the day wore on the weather deteriorated and even as the German Army entered Bohemia snow and poor visibility grounded the *Luftwaffe*.

Göring was not enthusiastic about the annexation and had been on sick leave in Italy when he was suddenly summoned to Berlin on 12 March. As Hácha resisted, the *Reichsmarschall* became more nervous, combining persuasion with threats by talking of destroying 'the beautiful city of Prague'. The old man fainted and, as he recovered consciousness, Göring bellowed 'Think of Prague!' Reluctantly the Czech president ordered the Army not to resist the invaders, and in the early hours of 15 March he signed a document granting Slovakia servile independence and reducing the rump to the Protectorate of Bohemia and Moravia.

The occupation was supported by 500 aircraft of *Luftflotte 3*'s *JG 233* and *333, KG 155, 255* and *355, StG 165* and *AufklGr (F) 123*. *Luftwaffe* activity was confined to demonstrations and leaflet drops over Prague, where a delighted, teetotal *Führer* had a glass of wine. He had reason to celebrate, for the excellent equipment of the former Czech Army allowed him to expand his own Army by a third, including several mechanized divisions, but there was little loot for the *Luftwaffe*, which lost two aircraft and seven men but gained only bases in the heart of Europe.

Of 1,500 Czech aircraft (including 600 trainers and 100 liaison aircraft), only some 150 combat aircraft and 250 trainers were initially discovered. Some 120 were transferred to the new Slovak Air Force (*Slovenské vzdusné zbrane*), while 205 were bartered to the Bulgarians at 40 per cent of cost.[42] A few Avia B.534s served briefly with *3./JG 71* but most requisitioned aircraft were flown only on second-line duties. There was a small aircraft industry with 10,500 people (which dropped to 7,100 within a year), but of greater value were the Czech foreign currency reserves, including £25 million ($115,750,000) which could be used to import raw materials and machine tools. The Czech aircraft industry was soon producing the Fi 156 Storch and DFS 230 glider as well as the As 10c engine for the Ar 96 advanced trainer and Fi 156. Preparations were also made to produce the new Fi 189 corps aircraft to replace the Hs 126.[43]

Hitler's flagrant violation of the Munich Agreement, followed three days later by the occupation of the former German port of Memel (Klaipeda) in Lithuania, steeled the hearts of the British and French Governments. They pledged military support to Poland, and on 31 March Chamberlain announced that aggression against Poland would lead to war. Göring, who was enjoying the fruits of peace despite frequent bellicose utterances, was to spend much of the summer using a Swedish emissary in secret attempts to avoid war while encouraging further appeasement.

The prospect of a two-front war was studied by the *Luftwaffe* staff from January 1939 and benefited from a major reorganization of *Luftwaffe* Intelligence completed the previous year. The *Luftwaffenführungsstab* acquired an Intelligence Section only in June 1935, but it consisted of reservists and civilians using foreign periodicals. Only in January 1938 did its status increase, when a regular officer, *Major* Josef 'Boy' (Beppo) Schmid, replaced the reservist *Oberst Freiherr* von Bülow. Schmid, aged 38, was a shrewd and ambitious man who had seen action with Bavarian *Freikorps* and participated in the Beer Hall *Putsch*. He had joined the *Reichsheer* as a private and quickly won a commission, but slow promotion led to his transfer to the *Luftwaffe* in 1935. His new role owed much to his friendship with Göring, for Schmid was no airman, spoke no foreign languages and had no experience of Intelligence, although he brought a professionalism previously lacking. It is a measure of the importance of his position that he was a *Major* while in the RAF an officer equivalent to a *Generalmajor* held a similar position. Objectives were set and Schmid widened his net to acquire data from *Abwehr* and from Martini's communications intelligence organization, which had seven interception stations.

Schmid used this data, his own resources, Rowehl's photographs and industrial inputs from Milch but it was not until the spring of 1939 that he acquired authority over the Air Attachés, including Wenniger, now a *Generalleutnant*. The greatest weakness was the lack of a technical intelligence capability, but his masters were sufficiently impressed with his work to promote him to *Oberstleutnant* by the summer of 1939. Schmid's work, supplemented by briefings from specialists and academics, was used in three detailed reports by Sperrle, Felmy and Kesselring on enemy air power in France (*Studie 'Rot'*), Britain (*Studie 'Blau'*) and Poland (*Studie 'Grün'*), with *Studie 'Blau'* remaining the *Luftwaffe*'s basic reference work for Great Britain until the end of the Second World War.

Greater attention was now being paid to the British, although Felmy's appreciation, *'Planstudie 39'*, so infuriated Göring with its pessimism that he demanded that it be either withdrawn or destroyed. In preparation for attacks upon British shipping and the Royal Navy, *Generalmajor* Coeler became *FdL* in May 1939 with responsibility for minelaying while *Generalmajor* (later *Generalleutnant*) Hans Ferdinand Geisler, once Wenniger's deputy in the *Reichsmarine* and later a member of LA, became *General zbV der Luftflotte 2* with responsibility for anti-shipping operations. During the early summer Geisler's *Staffeln* conducted practice missions across the North Sea, leading to a full-scale British air alert when radar spotted a *Gruppe*-size mission 50 miles (80km) off Norfolk. It approached to within six miles (10km) of the coast before turning back. The Germans were aware of British radar experiments and three signals intelligence sorties were flown off Britain between May and August 1939 using the retired commercial airship LZ 127 *Graf Zeppelin I*. The 12m transmissions intercepted were regarded as so crude compared with the centimetre wavelengths used by German radars that the *Luftwaffe* concluded that British radar work was at an experimental stage, although Martini had his doubts. In fact, the signals were from the Chain Home system's Type 1 radars, which detected, located and tracked the airship without difficulty. The *Luftwaffe* remained in ignorance of the radar system for more than a year—with disastrous results.[44]

At that time only one German air search radar, a development of the naval *Freya* surface search system, was operational at Potsdam, although at the outbreak of war two more, designated *FuMG 39G (gB) Dete 1 Freya*, were on order. In addition, a fire control radar under development since 1936 was also ordered in 1939 as the *FuMG 62A Würzburg*. The German Navy led the way in radar development from 1934 and its progress was monitored by Martini, who used an experimental naval *Freya* to detect aircraft at distances of 80–100km (50–60 miles) during the *Wehrmacht*'s autumn manoeuvres of 1937. However, he does not appear to have appreciated the use of radars as the basis of a strategic air defence system, and at the outbreak of war all that existed was a chain of naval *Freya*s for the coastal defence of the German Bight.

If Hitler knew this it did not undermine his confidence in the *Luftwaffe* as the Polish Crisis deepened, and he publicly stated that air power would help destroy Poland quickly and then neutralize the Western Powers. On 12 August he informed Ribbentrop and the Italian Foreign Minister Count Galeazzo Ciano that, while the British were making progress on air rearmament, it would take a year or two before their efforts became apparent. Only ten days later he emphasized Britain's vulnerability to air attack while he was addressing his military leaders.

There was good reason for this confidence, for the expenditure of some RM17,200 million ($6,680 million)—a quarter of the military budget—had produced the best equipped air force in Europe. On 1 September the *Luftwaffe* had a first-line strength of 4,093 aircraft, of which only 268 (7.5 per cent) were of the first generation. However, the supporting structure of spares, repair facilities, reserves and training facilities were inadequate for a prolonged war, as was the fuel reserve of 400,000 tonnes.

The bomber force, 1,176 strong, was formidable in quantitative and qualitative terms. In the *X*- and *Y-Geräte* it was unique in possessing radio-beam navigation and bombing systems for attacks in poor weather during daylight and at night. The fighter arm was well equipped, although many *Zerstörergruppen* retained the single-seat Bf 109, but strategic defence remained the realm of the *Flakwaffe*, using guns and observers with optical and audio sensors. Night fighter operations had been practised from May 1936 and a system involving co-operation with searchlight belts had been developed, although only a handful of light fighter *Staffeln* were available for this role.

In June 1939 the *Luftwaffe* conducted its last major pre-war exercises, '*Generalstabs-reise 1939*', to evaluate air operations against Poland. This emphasized the need to destroy enemy air power within the first few hours of an attack, while Jeschonnek demanded improved tactics to provide a fundamental advantage over fleeting technological superiority. He was confident of destroying the Poles but hoped that Hitler would ensure that the *Luftwaffe* did not fight a multi-front war for which it was not quite ready. Nevertheless, Schmid's reports gave him grounds for optimism as the *Staffeln* began to move eastwards.

One summer evening in 1939 Hoffmann von Waldau was driving *Oberst* Hans-George von Seidel, the *Luftwaffe* Quartermaster, across the Grosser Stern Square in Berlin. Six years before the Russian occupation (after which it was in the Western Zone), the perceptive Operations Chief suddenly asked Seidel, 'Do you think we will live to see this called Red Army Square?'.

NOTES TO CHAPTER 6

1. The Boeing 299 (B-17) prototype was lost to the same cause in October 1935.
2. Macksey, pp.52–4.
3. See Irving, *Milch*, pp.55, 366/n.192.
4. Deichmann, p.87.
5. Taylor, *Munich*, pp.603–42.
6. See Ishoven, pp.358–61; Irving, *Milch*, pp.57–9; and Malcolm Smith, pp.177–8.
7. See Taylor, *Sword and Swastika*, p.172, and *Munich*, pp.324–6; and Völker, *Luftwaffe*, pp.96–7.
8. For Jeschonnek see Mitcham, pp.27–9; Irving, *Milch*, pp.68–9; and *Landser Magazin*. I am also indebted to *Admiral* Gert Jeschonnek for personal information.
9. See James, pp.188–95; and Overy, pp.178–9; and also *Dr* Horst Boog's *Luftwaffen-führung* and his article in the Royal Air Force Historical Society Proceedings (to which my attention was drawn through the kindness of Mr Ken Munson).
10. Ishoven, pp.332–3.
11. Morrow, pp.126–8.
12. Ishoven, pp.365–6, 369; Völker, *Luftwaffe*, p.81.
13. Ishoven, pp.391–3.
14. Green, *Warplanes*, p.198.
15. For the fuel situation see Homze, pp.146–8.
16. For production and financial problems in 1936–37 see Homze, pp.144–6, 149, 151–7, 160–2.
17. Hermann, pp.44–5.
18. For the Ju 88 programme see Homze, pp.163–5, 194; and Irving, *Milch*, pp.65–7, 75–6.
19. Deichmann, p.88.
20. PRO Air 2/2826.
21. Goulding and Moyes, pp.50–9.
22. For the heavy bomber see Irving, *Milch*, pp.54, 65; Green, *Warplanes*, pp 336–48, 359–61; Homze, pp.166–8, 231; Völker, *Luftwaffe*, pp.32–4; and Griehl and Dressel.
23. Steinhilper, pp.108–9.
24. For examples and *Luftwaffe* data see Ries, *Luftwaffen Story*, pp.210–29.
25. Haubner, pp.63–87, 106.
26. Tuider, p.4.
27. Haubner, *Teil 1*, pp.132–3, and *Teil 2*, p.86.
28. See the report of 29 March by the British Air Attaché to Berlin in PRO FO 371 21710 C 2354, and also a report from the British Consul in Munich on a debriefing conference by the Commander of *I/Flak Regiment 5*, *Oberstleutnant* Karl Veith, in PRO FO 371 21710 C 3961.
29. This information has kindly been provided by *Dr* Erich Gabriel of the *Heeresgeschichtliches Museum*.
30. *SHAA* File 2B97; Taylor, *Munich*, p.463.
31. Quoted in Taylor, *op. cit.*, p.684.

32. Jones, pp.129–30, quoting Air Ministry Planning Conference, PRO Air 2/2731 and PRO Air 2/2961.
33. For the British situation see PRO Air 9/90: Group Captain Slessor's collection of special papers in connection with the German emergency (hereafter Slessor Report).
34. PRO CAB 24/279, CP 218.
35. See *Notice sur l'Armée de l'Air Allemande, Janvier 1938*, *SHAA* File 2B61.
36. Taylor, *Munich*, pp.756–65.
37. Christienne and Lissarague, p.300.
38. For Vuillemin's full report see Patric Façon's article.
39. Nemecek, p.147; Titz, p.4.
40. Slessor Report, appreciation of 29 August with map.
41. For details see Völker, *Luftwaffe*, p.170/n.333.
42. Green and Swanborough, 'Balkan Interlude'.
43. For the Czech aircraft industry see Homze, pp.232–3.
44. See Pritchard, pp.55–6; and Wood and Dempster, pp.17–21.

CHAPTER SEVEN

BLITZKRIEG/SITZKRIEG

September 1939 to May 1940

By the summer of 1939 Poland was the only reminder of the Versailles *Diktat*, but German plans were well advanced to remove this blemish on the Fatherland's prestige. Preparations for *Fall 'Weiss'* began on 3 April 1939 when Hitler issued a directive for operations against Poland, ostensibly as a contingency plan. Göring knew nothing of *'Weiss'* until he returned to Berlin from a holiday in Italy on 18 April and dined with Hitler. On learning the news he protested, and Hitler reportedly called him an 'old woman'.[1]

Göring's distress was a combination of an awareness of German economic weakness and personal setback, for he had used mutual hunting interests to maintain good relations with Poland's leaders. He may also have feared the Anglo-French reaction in the wake of the occupation of Prague and Memel, for the day the latter was occupied German Foreign Minister Joachim Ribbentrop had uttered his first threats to Poland. Ten days later Chamberlain responded with his public pledge that a German attack on Poland would lead to war with both Britain and France. In the following months, while Goebbels' propaganda machine fanned the flames, Göring used personal intermediaries to obtain British acquiescence for Hitler's plans and thus ensure another 'Flower War'.[2] It was Ribbentrop's triumph in accomplishing the Russo–German Pact, signed on 25 August, which appears to have changed Göring's mind, although he continued to send peace feelers out to Britain even after the Second World War began.

The *Wehrmacht* began outline planning for the invasion of Poland at Bad Salzbrunn in *'Generalstabsübung Schlesien'* early in May, although Jeschonnek gave Göring some form of briefing on 25 April. The main thrust was to be launched in the south from Silesia and Slovakia by *Generaloberst* Gerd von Rundstedt's *Heeresgruppe Süd* (8., 10. and 14.*Armee*), which was to destroy the bulk of the Polish Army. Its spearheads would be the mechanized forces of *XIV., XV.* and *XVI.Armee Korps (Motorisiert)*, which would strike north-eastwards along the boundary of the Polish Armies Lodz and Kraków towards Warsaw. *Generaloberst* Fedor von Bock's *Heeresgruppe Nord* would have a subsidiary role, with 4.*Armee* cutting the Polish Corridor to Danzig from Pomerania while 3.*Armee* advanced on Warsaw from East Prussia. *Luftwaffe* operations were moulded around these plans, with Kesselring's *Luftflotte 1* supporting Bock while Löhr's *Luftflotte 4* supported Rundstedt. As with Czechoslovakia a year earlier, virtually all the *Luftwaffe*'s strike force was concentrated in the east.

Intelligence activity against the Polish Air Force (*Polskie Lotnictwo Wojskowe*, or *PLW*) intensified from mid-April using agents and 74 radio interception stations. From July the *Fernaufklärungsgruppen* flew up to 100 miles (160km) inside Polish territory to produce a detailed picture of the communications and supply systems, the industrial base and, later, the dispositions of the Polish Army. It was estimated that Poland had 800 combat aircraft (it was actually 699), including 130 bombers (156), distributed at a dozen air bases many within *Stuka* range. However, according to Speidel, the *Luftwaffe* lost track of most Polish squadrons when Poland mobilized on 31 August.[3]

The original *Luftwaffe* plan envisaged a massive blow to annihilate the *PLW* on the ground and another (*Unternehmen 'Wasserkante'*) on Warsaw, both to destroy the Poles' industrial base and to erode their nation's political will. *ObdL* believed that these operations would win air superiority and allow them to concentrate exclusively upon supporting the Army. The generals did not agree, and their complaints to Hitler about the absence of air support in the crucial first hours of the campaign led him to demand major planning revisions by the *Luftwaffe* in August in the Army's favour. This meant reducing the weight of the blow aimed at enemy air bases and postponing *'Wasserkante'* to the latter part of A-Day, followed by precision night bombing using the *X-Gerät* of *Ln Abt 100*.

The support of *10.Armee* was already the prime task of Richthofen's *Fliegerführer zbV* within Loerzer's *Fliegerdivision 2* (*Luftflotte 4*), but with *Kampfstaffeln* diverted to targets which would ease the Army's burden Loerzer had only enough bombers to attack three bases. Many of Kesselring's *Kampfstaffeln* were also given new orders, and his cheerful disposition must have been further strained when he had to divert about half of Grauert's *Fliegerdivision 1* to neutralize the minuscule Polish Navy and its base at Hela. As compensation *ObdL* arranged for him to receive *Oberst* Behrendt's *KG 27* from *Luftflotte 2* immediately after hostilities began. Contingency plans were also drawn up for *Fliegerdivision 7* to open the Polish Corridor from the east by taking key bridges at Tczew (Dirschau) and Grudziadz (Graudenz), although these were replaced in August by precision air attacks.

The summer saw a steady increase in *Luftwaffe* activity as supplies were moved eastwards and training was intensified. There was a substantial (nearly 30 per cent) increase in the fighter force, with 26 *Staffeln* (eleven day, twelve night, three heavy) activated between July and August, although the *Jagdstaffeln* tended to use their 'Doras' (Bf 109Ds) rather than their 'Emils' (Bf 109Es) for training.[4] Because production of the Bf 110 remained low, 'Doras' equipped seven of the ten *Zerstörergruppen*, and apart from *I* and *II/ZG 26* these were given temporary *Jagdgruppe* designations—*JGr 101* (*II/ZG 1*), *JGr 102* (*I/ZG 2*), *JGr 126* (*III/ZG 26*), *JGr 152* (*I/ZG 52*) *and JGr 176* (*II/ZG 76*). Preparations intensified during a hot and sunny August as 155 airfields and airstrips were laid out and stocked, while with mobilization a substantial part of Lufthansa's fleet of 147 aircraft was requisitioned.

Senior *Luftwaffe* officers met their Army colleagues to complete details of tactical and operational-level air support, one of the most important meetings being that involving Richthofen, the *10.Armee* Commander *General der Infanterie* Walther von Reichenau and the latter's Chief of Staff *Oberst* Friedrich Paulus. From the evidence of Richthofen's diary, Paulus appears to have made little personal impression, but the meeting established a professional relationship which lasted until the defeat at Stalingrad. The most serious setback occurred on 15 August when Richthofen's *I/ StG 76* conducted a demonstration at Neuhammer in dense cloud mixed with ground fog which caused all nine *Stuka*s of *2./StG 76* and two each from the other *Staffeln* to crash, with the loss of 26 dead.[5] The leader of *1./StG 76, Oberleutnant* Dieter Peltz, escaped to become Inspector of Bombers in 1942 and Commander of *Fliegerkorps IX (J)* before falling into Russian hands. He returned to industry after his repatriation.

The incident added to Richthofen's problems, for he was worried about the ability of the *Luftwaffe* supply organization to support him, and especially his forward units, during a fluid campaign. His superior in *Fliegerdivision 2*, Loerzer, did not seem to have a clue, so Richthofen raised the matter with Jeschonnek on 28 August without result. Also unresolved was the problem of who would authorize air support, the Army or the *Luftwaffe*, and on his own initiative Richthofen assigned a liaison officer from each *Stukageschwader* to Reichenau's spearhead, *XV.Korps*. On 1 July the four *Luftgauen* bordering Poland (*I, III, IV* and *VIII*) began to create *Luftgauenstaben zbV* to organize such support, *Luftgau zbV 16* under *General der Flieger* Pflugbeil (previously the Inspector of Bombers, Dive Bombers and Reconnaissance Aircraft) being attached to Loerzer.

On 28 August *ObdL* moved to their wartime headquarters at Wildpark-Werder, Potsdam, and the following day Hitler held a final conference at Obersalzburg. Afterwards senior *Luftwaffe* commanders went to their command posts while some of Wimmer's *Luftwaffekommando Ost Preussen's Nahaufklärungsstaffeln* flew into forward airstrips, although most *Gruppen* remained at their peacetime bases. While the main Army effort was in the south, the majority of the *Gruppen* were under Kesselring's *Luftflotte 1*, which had 1,105 aircraft, including 526 bombers. The strike force was assigned to Wimmer, Grauert's *Fliegerdivision 1* and Förster's *Lehr Division*, the last two having 77 Bf 110s which were making their combat début while Grauert had 70 per cent of Kesselring's 172 *Stuka*s. Most of the Bf 109s (222, or 82 per cent) were assigned to defensive duties with the *Luftgauen*, which also had 337 *Flak* batteries (including 44 searchlight). Löhr had 729 aircraft, including 303 bombers and 209 *Stukas/Schlachtflieger*. Apart from a home defence force of 99 fighters and 57 *Flak* batteries (six searchlight) with *Luftgau VIII* (Breslau), most of Löhr's forces were under Loerzer's command, including Richthofen (224 aircraft), whose enthusiasm for the arrangement was markedly muted. He did not like being shackled to the Army command post because there might be no sympathy for, or appreciation of, the *Luftwaffe*'s problems and capabilities.[6] The two Services also had their own organic air support, Bock having 94 reconnaissance aircraft and Rundstedt 168, while *Generalmajor* Coeler as *FdL Ost* had 56 floatplanes. (For the *Luftwaffe*'s order of battle on 1 September, see Appendix 9.)

Facing this force of 2,152 German and 30 Slovak aircraft, the *PLW* under Brigadier Jozef Zajac had a total first-line strength of 1,900, including 650 in training units and 713 in reserve or under repair;[7] in addition, the Naval Aviation Wing (*Dyon*) had 25 aircraft, including twelve combat floatplanes. From 1935 to 1939 Poland's total defence expenditure was equivalent to $750 million (compared with $24,000 million for Germany), the majority of the money going to the Army. The 1938–39 budget was for 797.5 million *zloty* (($150 million), of which 46.3 million *zloty* ($8,730,000), or 6 per cent, was allocated to the *PLW*.[8] The aircraft included the gull-wing PZL P.7 and P.11 fighters, the monoplane PZL P.23 *Karas* fixed-undercarriage reconnaissance-bomber and various corps machines. New aircraft were being developed, including the PZL P.50 *Jastrzab* (Hawk) and RWD-25 *Sokol* fighters, the P.48 *Lampart* (Leopard) two-seat fighter bomber, the LWS *Mewa* (Gull) light bomber and the PZL P.46 *Sum* (Sailfish) corps aircraft. But only the twin-engine PZL P.37 *Los* (Elk) medium bomber had been delivered, and although a mission dispatched to Britain and France purchased 160 MS 406s and fourteen Hurricane fighters together with 100 Battle light bombers, none had been delivered by September.

Polish views of air power were always conservative; indeed, the dictator Marshal Josef Pilsudski demanded that the *PLW* be a purely reconnaissance arm. The demand was ignored, although during Pilsudski's rule (1926–35) the *PLW* had no bomber arm. An Air Staff was created only in 1937, following the adoption in October 1936 of an ambitious plan calling for 78 squadrons (688 aircraft) by March 1942. The Staff worked in splendid isolation from the Aviation Commander, the *PLW*'s administrative head and wartime commander, and the appointment of Zajac to this position in March 1939 was last nail in the Service's coffin. A former infantry officer and previously Inspector of Anti-Aircraft Defences, Zajac was a reactionary who promptly cancelled production of the *Los*.

His conservatism meant that Polish air power was fragmented in accordance with traditional beliefs, and three to five squadrons of fighter, *Karas* and corps aircraft were distributed to each of the reinforced corps (designated 'armies')—Armies Modlin (East Prussia), Pomorze and Poznan (Pomerania) facing *Heeresgruppe Nord* as well as Armies Lodz, Kraków (Silesia) and Karpaty (Slovakia) opposite *Heeresgruppe Süd*. Zajac retained a third (234 aircraft) as the Tactical Air Force with a brigade each of fighters and bombers.[9] Sensibly, upon mobilization Zajac dispersed his squadrons to camouflaged airstrips to shield them from attack, but the absence of a robust signals network meant that he would soon lose touch with them.

On 22 August Hitler ordered '*Weiss*' to begin within six days, only to postpone the operation on 25 August when his prospective ally Mussolini suddenly developed cold feet. Shortly afterwards Hitler finally recognized that the British would not accept another Munich-style agreement as he had believed and that the prospect of a two-front war was very real. On 25 August Kesselring and Löhr received the code '*Ostmarkflug 26.August 04.30 Uhr*' to begin operations the next day, but during the evening this was hastily cancelled. Unfortunately the change of plan did not reach Reichenau, who continued his preparations until *OKW* relayed a message through Richthofen.[10] The *Staffeln* were hastily stood down, although there was some concern when a Slovak Army mutiny threatened Zipser Neudorf airfield (where *Stab* and *II/StG 77* were based) on 28 August and a paratroop battalion was sent to secure it. Meanwhile there was a last-minute flurry of diplomatic activity in which Göring played a significant role, but Hitler demanded Polish capitulation and this Warsaw would not accept. After pondering the situation for a few days Hitler decided to risk Anglo-French intervention and on 31 August ordered that '*Weiss*' commence the next day, the *SS* staging an incident which provided an excuse for war. At 1815 the eastern *Luftflotten* were informed that '*Ostmarkflug*' would begin the next day at 0445.

In fact, the first shots of the campaign had been fired the previous day when *Luftflotte 4*'s *Flakgruppe 18* warned off an RWD-14 *Czapla* (Heron) corps aircraft.[11]

However, even as the *Staffeln* were alerted, the airfields were covered by a clammy blanket of fog and mist. Only four *Kampfgruppen* took off on schedule, two of them (*I/KG 1* and *III/KG 3* in *Luftflotte 1*) exploiting the facilities of a peacetime base. The fog cleared first from *Luftflotte 4*'s airfields so that later in the morning *I* and *III/KG 4*, together with *KG 76*, were able to get airborne, although cloud forced the last to abort its mission. *Luftflotte 1*'s airfields were blanketed for most of the morning and by midday only five or six *Gruppen* had completed their missions. The first unit into the air was *Oberleutnant* Bruno Dilley's *Kette* of *3./StG 1*, which successfully attacked a blockhouse controlling demolitions at the Tczew railway bridge. Dilley's men flew in at tree-top height to drop their bombs accurately ten minutes before the official outbreak of the war, but it was all in vain because the Poles repaired the wires to the explosives, despite a further attack by *III/KG 3*, and two hours later the bridge was blown up.

Achieving air superiority was the very core of the *Luftwaffe*'s creed, and within the first week of operations it appeared to achieve this aim through relentless attacks upon *PLW* bases. These became more intense on 1 September as the fog cleared over Germany, *Luftflotte 1* alone attacking five bases in the morning and thirteen during the afternoon (some of them receiving their second visit of the day), involving 56 per cent of its 1,100 offensive sorties. *Luftflotte 4*, which flew a similar number of sorties, directed 50 per cent of them against the *PLW* with operations by *I/StG 76*, *I/KG 76* and *II/KG 77*, while *KG 4* dropped 200 tonnes of bombs upon Cracow's airfields in missions involving 150 sorties. It was while returning from a morning attack upon Cracow that *Major* Oscar Dinort's *I/StG 2* claimed the first *Luftwaffe* victory when *Oberleutnant* Frank Neubert shot down Captain Medwecki's P.11 of No 122 Squadron, only for Medwecki's wing man, Second Lieutenant Wladyslav Gnys, to shoot down one of Dinort's *Stukas*.

So seriously did Löhr regard the destruction of the *PLW* that on the first day he ordered Richthofen to use half his force against Cracow, leaving a *Jagdgruppe*, a *Schlachtgruppe* and a *Stukagruppe* to support Reichenau. That afternoon, shielded by *I/ZG 76*, there were co-ordinated strikes involving four *Kampfgruppen* (*I* and *III/KG 4* and *I* and *III/KG 77*), with *KG 77*'s bombers flying so low to find targets in the dust and smoke that they were damaged by splinters from their own bombs. The *Luftwaffe* was able to strike deep into Polish territory, thanks to support from the *Zerstörergruppen*, at a ratio of one Bf 110 *Rotte* for every *Kampfstaffel*, flying above and behind the bombers. But there were barely 100 Bf 110s available, and spares may have been in short supply, for *Major* Walter Grabmann, *Kommandeur* of *I(Z)/LG 1* and former *Kommandeur* of *J/88*, noted in his after-action report that many *Staffeln* had only seven operational aircraft.[12] Nevertheless, Grabmann's *Gruppe* claimed 30 victories, four by *Leutnant* Werner Methfessel (four more in later operations), to become the top-scoring ace of the campaign, while *I/ZG 76* was officially credited with nineteen, with score-books being opened by three future twin-engine *Experten*, Wolfgang Falcke, Helmuth Lent and Gordon Gollob. The success was shared by the nominal *Zerstörergruppen* with their Bf 109Ds, and *Hauptmann* Hannes Gentzen's *JGr 102* (*I/ZG 2*) claimed 78 victories including 50 on the ground.

Although Göring quickly postponed 'Wasserkante' until the second day, Warsaw's airfields were struck twice on A-Day, by II(K)/LG 1 in the morning and by KG 27 in the afternoon, the latter flying 450 miles (725km) from the west before landing at Pomeranian bases. Both missions were attacked by Colonel Stefan Pawlikowski's Pursuit Brigade, but Grabmann's I(Z)/LG 1 intervened and bomber losses were light while the Pursuit Brigade suffered severely. By the end of the day it had lost ten fighters with another 24 unserviceable—nearly 62 per cent of its strength. The first day's operations cost the Luftwaffe 25 aircraft,[13] and their apparent success led ObdL increasingly to concentrate Luftwaffe resources away from counter-air to battlefield support operations. In fact, the counter-air campaign was a wasted effort, and while 'several dozen derelict or obsolete aircraft'[14] were destroyed, including 28 unserviceable trainers at Cracow-Rakowice,[15] the bombs left the PLW's first line unscratched. Post-war analysis indicates that the greatest Luftwaffe success against the PLW came towards the end of the campaign when seventeen Karas were destroyed on Hutniki airfield on 14–15 September.

One reason for the Luftwaffe's delusion was the absence of Polish bombers, which were being held back pending the discovery of good targets. Otherwise the PLW reacted vigorously but without co-ordination, and many of the 221 German aircraft claimed by the Poles were awarded to airmen. On the second day of the campaign Göring still hoped to carry out 'Wasserkante' and there were plans to attack Poznan (Posen), but the need to support the Army and to hold back part of the Kampfgruppen until AufKlGr ObdL discovered the enemy bombers' bases led to another postpone-ment. During 2 September only 364 bomber sorties out of 1,247 were directed against the PLW (an average of 28 per cent of sorties), although 36 per cent of Luftflotte 4's sorties were counter-air missions. During the day KG 4 conducted a sustained attack involving thirteen Staffel missions against three Deblin airfields and dropped 180 tonnes of bombs, then I/ZG 76, led by future night fighter ace Leutnant Helmut Lent, claimed eleven aircraft on the ground, but these were all from training units.

The day also saw attacks upon Poznan's and Warsaw's airfields, but the Luftwaffe's post-campaign analysis of these and other attacks concluded that they did not deny the PLW the use of its air bases because many of the runways were still intact.[16] The campaign's second day also saw the PLW sword unsheathed when 24 Karas attacked Reichenau's armoured spearhead. Flak displayed the efficiency it was to retain until the end of the Second World War by shooting down four bombers (a fifth fell to 'friendly' fire), while two badly damaged Karas were wrecked on landing. There was another German success when nine of Army Lodz's fighters were claimed by I/ZG 76, which in turn lost three aircraft during the day. Army Lodz was the prime barrier to the German advance on Warsaw and the destruction of its shield was completed on 4 September when nine Bf 109Ds from Gentzen's JGr 102 bounced the survivors and destroyed five. The transfer of the remainder to the Pursuit Brigade gave the Luftwaffe a free hand to support Reichenau. The Luftwaffe's losses on 2 September were relatively light, Luftflotte 4 reporting a total of six.

With the PLW clearly incapable of affecting the air situation, ObdL demanded that the maximum effort be committed to supporting the Army on 3 September but hedged its bets by ordering attacks upon the Polish aircraft industry. Counter-air operations were to be conducted only when reconnaissance discovered occupied enemy airfields, but the two *Luftflotten* were ordered to conserve the *Zerstörergruppen*. During the afternoon there were renewed attacks on Warsaw air bases and also on the PZL factory, leading to another major air battle again won by Grabmann's *Zerstörer*, which shot down three of 40 Polish fighters for the loss of one of their own.

ObdL's caution proved well founded, for the same day the *PLW*'s Bomber Brigade (Colonel Wladyslaw Heller) struck in force against Reichenau's armoured spearhead. Heller's pilots flew 40 low-level sorties into a maelstrom of ground fire which shot down five bombers and consigned another seven to scrap. However, the *Luftwaffe* had earlier reduced the odds of a repeat attack by wrecking the *Los* fitting-out depot at Malaszewicze. Despite the setback, Heller's bombers returned on 4 September to make their supreme effort against the same target, but there was no change in their fortunes. The Poles flew 43 sorties, almost all with *Los* medium bombers, and lost nine aircraft (two to *JGr 102*), and they lost another two during the afternoon when their base at Kuciny was bombed.

The *PLW* threat had now been neutralized, although Polish fighters remained active and took a steady toll of corps aircraft. The *Jagdgruppen* conducted occasional sweeps and the *Kampfgruppen* made sporadic attacks upon Polish airfields (the last being on 18 September by *Luftflotte 4*), but by 4 September the *Luftwaffe* could virtually ignore the *PLW* and concentrate upon supporting the Army, both *Jagd-* and *Zerstörerstaffeln* being increasingly committed to ground attack operations. The *Luftwaffe* was able to transfer *Flak* batteries to the front to provide the troops with direct fire support, and by 7 September *Luftflotte 4* alone had fifteen, many stripped from the Breslau area defences of *Luftgau VIII*. Other batteries moved westwards, where the growing British air threat to the coast led to *Luftflotte 1* transferring many units to Felmy's *Luftflotte 2*.

Yet the victory over the Polish Air Force owed less to the direct actions of the *Luftwaffe* and more to the combination of attacks upon the enemy rear and the speed of the German advance. The bombing of Polish communications and their supply system forced the *PLW* to concede the skies by 6 September as their supplies ran out, while the speed of the German advance destroyed the early warning system and forced the Pursuit Brigade to spend most of its time hopping from airfield to airfield, abandoning aircraft at every one. By 6 September it was down to sixteen serviceable aircraft and within four days it had absorbed most of the 24 surviving Army fighters (which had lost 41 aircraft) before moving to Lublin. There it was joined by Heller's bombers, which had flown 159 sorties (40 reconnaissance) and lost 38 aircraft, a loss rate of nearly 24 per cent. In the same period the armies had lost half their *Karas* (34 aircraft) and 60 per cent (37) of the corps aircraft, many of the latter to Polish ground fire.[17] The *PLW* continued to spit defiance with decreasing effect. The Bomber Brigade flew 147 sorties from 7 to 16 September and dropped 75 tonnes of bombs, but it lost 49 bombers, while the Pursuit Brigade was down to 54 fighters by 14 September.

While the battle for air superiority raged, the *Luftwaffe* increasingly committed resources in support of the land campaign, although *Luftflotte 4* flew only seventeen such bomber missions and *Luftflotte 1* eight on the first day of the campaign. Some of the first attacks were by the 'Ein-Zwei-Drei' (Hs 123) of *Major* Werner Spielvogel's

II(Schl)/LG 2, which provided continuous close air support for the armoured *XVI.Korps*, but the course of the campaign was shaped by the operational-level attacks. The bombing of the rail network, crossroads and troop concentrations played havoc with Polish mobilization, while attacks upon targets in towns and cities also disrupted command and control by wrecking the antiquated Polish signal network. Because Polish divisions had only nineteen radios (compared with 133 in a German division) they relied extensively upon the civil telephone network, but with switchboards wrecked and lines cut Polish communications became 'marginal at best'.[18]

Much of *Luftflotte 4*'s attention was focused upon the left of Army Lodz opposite Reichenau's armoured spearhead, where the first day saw Dinort's *Stukagruppe*, flying their third mission of the day, decimate Cavalry Brigade Wolynska, which was further pounded by *I/StG 77* and by *I/KG 77*. In an after-action report, *Hauptmann* Paul-Werner Hozzel (acting *Kommandeur* of *I/StG 1*) noted that 500kg bombs proved especially effective against even the strongest bridges and that one near miss wrecked an armoured train, one of four (out of ten) lost to air attack.[19]

Kesselring's airmen also had to support the Navy, for whom the reinforced *Geschwader Kessler (KG 1, II* and *III/StG 2, IV (St)/LG 1* and *4. (St)/TrGr 186)*, with 203 aircraft, flew five missions. They destroyed the Naval Aviation *Dyon*, the torpedo boat *Mazur* and the tender *Nurek* as well as damaging the destroyer *Wicher* and the minelayer *Gryf*, which were finished off by *Hauptmann* Blattner's *4. (St)/TrGr 186* and *3./KüFlGr 706* two days later. Tactical air support was also provided to *Grenzschutz Abschnittkommando 1* under *General der Flieger* Leonhard Kaupisch, the former *Luftkreiskommando II* commander who had been recalled to the Army's colours. Despite Kessler's support, the soldiers made slow progress and the Poles in the Hel Peninsula did not surrender until 2 October.

Army support assumed greater importance during the second day of the campaign, although the majority of 'missions' in both were at tactical rather than operational level. Of 575 flown by *Luftflotte 1* on 2 September there were 325 tactical ones (56 per cent) and 45 at operational level, while 43 per cent (295) of *Luftflotte 4*'s 674 'missions' were tactical and 125 were operational-level as Loerzer's bombers disrupted the rail network supporting Army Lodz and Army Kraków while preventing the transfer of reserves from Warsaw. Several *Kampfgeschwader*, including *KG 4*, had rapid reaction forces of three to six aircraft ready to respond promptly to radio reports by reconnaissance aircraft of potential targets. Attacks upon the railways between 2 and 5 September inflicted heavy losses on the 13th Polish Division en route to Army Prusy from Army Pomorze reserve and upon Army Lodz's Cavalry Brigade Kresowa as it detrained.[20]

With the *PLW* effectively eliminated by 3 September and bombers able to operate in *Staffel*-size or smaller formations, *ObdL* demanded the maximum effort to support the Army and set an example by using its own reserve, *KG 27*, to interdict rail traffic east of Warsaw. During the day Kesselring's *Luftflotte 1* focused upon the Polish Corridor, strengthening the support of *4.Armee* with two *Stukagruppen* released by Kessler, and helped to drive Army Pomorze southwards. Meanwhile Wimmer in East Prussia concentrated fifteen *Staffeln* of *LG 1, KG 2* and *I/StG 1* around Mlawa, where *3.Armee*'s advance was delayed by Army Modlin's fortifications. *Luftwaffe* attacks proved extremely accurate, bombs released from 500m (1,600ft) striking within five metres of bunkers. While the bunkers usually remained intact, the concussion demoralized the defenders (as noted also at Sedan in May 1940) and helped rout the Polish forces, leaving German armour barely 35 miles (60km) north of Warsaw. The harassment of the retreating Poles was continued by the use of reconnaissance aircraft to locate their columns and then radio the locations to the *Kampfgruppen*, which then made low-level bombing runs.[21] Inevitably civilians were killed, for a well-founded fear of German reprisals led thousands to flee the invader, the refugees and troops becoming mixed on the roads and suffering terribly.

On the fourth day of the campaign the operational-level effort concentrated upon rail interdiction, with *Luftflotte 1* dispatching seventeen 'missions' (involving 39 *Staffeln*), a third of them in support of *4.Armee*, compared with nine tactical ones (twenty *Staffeln*). Löhr used *KG 4, 55, 76* and *77* against rail targets to cause congestion at railheads, while attacks upon road traffic were made using a medium-altitude approach—2,000m (6,500ft) and above—followed by 10°–30° dive attacks. So effective were attacks at operational level that from 5 September the shortage of targets led to a change of emphasis to tactical operations. During the day the *Luftwaffe* began to conserve its resources, each *Luftflotte* holding back a third of its aircraft daily, with the result that *Fliegerdivision 2* bomber sorties dropped from 590 on 4 September to 414 on 6 September.[22]

The most dramatic display of tactical air power was by Richthofen's *Stuka*s and *Schlachtflieger*, supporting Reichenau's armoured spearhead (*XV.* and *XVI.Korps*) as it drove rapidly north-eastwards towards Warsaw, splitting Army Lodz and Army Kraków. On 3 September Richthofen's airmen destroyed the 7th Polish Division which anchored Army Kraków's right, but the fluid battle created problems and Richthofen commented on that day that no one in the Army command posts ever knew where the forward line was;[23] five days later, just as *1.Panzer Division* was about to seize a bridgehead over the Vistula, his *Stuka*s destroyed the key bridge.

The failure to establish a single command and communications network for Army support was the cause of the problem, but at its heart was an unresolved inter-service dispute. The *Koluft*—Army staff officers who advised *Heeresgruppen* and *Armee Oberkommandos* and acted as their liaison officer with *Luftflotte* and *Fliegerdivision* headquarters—were supposed to co-ordinate air–ground operations using data from the military staff and their own air reconnaissance units, the *Nahaufklärungsstaffeln* and *Fernaufklärungsstaffel (Heer)*.[24] As they had little knowledge of air operations and no control over the strike forces, they were dependent upon visits from the *Luftwaffe*'s *Flivo* to organize bomber and fighter support. The frequent *Koluft* demands for bombing missions at suicidally low altitudes (sometimes 20–50m, or 65–165ft) angered *Luftwaffe* leaders, and Richthofen severely restricted such missions from 1 September after suffering what he claimed to be 20 per cent of his fighter and *Zerstörer* losses.[25] On 6 September *4./KG 26* lost three aircraft on a low-level mission—to fragments from their own bombs according to the *Kommodore*, *Generalmajor* Hans Siburg, although the *Geschwader* War Diary blamed fighters.[26]

The cumbersome system might have worked more smoothly had there been a common radio frequency between the two Services. Not only was this lacking, but the *Luftwaffe*'s communications system itself was brought to the verge of collapse. Each *Fliegerdivision* had a telephone company, five liaison platoons and two radio aircraft, reinforced by telephone/teleprinter units, and the system was supposed to expand at the rate of 7km (4.3 miles) a day. However, the *panzer* divisions were advancing 40km (25 miles) a day. Land-line communications failed completely, while radio messages were taking three hours to get through.[27]

In a welter of mutual recriminations, the *Luftwaffe* leaders sought to salvage something and many followed the example of Richthofen, who took to flying reconnaissance sorties in a Fieseler Fi 156 from A-Day and personally co-ordinating operations at Army command posts, although his reports were sometimes disbelieved. These missions were a waste of a senior officer's time and exposed him needlessly to danger; indeed, while following his superior's example on 9 September, Richthofen's *Schlachtgruppe* Commander, Spielvogel, was shot down and killed.

The creaking supply system was a further headache for the *Luftwaffe* commanders, especially Richthofen, whose close support and fighter forces moved ever further from the depots. A single transport *Gruppe (I/KGzbV2)* was allocated to support both Reichenau and Richthofen, but the former often had priority; for example, as early as 3 September Löhr allocated *4./KGzbV2* to fly 30 tonnes of fuel to *1. Panzer Division* and 74 tonnes two days later. By 8 September Richthofen was expressing his concern about supplies to Löhr,[28] and within three days fuel shortages reduced the *Stuka*s to one sortie a day, with the helpless Pflugbeil blaming the bureaucrats in the rear.[29] Milch roved the battlefield helping where he could, but there was little he could do and Richthofen's situation did not ease until the arrival of substantial transport reinforcements from 13 September. Within two days 21 *Staffeln (II* and *IV/KGzbV 1* and *I, III* and *IV/KGzbV2*, plus *Transportstaffel 172)* were supporting *Luftflotte 4* and they were all subordinated to *Geschwader zbV Ahlefeld.* The situation would have been worse had the *Luftwaffe*'s twin-engine units moved forward, but bases in the *Reich* proved to be adequate during the campaign.

Despite all these problems, Reichenau's armour drove remorselessly onwards, supported by Richthofen's aircraft. On 4 September it destroyed the right of Army Kraków, drove deep into the rear of Army Lodz (which quickly disintegrated) and pushed aside Army Prusy to advance on Warsaw, the southern suburbs of which it reached barely eight days after the invasion. Three Polish armies (Prusy, Lodz, Pomorze) were isolated west of Warsaw, which had been abandoned by the Polish High Command 24 hours earlier. The Polish leaders decided to fight behind the Vistula and moved 125 miles (200km) eastwards to Brest-Litovsk (Brzesc-nad-Bugiem)—where the poor communications system further fragmented the Polish defence.

The German success accelerated a major reorganization of *Luftwaffe* forces. Richthofen had been given independence from Loerzer in the afternoon of 4 September after complaining to Löhr that Göring's friend was often out of touch; indeed, Loerzer admitted after the war that he had no contact with Reichenau's headquarters. But on 8 September he had his revenge by palming off on Richthofen as reinforcements *KG 77* and its headstrong *Kommodore, Oberst* Seywald, whom Richthofen icily described on 9 September as 'apparently a comic man'.[30] In the north Kesselring's *Luftflotte 1* was gradually concentrated in East Prussia to support *3.Armee*'s drive down the Vistula to isolate Warsaw from the east. However, the limited number of *Luftwaffe* bases in East Prussia were swamped, while to facilitate command the forces were divided on 6 September between Wimmer's *Luftwaffen-kommando Ostpreussen* as *Gruppe West* and Förster's *Lehr Division* as *Gruppe Ost*, Kesselring moving his headquarters to Königsberg on or about 10 September. Wimmer had two *Gruppen* each of *KG 3* and *27* together with *I/StG 1*, while Förster had two *Gruppen* each of *LG 1* and *KG 2* together with *III/KG 27* and *I(Z)/LG 1*. Grauert's *Fliegerdivision 1*, which had supported *General der Artillerie* Günther Kluge's *4.Armee*'s pursuit of Army Poznan for two days, was assigned to *Luftflotte 4* on 6 September and began transferring southwards.

Its arrival was fortuitous, for by 8 September, as the Germans began to assault Warsaw, a crisis developed in their rear. The city itself attracted much *Luftwaffe* attention on 8 September, with *Luftflotte 1* beginning daily attacks on bridges over the Vistula in the city centre. In the southern suburbs the *Luftwaffe* sought to help Reichenau's advance in the face of fierce resistance, flying 140 *Stuka* sorties (*StG 77* and *III/StG 51*) on the first day alone. Meanwhile *KG 77* struck railways east of Warsaw and during the night of 10/11 September *LnAbt 100* conducted the last of four precision night missions using the *X-Gerät* navigation system. The first attack had taken place on 3/4 September and this last one destroyed a munitions dump.

West of Reichenau the three Polish armies sought to break out of encirclement and *General der Infanterie* Johannes Blaskowitz's *8.Armee* was too overstretched to hold them. On 8 September reconnaissance aircraft spotted Army Prusy assembling in the forests around Ilza, south of Radom, which attracted a massive response from both Richthofen and Loerzer: three *Kampfgruppen* (*I* and *II/KG 55* and *I/KG 77*) and four *Stukagruppen* (*I/StG 1, I/StG 2, I/StG 77* and *III/StG 51*) as well as *II(Schl)/LG 2* were used and by 10 September they had helped to annihilate the Poles. To disrupt any relief operation from Lublin, Göring used three *Kampfgruppen* to attack communications around the city on 9 September. During the battle *I/Flak Regiment 22* (*Major* Weisser) distinguished itself in ground support operations with *2.Leicht Division*, breaking up Polish ground attacks, although Weisser was killed in the fighting. Grauert's *Fliegerdivision 1* was also committed to the Ilza encirclement using four *Kampfgruppen* (*Stab* and *II/KG 26, I/KG 53, I/KG 1* and *I/KG 2*) and *III/StG 2* together with *I/ZG 76* (borrowed from Loerzer), and despite Löhr's objections these attacks were made with incendiaries rather than high explosive. Grauert found himself involved in a more serious crisis on 9 September when Army Poznan struck southwards across the River Bzura and into the northern flank of Blaskowitz's *8.Armee*. The Germans were over-extended and over-confident and by the next day they were retreating, leaving 1,500 prisoners in Polish hands. So severely stretched was the German Army that *III/FJR 1, I/Flak Regiment 23* and two *FBK*s as well as groundcrews found themselves containing the desperate Poles. By skilful manoeuvring Rundstedt held the Poles by 16 September and during the next four days reduced the pocket where 100,000 men (a quarter of the Polish Army) were resisting, thereby breaking the back of Polish opposition.

On his own initiative Grauert helped Blaskowitz's hard-pressed troops, and from 11 September *ObdL Direktive Nr 12* ordered him to concentrate on this task while Richthofen supported Reichenau. The clear, sunny weather offered the *Luftwaffe* a perfect opportunity to demonstrate the awesome effects of air power, and it first isolated the battlefield by cutting the bridges across the Bzura. This slowed the Polish advance, and as troops reached the river bank they created huge traffic jams which attracted *Luftwaffe* bombers. The Poles were deluged with 50kg (110lb) bombs, and the attacks became more intense as the defenders' 40mm AA guns ran out of ammunition. In the later part of the battle Richthofen joined in, and the *Staffeln* were allowed precisely ten minutes over the battlefield, with orders to expend all their ammunition. The Poles sought refuge in the forest but Richthofen's men smoked them out by dropping 100kg incendiary bombs.[31] On 12 September air–land co-operation was disrupted when *Fliegerdivision 1* lost contact with their *Flivo* at Blaskowitz's headquarters and a replacement was seriously injured when his Storch crashed. Only through the intervention of Korten were communications with *8.Armee* restored. Immediately after the break-out had been contained thick cloud covered the battlefield, to be followed by heavy rain, but there was little respite for the Poles and Richthofen's men alone flew 750 sorties on 17 September. As the size of the success became apparent Göring told his airmen, 'If you perform in the west as you did against Poland, then the British, too, will run.'

Yet Germany's leaders were shocked by the brief Polish success and demanded retaliation. *ObdL* considered launching '*Wasserkante*' on 10 September, justifying an attack on Warsaw as retaliation for unspecified crimes against German soldiers—presumably defeat by the Polish *Untermensch*. The following day Hitler demanded that Warsaw be bombed on 12 September, the assignment being passed to Richthofen and Grauert. Richthofen dashed to meet Hitler and protested that he was not ready for such an attack, but all he received was a couple of days' grace.

Early on 13 September Jeschonnek telephoned Löhr with orders for an incendiary attack upon northern Warsaw, the Ghetto (due north of the Danzig Railway Station) possibly being included in the target folder.[32] That afternoon two *Staffeln* of *Fliegerdivision 1*'s *KG 4* dropped a 50:50 load of high explosive and incendiaries, and the *Kommodore*, *Oberst* Martin Fiebig, reported that the 7,000 incendiaries had helped set ablaze the Ghetto. With his forces stretched to the limit, Richthofen was able to put only 183 sorties (involving six or eight *Gruppen*) over the target towards evening.[33] The operation was a shambles: none of the units attacked on schedule, there were several near misses and smoke prevented any assessment of the damage.

Richthofen was furious and when he met Göring at Radom he demanded a single air commander if the operation were renewed. The meeting was also an opportunity to rid himself of Seywald, whom Richthofen accused of poor leadership and failing to attend meetings. Loerzer was also present and added that he had been glad to get rid of *KG 77* because he could not cope with Seywald either! According to Boog, Richthofen had salt rubbed into his wound when Seywald and his *Kommandeure* on their own initiative ignored the selected targets in favour of purely military ones. Seywald was relieved at 2200 hours by Richthofen's choice, *Oberst* Wolf von Stutterheim, *Kommandeur* of *III/KG 77*.[34]

The fighting in central Poland overshadowed operations in the south, where Loerzer, reinforced by *KG 1*, *KG 26* and *KG 55* (bringing his force to eleven *Kampfgruppen*) supported *14.Armee* as it swept through the south of the country to isolate the remnants of the Polish Government from their escape route to Romania. On 14 September *ObdL Direktive Nr 15* resumed rail interdiction missions, after a week's lapse, on lines leading into Romania and eastern Poland. Attacks on the lines into Romania were conducted by individual aircraft or *Rotte* of the Slovakia-based *KG 76*, which flew low-level sorties against choke points such as cuttings and bridges. By 20 September the defenders in southern Poland had been destroyed, the nation's death agonies having increased on 17 September with the Russian invasion in accordance with a secret protocol in the Russo–German Non-Aggression Pact. The *Luftflotten* were notified by teletype at midnight on 16/17 September and immediately operations in central and southern Poland were restricted. With a Parthian shot on 17 September, *Luftflotte 1* ended its attacks upon enemy troop columns to avoid bombing friendly troops. The dictators divided Poland along the Rivers Bug and Narev on 22 September, but the Russians' task had been eased by the *Wehrmacht*'s victories. The surviving *PLW* aircraft were ordered to neutral Romania, Hungary and Latvia, while pockets of Polish troops fought on until early October and those who avoided enemy prison camps became guerrillas.

Long before that the *Luftwaffe* had begun to withdraw its squadrons, moves anticipated by *OKW Direktive Nr 3* (9 September) and *Nr 4* (25 September). The move westwards began on 11 September with *ObdL Direktive Nr 12*, which transferred *Stab LG 1* (*II* and *III*) and *Stab KG 26* (*II*) to *Luftflotte 2* and *I/KG 53* to *Luftflotte 3*; two days later Grauert was informed by Korten that *Fliegerdivision 1* would follow, although the transfer began only on 20 September. The new assignments were announced on 16 September and soon the *Staffeln* were flooding westwards so that by 24 September only the *Lehr Division*, fifteen *Gruppen* and a few reconnaissance *Staffeln* were left in the east, the majority with *Luftflotte 4*.

By then Polish resistance was confined to the Hel Peninsula, Warsaw and Modlin. Warsaw itself was a fortress garrisoned by some 150,000 men and therefore no longer an 'open city', so under the international rules of warfare it could be subjected to wholesale bombardment. There clearly remained a desire on the part of the Germans to avenge the Polish slights of the previous two decades. On 22 September Richthofen signalled *ObdL*, 'Urgently request exploitation of last opportunity for large-scale experiment as devastation and terror raid . . .' He added that, if he was assigned the task, 'every effort will be made to completely eradicate Warsaw.' Although this request was rejected, the bombing nine days earlier had clearly been a terror attack, while on 14 September *ObdL Direktive Nr 15* contemplated a follow-up operation. On 18 September leaflet drops began to demand Warsaw's surrender, although bad weather restricted air operations over the city. But on 20 September Richthofen, on his own initiative, flew 620 sorties to sever the last links between Warsaw and Modlin, while the *Nahaufklärungsstaffeln* located targets and ranged the 1,000 guns assembled to support the assault.

Luftwaffe Direktive Nr 18 (21 September) gave Richthofen the responsibility for directing air operations, and the following day *Direktive Nr 20* assigned targets in northern Warsaw to *Luftflotte 1* and those in the west and south to *Luftflotte 4*, which was reinforced by three *Stukagruppen* from Kesselring. Eastern Warsaw was untouched because it was scheduled to be handed to the Russians. Richthofen's preparations were hampered by *ObdL* restrictions: the He 111 *Kampfgeschwader* he requested was replaced by Ju 52 transports of *IV/KG zbV 1* to conserve the *Kampfgruppen* for the western campaign, and for this reason the bombers were to carry only 50kg (110lb) high explosive and incendiary bombs; *ObdL* even withdrew permission granted on 16 September to equip *Stukas* with 250kg (550lb) and 500kg (1,100lb) bombs. The air and artillery bombardment began on 23 September but low cloud restricted air operations to harassing attacks and the destruction of city utilities such as gas, water and electricity. The full horror came when the weather cleared two days later on what the Poles came to call Black Monday. From 0800 onwards a force of some 400 bombers, including five *Stukagruppen* and a *Kampfstaffel*, flew 1,150 sorties, sweating airmen literally shovelling out incendiary bombs from the open doors of Ju 52s—a method described by Speidel as 'worse than primitive'.[35] By dusk 560 tonnes of high explosive bombs and 72 tonnes of incendiaries had been dropped, many of the latter being snatched by a strong north-easterly wind away from their targets and some hitting German positions, causing casualties. The bombardment killed 40,000 Poles and, together with the 13 September raid, totally destroyed 10 per cent of Warsaw's buildings and heavily damaged another 40 per cent. The *Luftwaffe* lost three aircraft, two of them Ju 52s.

The attack's inaccuracy and the smoke which rose to 10,000ft (3,000m) obscured targets for the gunners and infuriated Blaskowitz, whose troops were preparing to assault the city. He demanded an end to the air attacks, but when the dispute went to Hitler the latter told Richthofen to carry on. The following day Richthofen informed Göring that he would provide support only for those missions which Blaskowitz had specifically requested. In fact, dense cloud and smoke over Warsaw limited air operations, and Richthofen flew 450 sorties against Modlin. Without air support, Blaskowitz's men made slow progress and in places lost ground, leading the *Luftwaffe* to prepare another massive air attack for the next day. However, during the evening the Polish commander, General Juliusz Rómmel, exploited his success to seek the best terms he could and surrendered during the afternoon of 27 September. The same day the Modlin garrison also surrendered after Richthofen's airmen had made a sustained attack involving 550 sorties against the city, dropping 318 tonnes of bombs in two days.

Within a few days Richthofen and his men were westward bound, he and his staff flying to Berlin on 30 September while the battered *Gruppen* returned to their peacetime bases to receive replacements. *Luftwaffe* casualties had been relatively light: only 285 aircraft had been lost and another 279 had suffered damage in excess of 10 per cent, while personnel casualties were 759 (including 317 dead and missing), of which 539 were aircrew (413 dead and missing). This was a small price to pay when the German Army had lost 16,000 dead and 32,000 wounded and the Poles had lost 66,000 dead, 133,000 wounded and 587,000 taken prisoner by the Germans and 200,000 by the Russians. The *PLW* had lost between 327 and 335 aircraft,[36] two-thirds to *Luftwaffe* action, which indirectly accounted for many of the 140 aircraft which crashed on landing. About a third of the Polish losses were due to *Flak*, while 'friendly' fire accounted for 33. Typical of *Luftwaffe* activity was that of *Stab* and *II/KG 26*, which flew 207 sorties while the attached *I/KG 53* flew 223: the former unit lost four bombers with twelve damaged while the latter had six aircraft damaged.[37] *KG 26* was especially unlucky, losing two aircraft with seven damaged in a bomb accident at Gabert airfield on 4 September. This was one of a series of such mishaps during September, possibly caused by faulty fuzes. Eleven men were killed at Neuhausen, the *KG 51* base in the west, on 27 September, while during the Poland

Campaign *II/KG 3*, *III/KG 4* and *I/KG 152* each lost a bomber destroyed in mid-air between 3 and 6 September.

Despite the successes in Poland, it was clear that the *Luftwaffe* had important lessons to learn. The ground organization needed overhauling if it were to support an advance in the west, and steps were taken to improve its mobility and flexibility. To improve air-to-ground co-operation several *Kommodoren*, including Kessler of *KG 1* and Fink of *KG 2*, recommended providing each *Koluft* with a standard *Luftwaffe FuG III* morse radio or the replacement of the *Koluft* with the *Flivo*. However, the problem remained unresolved and air-to-ground co-operation still depended upon the initiative of the local *Luftwaffe* commanders, although Richthofen created mobile signal teams which could move with the motorized corps.[38] While *Flak* had proved extremely successful, the campaign had shown the need for units accompanying the ground forces to be under mobile regional commands, and by the end of the month the western *Luftflotten* had each created a *Flakkorps*.

The success of Richthofen's command earned a just reward on 3 October when it became *Fliegerkorps VIII*, but the *Lehr Division* had clearly outlived its usefulness and the following day it was broken up, Förster becoming *General zbV* attached to Milch's office. The success of *Aufklärungsgruppe ObdL* was also rewarded with a new *4.* raised, while *2.* and *3.* were the former *2. (F)/121* from *Fliegerdivision 1* and *8. (F)/LG 2*, which was already attached to the *Gruppe*. The other *LG 2* unit previously attached to *AufKlGr ObdL*, *10. (F)* under *Major* Wolff, appears to have become *1. (F)/124*.

In contrast to the savage struggle in the east, the Western Front seemed almost a haven of peace. True, the French made a token military thrust towards the *West Wall* from 7 September, but while it may have stretched the nerves of *Generaloberst* Wilhelm *Ritter* von Leeb's *Heeresgruppe C*, it was not a serious threat and it ended on 12 September at the cost of some 3,500 casualties. The land war declined into skirmishes and exchanges of shells, but its curiously unreal nature led to the descriptions 'Phoney War' or *'Sitzkrieg'*. Neither expression could be applied to the war at sea or that in the air, where some 460 *Luftwaffe* aircrew from first-line units were killed or went missing on operations between 3 September 1939 and 9 May 1940, of whom 180 were lost in the Scandinavian campaign, *Unternehmen 'Weserübung'*. In addition, another 547 men in front-line units were killed and 202 were injured in accidents.[39]

To universal relief, the air war failed to begin with the dreaded holocaust of aerial bombardment, partly because neither side wished to earn international condemnation and, more practically, because neither was in a position to launch such attacks. The option remained, however, and the RAF's AASF planned to attack the Ruhr if Germany invaded the Low Countries, while for his part Hitler told a secret meeting of Nazi leaders on 21 October 1939 that he would attack Great Britain 'with all available means'. However, he was also worried that if bombs fell upon Berlin they might spark an uprising among the poor, who were still Socialists and Communists at heart.[40]

The rump of the *Luftwaffe* in the west was immediately placed on the defensive by both Hitler's *OKW Direktive Nr 1* and *Luftwaffe Direktive Nr 1*, of 31 August. Felmy's *Luftflotte 2* (557 aircraft) and Sperrle's *Luftflotte 3* (579) were to shield the Ruhr and *Heeresgruppe C* respectively, and while they were to observe the neutrality of the Low

Countries and Switzerland, large enemy air formations might be engaged over neutral territory.[41] Felmy was also to disrupt British imports and armaments production, but Hitler reserved for himself the decision to strike London. Attacks upon the Royal Navy and upon troop transports were originally permitted, but following the Allies' declaration of war *OKW Direktive Nr 2* and *Luftwaffe Direktive Nr 2* took a more cautious line. Hitler's directive made no mention of strategic bomber raids, while attacks upon naval forces were permitted only if the enemy bombed Germany (a restriction lifted on 6 September, when *FdL West* was allowed free reign in the German Bight). By contrast, Göring's directive permitted un-restricted attacks upon warships anywhere as well as upon troop transports at sea.[42] Restrictions were also placed upon operations over France in Hitler's second directive, which noted that 'The guiding principle must be not to provoke the initiation of aerial warfare on the part of Germany.'

The *Luftwaffe* recognized that it was leading the race for aerial superiority, and Schmid's assessment of 2 May 1939 that the British and French air forces were still 'much out of date' remained accurate. He correctly estimated RAF Bomber Com-mand's strength to be 500 modern aircraft and that the French bomber force was obsolete—indeed, of the latter's 360 aircraft only eight (five LeO 451 B4s and three Potez 633 B2s) were modern, forcing Vuillemin on 25 August to restrict them to night operations except in an emergency.

Like *Luftflotte 2*, RAF Bomber Command was also restricted to naval targets, and the North Sea became the bombers' arena until the beginning of '*Weserübung*'. The RAF struck the German coast several times in September and December to attack naval bases or to seek naval targets of opportunity in armed reconnaissance. It was fighters rather than *Flak* which thwarted them, and RAF formations lost between 19 and 58 per cent of their strength, the first falling to *Feldwebel* Alfred Held and Hans Troitsch of *II/JG 77*, who dispatched Wellingtons of No 9 Squadron on 4 Septem-ber.[43] Nevertheless, these attacks led to the strengthening of Germany's air defences during the autumn. *Luftflotte 2* received *Flak* units from less exposed areas, *Luftgau XVII* (Vienna), for example, providing it with four gun and one searchlight batteries on 4 September. Fighter forces in the Heligoland Bight were reorganized on 12 November when *Major* Carl Schumacher, *Kommandeur* of *II/JG 77*, was promoted to *Oberstleutnant* and given the new *Stab JG 1* with *II(J)/TrGr 186*, *JGr 101* and *10. (N)/JG 26*. By now the Navy had a coastal defence early warning system with eight *Freya* radars. Schumacher recognized the value of these for air defence purposes and had *JG 1* plugged into the system.

Thanks to this initiative, heavy casualties were inflicted upon two Wellington formations on 14 and 18 December. The second action, dubbed the Battle of Heligoland Bight, was a special morale-booster as it came a day after the *Panzerschiff* ('pocket battleship') *Admiral Graf Spee* was scuttled in the River Plate. Out of 24 bombers (German radar reported 52), fourteen were shot down, many by Bf 110s whose 20mm cannon permitted attacks beyond the defenders' fire and tore apart aircraft lacking armour or even self-sealing tanks.[44] Yet the British proved as reluctant to accept the collapse of their theories about daylight operations as the *Luftwaffe* was to recognize that fighters rather than AA guns were the new basis of strategic air defence. Nevertheless, the value of fighter commands for defensive and offensive purposes was understood, and on 1 January *ObdL/RdL Ac 16/13* created three *Jagdfliegerführer* (*Jafü*). Schumacher became *Jafü 1* (also *Jafü Deutsche Bucht*), *Generalmajor* Hans von Dörning was *Jafü 2* (with his headquarters at Dortmund) and *Generalmajor* Hans Klein was *Jafü 3* (at Wiesbaden), but these were essentially co-ordinating staffs on the lines of First World War *Jagdgruppen*, and tactical command remained with the *Kommodoren* and *Kommandeure*.

The threat to the German Navy declined, although RAF Bomber Command continued to make anti-shipping sweeps across the North Sea, flying nearly 500 sorties between January and April 1940 for the loss of six aircraft. As preparations for *'Weserübung'* neared completion, the German Navy feared that the RAF might disrupt them and on 31 March requested reinforcement for *Jafü 1*. However, the opening of the Scandinavian campaign ended the threat to the German coast, apart from occasional RAF Coastal Command anti-shipping missions, which were easily contained.

Defending German territory at night proved more difficult, but fortunately both sides' bombers dropped only propaganda leaflets. RAF Bomber Command opened the leaflet campaign on the first night of the war with missions usually flown from British bases, although the bombers sometimes staged through French airfields. The French themselves joined the campaign from 10 October, and by 10 May 1940 the Allies had flown some 500 leaflet sorties, of which 432 had been undertaken by the RAF (94 by the AASF from 18 March). Losses were extremely light, the British losing 4.16 per cent of their bombers (18), with *Flak* claiming the first of five victims during the night of 8/9 September and later the sole French aircraft to be lost.[45] Seven *Nachtjagdstaffeln* had been formed in February 1939, only to be absorbed by the day fighter arm during the summer, and three replacements had to be raised during the autumn as *10. (N)* of *JG 2* and *26* and *LG 2*. They were used for point defence operations during twilight and for this reason had Bf 109Ds, whose pilots removed the canopies partly for a better view and partly to prevent themselves being dazzled by searchlight beams. On 18 February 1940 all *Nachtjagdstaffeln* came under the administrative command of *IV(N)/JG 2* at Jever, but this brought no operational benefits and the first victory did not occur until 20/21 April, when *Oberfeldwebel* Willi Schmale shot down a Battle of No 218 Squadron (a second followed against minelayers four nights later).[46] However, the path to a true night-fighting capability was to be demonstrated by *I/ZG 1* in Denmark a few days later (as described in Chapter 8).

The *Luftwaffe* was slower to undertake its own leaflet sorties, which began on the night of 5/6 November with five by *KG 51* and *53*.[47] All were conducted over France, and *Luftwaffe* situation reports to 9 March suggest that they were infrequent and carried out only by small numbers of aircraft, rarely more than three bombers and sometimes only one. By 9 March no more than 100 sorties appear to have been flown, the majority by *Oberst* Stahl's *KG 53* (although the *Geschwader* history suggests that no more than six aircraft at one time were dispatched[48]), while *III/KG 51* flew only eighteen leaflet missions (32 sorties) to 19 June 1940.[49] Casualties were light but missions during the night of 16/17 November encountered high winds which blew a dozen bombers eastward, two crashing in Austria and northern Italy. There appears to have been an increase in sorties in the run-up to *'Gelb'*, and during the night of 20

April *KG 1* and *KG 2* apparently joined in. As with the Allies, losses were light—two aircraft in April, to mechanical failure rather than to the ineffective French night defences.

Luftflotte 2 does not appear to have flown leaflet raids, possibly because it was committed to the duel with British maritime power across the North Sea (see Chapter 8). Moreover, most of Felmy's time (and Sperrle's) was spent preparing to support the German Army's advance westwards, *Fall 'Gelb'*. As early as 9 September *OKW Direktive Nr 3* warned of plans to move the *Luftwaffe* westwards from Poland, although Hitler retained a veto on flights beyond Germany's western frontier. Operations west of that line were still subject to the earlier guidelines, and it was not until *OKW Direktive Nr 4* of 25 September that even reconnaissance flights were officially authorized.

Reconnaissance was now essential, because on 27 September Hitler told his military leaders that if the Allies continued the war he intended attacking before their armies were prepared and would push through the Low Countries (including the Netherlands) to the Channel.[50] The generals were appalled, for they had no confidence in victory until shortcomings in training and equipment had been overcome. They advocated a defensive strategy and pointed out that fog would hamper air operations, an argument which Göring also used in early October. However, the generals were willing to allow the Navy, Germany's weakest service, to carry the burden of offensive operations, with some *Luftwaffe* support.

Hitler was not to be dissuaded and, although he made a peace offer on 6 October, three days later he issued *OKW Direktive Nr 6*, which confirmed his intention of striking westward as soon as possible in order to win territory in the Low Countries, thus providing the *Luftwaffe* with the means to attack the British Isles more effectively. Nine days later, on 18 October, *OKW Direktive Nr 7* addressed the need for active reconnaissance and permitted the *Luftwaffe* to escort such flights across the western frontier if necessary. While attacks upon industrial targets in the Low Countries were forbidden because of the danger to the civilian population, the *Luftwaffe* was permitted to attack British naval forces at their bases.

Meanwhile the victorious troops began arriving back from Poland during early October. Bock's *Heeresgruppe Nord* became *Heeresgruppe B* on 10 October opposite the Netherlands and Belgium, with *4.Armee* facing central Belgium and *6.Armee* (Reichenau's former *10.Armee*) facing northern Belgium and the southern Netherlands. Leeb's *Heeresgruppe C* concentrated opposite the French border. On 24 October Rundstedt's *Heeresgruppe Süd* was redesignated *Heeresgruppe A* and inserted between Bock and Leeb facing Luxembourg, with *12.Armee* in the north and *16.Armee* in the south. The *Luftwaffe* also regrouped, Felmy and Sperrle being notified on 16 September which units they were to receive from the East. *Luftflotte 1* was to give Flemy *Fliegerdivision 1*, *KG 27* and *I(J)/LG 2*, while *Luftflotte 4* would provide *KG 1* and *KG 4*. Sperrle was to receive from *Luftflotte 4* the *Fliegerdivision 2* headquarters and Richthofen's *Fliegerführer zbV*, together with *KG 76*, *III/StG 51*, *III/StG 2*, *I(Z)/LG 1*, *I/JG 77* and *I/ZG 2*.[51] The *Fliegerdivisionen* were redesignated *Fliegerkorps* at the beginning of October, and on 5 October Richthofen's anonymous command was upgraded to *Fliegerkorps VIII*; three days later he and Sperrle discussed future operations. On 18 October his headquarters moved to Koblenz, by which time he had 46 *Staffeln* (27 *Stuka*) under *JG 53, LG 2, KG 77, StG 2* and *StG 77*.

Richthofen did not remain long under Sperrle's command. On 19 October the first draft of the *'Gelb'* operational plan was published and envisaged an advance through Belgium to the Channel coast with the main blow launched by Bock. His armour, under Reichenau, was to drive through the southern Netherlands into northern Belgium, shielded from an Allied counter-offensive from the south by Rundstedt with a motorized corps. As Sperrle would be supporting Rundstedt and Felmy would work with Bock, Richthofen moved to *Luftflotte 2* on 20 October in exchange for Grauert's *Fliegerkorps I* and within a week he had been reinforced by *JG 77* and another seven *Staffeln, II/StG 2* replacing *III/StG 51*, which was transferred to *Fliegerkorps I*.[52]

No date for *'Gelb'* was announced, but after presenting the *Ritterkreuz* to fourteen officers, including Jeschonnek, on 27 October Hitler announced that A-Day would be Sunday 12 November. No one was happy with the *'Gelb'* draft, which was amended on 29 October so that the only part of Dutch territory to be violated would be that around the vital Maastricht bridges. The next day Jeschonnek carried Göring's protest about the exclusion of the Netherlands to Hitler and warned that the British might take Dutch airfields. Hitler was unmoved, so Jeschonnek worked on *OKW* and insisted on 11 November that the Netherlands be occupied in order to protect the Ruhr from air attack, an argument *OKW* reluctantly accepted on 14 November. So anxious was Jeschonnek to occupy the Netherlands that he proposed to Jodl as late as 6 February 1940 guaranteeing Belgian neutrality but occupying only the Netherlands, Norway and Denmark. As a result of *ObdL* pressure Hitler issued *OKW Direktive Nr 8* on 20 November, which added the occupation of the Netherlands to Germany's war aims, although it was only at the end of January 1940 that the task was assigned to *General* Georg von Küchler's *18.Armee*.

The sharpest critic was Rundstedt's Chief of Staff, *Generalleutnant* von Manstein, who sought a more decisive conclusion by moving the armour from the right to the left flank. This was too strong for *OKH*'s stomach, but Hitler had proposed a month earlier using Rundstedt's *12.Armee* with some armour to cross the Meuse at Sedan and advance to Laon, to occupy France north of the Aisne. This was accepted by *OKH* on 3 November, but six days later Hitler allocated *12.Armee* a full mechanized corps, *General der Panzertruppe* Heinz Guderian's *XIX.Armee Korps (Motorisiert)*. These changes were incorporated in *OKW Direktive Nr 8* of 20 November which, stated that if Rundstedt made faster progress than Bock then this success would be exploited. The Directive also extended restrictions upon industrial targets in the Low Countries to population centres.

From the *Luftwaffe*'s viewpoint, *OKW Direktive Nr 8* created an unresolved anomaly through the corresponding *ObdL Direktive Nr 5* of 7 December. Its Paragraph 11c noted that, in order to support *4.* and *12.Armee*, from A+1 or A+2 *Fliegerkorps VIII* would transfer *Stab StG 1* and *JG 77* to *Fliegerkorps I* as well as *I(St)/TrGr 186* to *Fliegerkorps II*.[53] Richthofen had already lost six *Stukastaffeln* to these *Fliegerkorps* and the orders meant that he might quickly lose control of up to another nine (42 per cent of his *Stuka* force)—yet there is reference to the order neither in the *Fliegerkorps VIII* War Diary nor in Richthofen's personal diary.[54] Events during the Western Campaign five months later would suggest that there was a serious lapse in staff co-ordination somewhere in the *Luftwaffe*, but the source can no longer be located.

In the *OKW* directive Hitler also revealed that he had given verbal instructions for airborne operations to capture bridges across the Albert Canal west of Maastricht, but *Fliegerdivision 7*, reinforced by *22.Infanterie Division (Luftlande)*, was to be used only when the bridges had been captured. In fact, during a two-hour briefing with Hitler and Keitel on 27 October, Student was told that glider-borne assault forces would land in the fortified line to take the canal bridges at Veldwezelt, Vroenhoven and Canne, which commanded the roads fanning westwards from neighbouring Maastricht. Another force would take Fort Eben Emael south of Canne, whose guns covered the bridges. Although it was agreed that air support would be provided by a task force, *Fliegerführer 220* (formerly Putzier's *Fliegerdivision 3*), the choice of landing zone for the main operation remained in dispute into the New Year, the Commander of *22.Division*, *Generalleutnant* Hans *Graf* von Sponeck, opposing landings at Namur-Dinant and later Ghent for fear that his lightly armed troops would be caught by superior forces. Meanwhile *Hauptmann* Walter Koch formed a battalion-size battle group based on *1./FJR 1* to storm the Albert Canal. However, when Richthofen met Reichenau on 9 January to complete arrangements, the *General* was unhelpful and refused even to provide artillery support.[55]

As the date for the offensive drew near, preparations for what was officially described as a defensive battle (*Abwehrschlacht*) intensified between 2 and 7 November. Richthofen had received *JG 27* (which was now reinforced by *JGr 126*) and *JG 52*, while for liaison with the Army he received ten Fi 156s and a Ju 52 with a small signal liaison command (*Luftnachrichtenverbindungskommando*) under *Major* Wurm together with some armoured cars for forward observers. One *Stukagruppe* was transferred to Sperrle on 14 November but Richthofen later received *I(St)/TrGr 186* and *Stab StG 1* in compensation. But when Seidemann visited the forward airfields on 8 November he found them 'barely usable'. Meanwhile, to enhance the psychological impact of its *Stukas*, *Fliegerkorps VIII* fitted at Udet's suggestion a siren which made a terrifying wailing noise as the aircraft dived. It was demonstrated to Richthofen at Dedelsdorf on 22 December and soon equipped all of the *Fliegerkorps'* *Stukas*.[56]

Felmy's other element, *Fliegerkorps IV*, was to support Sperrle's *Fliegerkorps V* (*KG 27, 51, 55* and *76*) in attacks on enemy airfields in Belgium and northern France before engaging the BEF. French air strength was estimated at 680 aircraft (including 130 single-engine fighters and 60 bombers), while British strength was estimated at 300 aircraft (54 fighters and 160 Battles). Sperrle's *Fliegerkorps II* (formed from *Fliegerdivisionen 2* and *6*) would attack communications along the Franco–Belgian border with *KG 2, 3* and *53* to delay the Allied advance into Belgium and for this task received *I/StG 76* from Richthofen on 14 November. Even this did not ease Sperrle's fears about the absence of close support forces both for Bock's left and for Rundstedt, especially for *12.Armee*. The *4.Armee* was assigned support from *Flakkorps I* and *Fliegerkorps I* (at one time under *ObdL* control), the latter reinforced by *III/StG 51*.[57]

There was considerable but inconclusive discussion about the timing and duration of the air offensive, especially whether or not it should precede the ground offensive, and on 16 October Jodl and Jeschonnek discussed the feasibility of the *Luftwaffe's* pinning down the enemy in northern France for several days before the ground offensive began. On 7 November *Luftwaffe* air transport units were warned to be on 24 hours' notice to move from midday the following day.[58] However, bad weather and problems with rail transport resulted in numerous postponements, giving time for extra *Luftwaffe* training, reconnaissance and fighter patrols.[59] Hitler hoped for clear, cold weather with frozen ground and good flying conditions, and on 27 December A-Day was scheduled between 9 and 14 January, but the next day he told Jodl that in the event of bad weather in mid-January he would postpone the offensive until the spring.

On 10 January the *Luftwaffe*'s meteorological service forecast a fortnight's clear winter weather from 15 January and A-Day was set for the 17th, following four or five days of *Luftwaffe* attacks to destroy enemy air power. The latter decision was made by Hitler on 9 January and the schedule was agreed with the *Luftwaffe* the following day. But on 11 January Jeschonnek stated that the *Luftwaffe* could not begin its attack until 14 January, presumably because of poor meteorological forecasts. Hitler began to reconsider the difficulties of synchronization between the air and the ground campaign and on 13 January he informed Jodl that he wished to avoid 'A Verdun in the air'. That evening he decided that the air and ground campaigns would occur simultaneously, but by then the decision was irrelevant.

Preparations for the offensive had been hindered by delays in the overstretched railway network, and these problems led *Major* Helmut Reinberger, *Fliegerführer 220*'s liaison officer with *Luftflotte 2*, to accept an offer from 'Old Eagle' *Major* Erich Hönmanns, *Kommandant* of Loddenheide airfield, to fly him to Cologne in a Bf 108 in violation of security regulations banning the transport by air of classified documents. They set off during the morning of 10 January but Hönmanns got lost and made a forced landing at Mechelen-sur-Meuse, north of Maastricht, where the two were quickly captured. Although Reinberger twice tried to burn the papers, the Belgians recovered sufficient to realize that they referred to an airborne landing at Ghent.[60]

The news broke on 11 January, the eve of Göring's birthday, and Hitler's apoplectic and justified anger terrified the *Luftwaffe* commander. Surprisingly, no action was taken against Putzier but Felmy and his Chief of Staff, Kammhuber, were made the scapegoats and dismissed on 12 January. Kammhuber became *Kommodore* of *KG 51* but Felmy, who had been on bad terms with Göring since his critical reports of 1938 and 1939, was cashiered. His sons, too, were tarred with the brush, and to rehabilitate himself Felmy joined the Nazi Party in the autumn of 1940. He was recalled to active duty in May 1941 and commanded *Luftwaffe* units in Iraq, then Army corps in Russia and the Balkans until the end of the war—a waste of one of the *Luftwaffe*'s most experienced and clear-thinking officers. Milch sought to replace Felmy but was blocked by Jeschonnek, who favoured his former patron Kesselring, as did Göring. Kesselring brought with him the *Luftflotte 1* Chief of Staff, Speidel, and assumed command of *Luftflotte 2* on 15 January, being replaced at *Luftflotte 1* by Stumpff and *Generalmajor* Heinz-Hellmuth von Wühlisch.

Prudence dictated the abandonment of the existing plans and at 1435 on 11 January Felmy informed his *Fliegerkorps*, 'All orders cancelled. New orders will follow. Previous orders invalid.' The following day fighter units were ordered to transfer to new airfields by 15 January (later amended to 16 January). Although Wenniger, now only Air Attaché in Brussels/The Hague, reported on 13 January that the documents had been burned, Göring remained very anxious. He attempted psychic means to determine whether or not the documents had been destroyed, then burned his hand making practical experiments. A lasting result of the 'Mechelen Incident' was the introduction of a 'need-to-know' rule which Hitler exploited to restrict discussion of his plans.

Hitler still wished to execute *'Gelb'* using the original plan, but the weather became capricious and after more postponements he decided on 16 January to redraft it. It was a decision partly forced upon him by the onset of Europe's worst winter for half a century, with blizzards covering the roads and railways in up to three feet or more of snow and ice and blocking rivers and canals. Airfields were blocked and *Fliegerkorps VIII's* War Diary noted that on 31 January, when Seidemann was slightly injured after his car skidded and crashed outside Münster, the snow made even transport by truck to the headquarters difficult.

The weather, and the Mechelen Incident, acted as a catalyst upon German decision-making. It was correctly anticipated that the compromised *'Gelb'* plan would encourage the Allies to advance into central Belgium to stop the Germans along the River Dyle or the River Escaut, and the Germans saw in this strategy an opportunity to trap the enemy's mobile forces. A demonstration of the Manstein concept attended by the principal officers of *Heeresgruppe A* and *Luftflotte 3* was held in Koblenz on 7 February but the revised *'Gelb'* which appeared on 18 February and was accepted six days later proved even more radical by reversing Bock's and Rundstedt's roles. Bock (28 divisions) would now draw the Allies into Belgium using some armour and pin them down on their chosen defensive line. Simultaneously Rundstedt (44 divisions) would use most the mechanized forces under *General* Ewald von Kleist's *Panzergruppe* headquarters to drive through the Ardennes. They would emerge around Sedan and isolate the enemy in Belgium by an advance towards the Channel, although this part of the plan was left open. The new draft did demand the completion of the reorganization of forces by 7 March.[61]

The *Luftwaffe*'s dispositions were a mirror image of the Army's for the revised *'Gelb'*, with the concentration on the right rather than the left flank. This was because the *Luftwaffe* was no longer expected to delay the Allied advance into Belgium but rather to disrupt the transfer of enemy forces from the Allied right on the Maginot Line to Belgium, where German aircraft would help to destroy them. The *Luftwaffe* had already been strengthening its right to occupy the Netherlands using *Fliegerkorps IV*, which was reinforced by Behrendt's *KG 27* from Sperrle's *Fliegerkorps V*. This *Fliegerkorps* was also to strike targets in Belgium, while *Fliegerkorps I* would now support Bock's advance into northern Belgium and for this reason was reinforced by Froehlich's *KG 76*.

The decision to mount simultaneous air and ground offensives, originally made by Hitler on 13 January, was confirmed after Göring informed Jodl on 10 February that the bombers could achieve surprise if they took off before dawn. But there remained ambiguity over *Fliegerkorps VIII's* role, which Richthofen believed was only to support Reichenau's *6.Armee* as the spearhead of the thrust into northern Belgium. There is evidence neither that *ObdL* informed him that he would be expected to switch his support to *Panzergruppe Kleist* within a few days of A-Day nor that *Fliegerkorps VIII* and *Luftflotte 3* staffs met between October 1939 and A-Day.

The airborne landing at Ghent had been compromised at Mechelen and new roles were sought for *Fliegerdivision 7, 22.Division* and Putzier's command (now renamed *Fliegerkorps zbV 2*). Scandinavia was considered, but it was recognized that the airborne forces offered an ideal means of helping *18.Armee* to take the Netherlands. Even during the New Year Hitler was considering alternatives to Ghent, including Sedan, Walcheren Island and Flushing, but on 10 January he accepted a new plan involving landings at The Hague and Rotterdam. These would decapitate the Dutch leadership and secure the advance across rivers and inundations which shielded the redoubt of Fortress Holland. Kesselring was briefed soon after assuming command of *Luftflotte 2* and obtained from Bock, with whom he had become friendly in Poland, a guarantee of rapid relief for the airborne troops.

Meanwhile, with the bombers largely rendered impotent by political restraint, the air war during the *'Sitzkrieg'* appeared to take a traditional stance, with the emphasis upon fighters and reconnaissance aircraft. However, the appearance of radar was to make the task of the reconnaissance aircraft more difficult, while easing that of the fighter. The success of naval surface-search radars in detecting targets at ranges of 34km (21 miles) during the coastal air battles encouraged the installation of the *Luftwaffe*'s first air-search *Freya* at Wangerooge in December. The following month two mobile *FMG 39 Freya* systems were allocated to *Luftflotte 2* to cover the approaches to the Ruhr, while a third was assigned to *Luftflotte 3* (in March 1940 *Luftflotte 2* received a third set). These radars, with a range of 150km (95 miles) against formations and 70km (45 miles) against individual aircraft, were used to complement the network of observers and sound-locators but had no direct link with the *Jagdstaffeln*. Nevertheless, through the initiative of unit signal officers, *Jagdgruppen* such as *I/JG 52* were exploiting the observer network and directing interceptions using radios; *JG 52* had its first success on 6 October 1939 with a LeO 451.[62]

By these means the *Luftwaffe* took a steady toll of enemy reconnaissance aircraft. The victims were mostly the underpowered and under-armed Bloch MB 131 RB4, the Potez 637 A3 and the Bristol Blenheim IV, which proved to be easy meat for *Luftgau XI*'s *II/JG 77*, *Fliegerdivision 5*'s *JGr 152* and *Luftgau XII*'s *Stab, I* and *II/JG 53*. The *Luftflotte 3* units especially enjoyed rich pickings over the *West Wall* and in the week beginning on Monday 25 September shot down eleven Allied reconnaissance aircraft (a twelfth force-landed), including a complete formation of five Battles by *2./JG 53* on the Saturday. This disaster ended unescorted daylight missions by the Battle, but even with a fighter escort several French aircraft were shot down. As a result, the French began to replace the MB 131 with the Bloch MB 174 A3, but the British had to soldier on with the Blenheim.

Luftwaffe reconnaissance was largely conducted by the Do 17P and the He 111H, and the first sortie was flown over Scapa Flow by *1./ObdL* on 5 September, followed, the next day, by a meteorological flight by *Wekusta 26* along the eastern coast of Britain. Although much of western Europe's weather is 'generated' from the Atlantic, it was only in June 1938 that *Oberleutnant* von Rotberg's *Wekusta 26* was raised. It was to conduct regular missions throughout the *'Sitzkrieg'*, and RAF Fighter Command quickly recognized the mission profiles of these *'Zenit'* flights. Reconnaissance over

France may have begun on 4 September with a *I/KG 53* mission,[63] but it was not until 13 September that Göring permitted Sperrle to conduct 'unobtrusive, long-range reconnaissance at extreme altitudes' without formal *OKW* authority. The first unofficial missions over France appear to have been carried out on 21 September to photograph air bases at Thionville, Metz-Frescarty and Rouvres (north-east of Verdun), and there was a similar mission the next day; both were probably undertaken by *1./ObdL*. It was not until *OKW Direktive Nr 4* of 25 September that reconnaissance flights beyond Germany's borders were formally authorized, and on that day three Do 17Ps of *Hauptmann* von Normann's *2. (F)/123* flew over air bases in north-eastern France.[64]

These operations were confined to the *Luftflotten Fernaufklärungsgruppen*, but poor weather restricted activity during October. Improved weather in November and the imminence of *'Gelb'* led to a surge in activity during November, especially after the 5th, when *OKW* gave permission for large-scale, operational-level reconnaissance over France and the Low Countries as well as the use of Do 17s in the *Aufklärungsgruppen (Heer)*.[65] Detailed figures for German reconnaissance activity no longer survive, but estimates can be made by cross-checking *Luftwaffe* daily reports with French and British records (see Table 12). The figures indicate that each side flew a very similar number of sorties during the *'Sitzkrieg'* and despite the superiority of their equipment the *Luftwaffe* lost only slightly fewer aircraft than the Allies—a reflection of the inadequate performance and defensive armament of the Do 17P.

In 1939 the *Luftwaffe* averaged five or six reconnaissance sorties a day over France, although the weather tended to create its own surges in activity, especially late in November when the effort reached double figures on a number of days. The first aircraft to be lost came down on 21 September when a fighter escort failed to save a Do 17P of *3. (F)/22* from French fighters over Zweibrücken, but it was not until November that casualties became serious, averaging almost one for every day a reconnaissance mission was flown. Eight aircraft were lost on 23 November alone. On that day *Unteroffizier* Arno Frankenberger of *4. (F)/122* had his revenge when his aircraft was severely damaged by Hurricanes of the RAF's No 1 Squadron. After the rest of the crew bailed out he managed to shoot down one of the fighters before taking to his parachute. That evening he was entertained by the rueful victors before going to a prisoner-of-war camp.[66]

Missions over France were conducted by the *Fernaufklärungsstaffeln* of the *Luftflotte*, of Sperrle's *Fliegerkorps* and of Rundstedt's armies, together with some *Stäbe* of *Kampfgeschwader*. They located Allied formations and airfields, the latter correlated with communications intelligence, although Bomber Command's use of forward French bases led the *Luftwaffe* to overestimate the RAF presence in France as being some 560 aircraft.[67] In September 1939 the *Armée de l'Air* and RAF nominally had 1,420 aircraft in northern Metropolitan France, including 160 Battle III bombers of the AASF; these were joined in late September by the RAF's Air Component, whose 200 aircraft were to support the BEF. With the division of the *Armée de l'Air* on 1 October between ZOAN (*Général* François d'Astier de la Vigerie) and ZOAE (first under *Général* Roger Pennès and then *Général* Marcel Têtu), the Air Component was attached to the former and the AASF to the latter. The *Aufklärungsgruppen* helped OKH to make an accurate estimate of Allied strength and to notice that most mobile formations were on their left, ready to advance into Belgium. *AufKlGr ObdL* (expanded to four *Staffeln* with a *Gruppe Stab* on 10 October) probably also conducted very high altitude sorties over the Low Countries, because by 3 November the *Luftwaffe* possessed detailed information about the distribution of Belgian AA batteries.[68]

The object of *Luftwaffe* reconnaissance was to produce a detailed geographic and military picture as the basis for staff planning, and the emphasis was upon tactical and operational-level sorties. For photographic reconnaissance each *Fliegerkorps* had its own *Stabsbildabteilung* (*Stabia*), while each *Aufklärungsstaffel* had some twenty processors and interpreters in a *Staffelbildgruppe* who would conduct rapid initial interpretation either from prints or negatives then telephone tactically valuable data to the appropriate headquarters within 90 minutes. After eight hours the film and a written report would be passed to the *Stabia* for second-stage interpretation before being dispatched for further examination and filing to the tri-Service *Stabsbildabteilung-Generalstab* (*Stabia-Gens*) at Zossen. Strategic reconnaissance was less important to the *Luftwaffe*, whose interpreters made little attempt to assess changes which were discovered (this was left to Intelligence)—a major handicap when assembling target folders on airfields or industrial sites.[69]

Unknown to the *Luftwaffe*, reconnaissance of the British Isles was made potentially more hazardous by an air defence system which closely integrated the radars and observer network with a command, control, communications and intelligence organization which could direct fighters to their targets by means of high frequency radio. A more enquiring *Luftwaffe* staff might have wondered why reconnaissance sorties over Britain, usually conducted by *AufKlGr ObdL* and *AufKlGr (F) 122*, supplemented by *Stab KG 26* and *KGr 100*, suffered loss rates as high as 9 per cent during the 'Sitzkrieg'; indeed, on 22 February 1940 *AufKlGr ObdL* lost half of the He 111Ps dispatched. Attention was focused upon maritime and air activity in eastern Britain but missions were flown as far west as Liverpool (for example, on 17 October). *Oberstleutnant* Joachim Stollbrock's *KGr 100* (created on 18 November by renaming *Ln Abt 100*) also tested the effectiveness of the *X-Gerät* system in the west, the first mission taking place on 20 December when *Oberleutnant* Hermann Schmidt (*Staffelkapitän* of *1./KGr 100*) flew at an altitude of 7,000m (23,000ft) over London. Other missions followed, but on 13 February Stollbrock was shot down and killed over the Thames Estuary following a text-book interception.[70]

From January 1940 the British system was extended into France where the *DEM* had proved a failure. A radar screen using mobile GM radars (the British equivalent of *Würzburg*), each with a range of 145km (90 miles), was established between Calais and Strasbourg under No 14 Group. By May 1940 six GMs covered a sector from the coast at Calais to Le Cateau, east of Cambrai, and they reported to a filter room at Allonville, but the system did not become operational until late April. Although designed both to support Allied forces and to extend Fighter Command's shield, it proved a disappointment because of poorly sited or assembled radars. On 10 May the commander of the British Air Forces in France, Air Vice-Marshal Arthur Barratt, complained to the Air Ministry that his radars had failed to detect a *Luftwaffe* aircraft which flew over at 20,000ft (6,100m) although Fighter Command headquarters at Stanmore, outside London, easily tracked it with their CH system Type 1 radars (which also monitored the whole of western Belgium). As a result of these complaints Professor Robert Watson-Watt, the brains behind the British radar programme, was dispatched to France. On 17 May he concluded, somewhat academically, that none of the radars was of substantial operational use.[71]

Nevertheless, the growing efficiency of the air defences forced both sides to adopt a high-altitude reconnaissance solution. At the beginning of the war the *Luftwaffe* had taken the first steps by introducing the Do 17S-0, which was supplemented by the Do 215B and later the Ju 88D (a photographic reconnaissance version of the Ju 88A), while the spring of 1940 saw prototype Ju 86Ps for very high altitude (11,000m, or 36,000ft) operations. But even the high performance of the Ju 88 did not ensure the aircraft's survival and two were shot down over France on 19 April and 9 May. From November 1939 the British used stripped-down Spitfires, which flew fifteen sorties in two months to photograph the whole of the *West Wall*. Another thirteen sorties were flown from 11 April to 10 May over targets hitherto inaccessible and alerted the Allies to preparations for operations in Scandinavia. However, the Spitfires were vulnerable too: one fell to *I/JG 20* on 23 March and another two were shot down by Bf 109s of *II/JG 51* and *I/JG 1* by the beginning of *'Gelb'*, although these victories owed as much to Fate as to the determination of the pilots.[72]

Severe weather in early 1940 reduced the *Luftwaffe's* daily reconnaissance to an average of three sorties in January and February, but as the weather improved this rose to seven by April. To reduce casualties, and to erode the enemy's fighter strength, direct and indirect fighter support was provided, with close escorts and sweeps being flown ahead of the mission, initially using Bf 109s but from late March Bf 110s also. March saw an intensification of *Luftwaffe* reconnaissance activities which lasted until 9 May, occasionally supplemented by flights by Ju 52 communications intelligence aircraft of the *Fliegerkompanie, Luftnachrichtenregiment ObdL*. One of these aircraft, a veteran of the Spanish Civil War, was shot down by French fighters on 7 April.[73] In late April these efforts briefly led to a scare when photographs of Sedan appeared to show new fortifications, but closer examination established that these works were still under construction and would not be completed for some time.[74] *Luftwaffe* reconnaissance aircraft had often strayed accidentally into Belgian air space—on 2 March, for example, a Do 17 of *1. (F)/22* shot down a formation of three Belgian Hurricanes attempting to force it to land—but in late April RAF Fighter Command noticed that the Germans were clearly flying reconnaissance missions over the country. It could hardly have reassured the Belgians—who, like the Dutch, had feared a German invasion since 7 November—when Berlin frostily rejected the neutral countries' attempts to broker peace talks.

With so much reconnaissance activity, both sides' fighters were extremely busy over the Western Front trying to 'blind' the enemy. *Luftflotte 3* had 312 single-seat fighters to meet the threat (including 33 Ar 68 night fighters) in September 1939, the brunt of the effort being borne by the 90 Bf 109Es of *Oberst* Hans Klein's *JG 53*, part of *Generalmajor Dr* Weissmann's *Luftgau XII*. The French had implemented a large-scale re-equipment programme during the spring of 1939 and also had 312 modern fighters, while nine *escadrilles de chasse* with obsolete fighters were gradually re-equipped during the winter of 1939–40. Supplementing these were four RAF squadrons with 64 Hurricane Is.

Klein's pilots had spectacular success against enemy reconnaissance aircraft, aided by the latter's obsolescence. Their average monthly loss rate was about 10 per cent, mostly to fighters, but against daylight missions the *Luftwaffe* was even more successful and No 70 Wing of the Air Component suffered a 19 per cent loss rate.[75] This forced the Allies to operate more at night: for example, of 35 reconnaissance sorties flown by *ZOAE* in April 1940, 27 were nocturnal (see Table 12).[76] The first fighter clash occurred on 8 September between a *Schwarm* of *I/JG 53*'s Bf 109Es and Hawks of *GC II/4*, following which the Condor Legion ace *Oberleutnant* Werner Mölders force-landed with a damaged engine. But that evening a patrolling *Rotte* from *II/JG 52* opened the score-book on the Western Front with a Mureaux 115 A2 near Karlsruhe. The fighters skirmished between the Maginot Line and the *West Wall*, and while the Hawks, MS 405s and Hurricanes slightly outclassed the Bf 109Ds, they were inferior to the Bf 109E.

TABLE 12: RECONNAISSANCE IN THE WEST, SEPTEMBER 1939–APRIL 1940

Month	Front	Sorties (losses)*					
		French		RAF		*Luftwaffe*	
Sept 1939	Western	148	(11)	110	(9)	30	(1)
	North Sea	–		33	(1)	10	(–)
Oct 1939	Western	44	(5)	16	(4)	40	(2)
	North Sea	–		14	(2)	15	(4)
Nov 1939	Western	58	(4)	37	(2)	120	(19)
	North Sea	–		11	(2)	40	(4)
Dec 1939	Western	22	(3)	25	(–)	75	(–)
	North Sea	–		10	(1)	30	(3)
Jan 1940	Western	36	(1)	10	(2)	75	(3)
	North Sea	–		12	(1)	20	(1)
Feb 1940	Western	11	(–)	5	(1)	45	(–)
	North Sea	–		21	(2)	10	(3)
Mar 1940	Western	76	(2)	5 **(1)		130	(4)
	North Sea	–		47	(3)	30	(4)
Apr 1940	Western	160	(3)	25 **(4)		180	(8)
	North Sea	–		20	(–)	45	(–)
Totals		**555**	**(29)**	**401**	**(35)**	**895**	**(56)**

Note: * Includes those by bombers. Some French sorties officially described as 'night reconnaissance' were actually leaflet dropping missions but have been included here. ** Estimated figures (as are those for all *Luftwaffe* sorties).
Sources: *Lagebericht*, in IWM Ger/Misc/MCR/19, Reel 1A; *GQG Aérienne Journaux des Eventements*, SHAA File 1D44/1; Fighter Command Operational Record Book, PRO Air 25/507; Bomber Command Operations Record, PRO Air 24/200; Advanced Air Striking Force Operations Record, PRO Air 24/21; Air Component Diary of Operations, September 1939 to January 1940, PRO Air 35/284; Brookes pp.56–68. British figures exclude Coastal Command photo-reconnaissance sorties.

In November the *Luftwaffe* threw down the gauntlet in the battle for French air space with sweeps by Bf 109s, which were joined on 21 November by Bf 110s from *V(Z)/LG 1* escorting reconnaissance aircraft as far west as Rheims, 100 miles (160km) behind the French lines. On the same day *Hauptmann* Dietrich *Graf* von Pfeil und Klein Ellguth, *Kommandeur* of *I/JG 52*, was severely burned in another battle and never flew again. He was killed by French guerrillas in 1944. However, the biggest setback for the *Jagdgruppen* occurred on 6 November when 27 Bf 109Ds of Gentzen's *JGr 102* lost four aircraft (including two flown by *Staffelkapitäne*) and had another four badly damaged when they met nine Hawks of *GC II/5*. Bad weather then restricted activity for the next three months, although in the first clash with RAF fighters on 22 December *III/JG 53* shot down two of No 73 Squadron's Hurricanes, one falling to Mölders (who was now a *Hauptmann*).

The introduction of constant-speed propellers for the MS 405 and Hurricane during the New Year improved their climb rate and acceleration. Armour protection was also installed, but the Bf 109Es retained a margin of superiority and from March onwards they and the Bf 110s ranged ever deeper behind French lines. From now until the beginning of *'Gelb'* the *Jagdstaffeln* shot down twice as many Allied fighters as they themselves lost, while the *Zerstörerstaffeln* shot down only two fighters (a third force-landed) but lost seven Bf 110s. By the middle of March skirmishes were frequently taking place up to 55 miles (90km) inside France, and the month ended in a major victory for *II/JG 53* when it bounced eleven MS 406s of *GC III/7*, shot down three and forced another three to crash land.

JG 53 was the most successful *Luftwaffe* unit, with 71 claims confirmed out of 160 made by the *Luftwaffe* for the *'Sitzkrieg'*, and Mölders was its leading ace. Other *Experten* who opened their score-books alongside him included Josef Wurmheller (102 victories), Günther *Freiherr* von Maltzahn (68), Gerhard Michalski (73) and Wolf-Dietrich Wilcke (162), while the most notable of the minor pilots was 'Old Eagle' *Hauptmann Dr* Erich Mix of *I/JG 53*, who added two Second World War victories to the three he won in the Great War (and later became *Kommodore* of *JG 1*). Many other *Experten* opened their scores over the Western Front, the leaders being Heinz Bär (221), Anton Hackl (192), Max Stitz (189), Johannes Steinhoff (178), Joachim Müncheberg (135), Erich Leie (118) and Helmut Wick (56). But success was not bought cheaply and the *Luftwaffe* lost 56 fighters here (see Table 13). The majority of the losses were caused by Allied fighters, but there was a steady trickle to accidents and at least two aircraft were lost when the pilots' oxygen supply failed.

While some French fighter squadrons were extremely active, Kesselring's Chief of Staff, Speidel, commented after the war on the lethargic response of the *groupes de chasse*, which he compared unfavourably with the RAF fighter squadrons.[77] French and British records provide a degree of confirmation, for although the RAF contributed about 20 per cent of the Allied fighter strength it flew an average of 23.5 per cent of the fighter sorties between November 1939 and February 1940 and about one-third of all operational sorties in March (the loss of RAF records makes an accurate comparison for April impossible).

Although French single-seat fighter strength rose from 300 to 500 aircraft during the *'Sitzkrieg'*, the average daily effort was only 50 sorties. In *ZOAE* the fighters' average daily sortie rate was 33 in September 1939 and only 22 in April 1940.

TABLE 13: FIGHTER LOSSES OVER THE WESTERN FRONT, 3 SEPTEMBER 1939–9 MAY 1940

	Luftwaffe EA	Luftwaffe Op	Armée de l'Air EA	Armée de l'Air Op	RAF EA	RAF Op
Sept 1939	13	1	12	1	–	–
Oct 1939	–	2	–	–	–	1
Nov 1939	9	3	7	–	–	3
Dec 1939	1	–	1	–	2	–
Jan 1940	2	3	–	–	–	–
Feb 1940	–	–	–	–	–	–
Mar 1940	1 (1)	3	7	–	5	2
Apr 1940	12 (5)	5	8	–	3	–
1–9 May 1940	1 (1)	–	–	–	1	–
Totals	**39 (7)**	**17**	**35**	**1**	**11**	**6**

Note: EA = enemy action; Op = operational loss. Figures in parentheses are Bf 110s.
Sources: Shores, *Eagles*; Martin.

Between 1 March and 10 May 1940 there were only sixteen days when the total effort of the *groupes de chasse* exceeded 100 sorties. The maximum effort was on 20 April with 236 sorties, when three French aircraft were lost for one German;[78] by comparison, RAF Fighter Command, with some 400 fighters, flew 2,079 sorties just between 3 and 13 March, a daily average of 189.[79]

The courage and determination of French fighter pilots during both the *'Sitzkrieg'* and the *'Blitzkrieg'* are without question and the shortage of pilots was one reason for the poor effort. On 10 May there were 579 fighters (419 serviceable) but only 428 pilots, many of whom would be reservists whose skills would be rusty.[80] Among NCO aircrew the shortfall in radio operators was 60 per cent and in air gunners 31 per cent, while the *Armée de l'Air*'s manpower shortages were aggravated when many of its reservists were drafted into second rate (*Série B*) infantry divisions. On 18 April the Commander of *ZOAE* warned Vuillemin that he could not conduct a sustained effort in a crisis, yet it was only on 26 September that the *Armée de l'Air* began to expand air training and this failed to bear fruit before France collapsed. The shortcomings of pre-war training needed to be overcome, but despite the wholesale introduction of new aircraft, serviceability rates were superior to those of the *Luftwaffe* (French fighter serviceability on 10 May was 72.4 per cent compared with 70.8 per cent for *Luftwaffe* forces in western Germany).

While French industry was overcoming its pre-war problems during the *'Sitzkrieg'* to produce an average of 241 aircraft a month (supplemented by an average 21 American aircraft), it frequently lacked key items such as radios, bomb sights and weapons. Production of these was slow and the French supply organization fitted them only after the aircraft had completed flight testing. Congestion at the factories forced the *Armée de l'Air* to accept machines that were not airworthy, with the result that by 10 May only seven of the 33 *groupes de bombardement* were operational and

TABLE 14: ALLIED FIGHTER SORTIES IN FRANCE, SEPTEMBER 1939–APRIL 1940

Month	Armée de l'Air	RAF	RAF % of total
Sept 1939	1,157	–	–
Oct 1939	813	60	6.8
Nov 1939	1,548	512	24.8
Dec 1939	402	199	33.1
Jan 1940	882	223	20.2
Feb 1940	136	25	15.5
Mar 1940	1,510	915	37.7
Apr 1940	1,907	256 (+)	11.8% (+)

Note: No 85 Squadron appears to have flown no operational sorties between 2 and 30 April while the monthly reports for April of Nos 87, 607 and 615 Squadrons have been lost.

Sources: *GQGAé* War Diary, *SHAA* File 1D44/1; *Resumé des Opérations, SHAA* file 1D44/2; Air Component Diary of Operations, September 1939 to February 1940, PRO Air 35/284; Squadron operations records for Nos 1, 73, 85, 87, 607 and 615 Squadrons in PRO Air 27 series.

twenty were still converting to modern aircraft.[81] By contrast British production benefited from capital investment programmes in 1937 and 1938: in April 1940 1,081 aircraft (546 combat) were produced and by June the figure had risen to 1,591 (960 combat), but the omens had proved so good that the RAF raised eighteen fighter squadrons. It was fighters which were a burr in inter-Allied relations, for the French were constantly demanding aircraft from the British—who were willing in principle to supply them provided this did not compromise strategic defence. The RAF was especially contemptuous of the French early warning system, which Dowding, in a letter of 6 October 1939 to Chief of the Air Staff Sir Cyril Newall, described as being of 'pathetic inefficiency'. Two days earlier, when Dowding was informed that he might have to transfer more fighters to France, he reminded Newall that Fighter Command was 35 per cent below establishment and had already transferred 15 per cent of its first-line strength to the Continent. In December two more Hurricane squadrons went to France and during the next three months Dowding's average first-line strength was 411 aircraft, of which 95 were Blenheims or Gladiators. Even with an improved flow of aircraft from the factories, it was clear that he could not afford to be profligate with his fighters.[82]

The same problem faced the *Luftwaffe*. The war had aggravated Germany's material shortages at a time when the Services' demands were increasing: there was, for example, a 600,000-tonne shortfall in steel production. Rationalization was the only means available to Udet for squeezing more aircraft from the factories, and on 12 September he virtually abandoned the manufacture of four-engine designs, although the need for such aircraft compelled him to reinstate production of the FW 200. By concentrating upon proven designs and introducing a ten-hour working day,

Udet was able to overcome the worst aspects of material shortages and actually to expand production. Yet the emphasis remained upon supplying and expanding front-line units while neglecting both the creation of an adequate reserve and adequate supplies of spares. Consequently, while there was a 20.75 per cent increase (822 aircraft) in first-line strength between September 1939 and March 1940, when it was 4,787 aircraft, during the same period serviceability dropped from 80 per cent to 75 (in December) and then to 70 (in March)—a figure maintained to the start of 'Gelb' (see Table 15).

Munitions also posed a problem, for their consumption during the Poland Campaign had been far greater than anticipated. At the outbreak of war the *Luftwaffe* had 69,015 tonnes of bombs, but a month's operations saw 21,324 tonnes (nearly a third of the total) expended and only 2,864 tonnes produced. With steel in short supply, a concrete-cased bomb was developed: the first 12,000 were delivered in February 1940 but they appear to have been kept in reserve. By the beginning of March 34,756 tonnes of bombs had been produced, enabling the *Luftwaffe* to create a reserve for forthcoming operations.[83]

During the 'Sitzkrieg' the *Luftwaffe* raised 30 *Kampfstaffeln* and sixteen *Jagdstaffeln*. The *Zerstörerstaffeln* of *JGr 101, 102, 126, 152* and *176* were finally able to exchange their Bf 109s for the Bf 110 to become *II/ZG 1, I/ZG 2, III/ZG 26, I/ZG 52* and *II/ZG 76* respectively, and in February a *Zerstörerstaffel* (Z./KG 30) was raised with the new Ju 88C-2, another sub-type converted from the Ju 88A-1. Re-equipment was not confined to the fighter units and the Ju 88 was issued to seven *Kampfgruppen* (I, II and III/KG 30, II(K)/LG 1, III/KG 4, I/KG 51 and II/KG 51) and a number of *Fern-*

TABLE 15: *LUFTWAFFE* AIR STRENGTH, SEPTEMBER 1939–MARCH 1940

Type	30 September 1939			30 December 1939			30 March 1940		
Bf 109	1,174	1,125	870	966	1,022	769	1,449	1,258	817
Bf 110	168	194	141	438	410	299	369	325	222
Bomber	1,248	1,270	1,060	1,493	1,402	1,022	1,786	1,690	1,124
Ju 87	381	347	230	417	418	331	420	411	341
Hs 123	39	37	37	39	41	25	39	42	27
Recon*	302	403	333	330	436	315	373	478	357
Corps	366	328	284	339	303	260	373	337	280
Naval	270	261	221	233	213	154	233	246	172
First-line total	3,948	3,965	3,176	4,255	4,245	3,175	5,024	4,787	3,340
Transport	255	227	186	210	228	189	463	466	404
Liaison	133	135	119	144	132	121	133	118	97
Grand total	4,336	4,327	3,481	4,609	4,605	3,485	5,620	5,371	3,841

Notes: Statistics are establishments/numbers with units/serviceable aircraft.
Luftwaffe reconnaissance *Staffel* strengths include *Wekusta* aircraft.
Source: IWM *Luftwaffe* Activity Book I, Reel 192, Frames 1110, 1111.

aufklärungsstaffeln. Plans existed to create fifteen *Kampfgeschwader Stäbe* and 46 *Gruppen* by July 1940, and by 10 May the bomber force had fourteen *Stäbe* and 46 *Gruppen*, while two *Küstfliegergruppen* had replaced their floatplanes with bombers. In addition March saw eight *Kampfgruppen zbV* raised, largely from *Blindflug-* and *C-Schulen*, although *KGrzbV 105* had requisitioned Lufthansa four-engine aircraft and *KGrzbV 108* had floatplanes.

There was little problem finding the men for these new units as the *Luftwaffe*'s strength by the spring of 1940 had increased from 2 million to 2.2 million but still included 85,000 officers, officials and engineers. The *'Sitzkrieg'* also provided an opportunity to replace 'Old Eagles' *Kommodoren* with younger men who perhaps had combat experience. This was especially so in the *Kampfgeschwader*, of which seven (*KG 1, 26, 51, 55, 76* and *77* and *LG 1*) received young blood in a process completed in March when *Generalmajor* Wilhelm Süssmann (*KG 55*) was replaced by *Oberst* Alois Stoeckel and Kammhuber relieved *Oberst Dr* Johann-Volkmar Fisser (*KG 51*). Because only *JG 26* and *JG 53* had *Kommodoren* when war broke out, it was possible to raise new *Geschwaderstäbe* with young officers. *Major* Hugo Witt, a scarred survivor of the *Hindenburg* disaster, relieved Great War air ace *Ritter* von Schleich of *JG 26* in December 1939, while *JG 53*'s Klein was promoted to *Jafü 3* in January 1940 and was ultimately replaced by *Oberst* Hans-Jürgen von Cramon-Taubadel. Many of the former *Kommodoren* were given administrative positions: Süssmann, for example, headed a new *Luftgaustab zbV* which was to move to Norway.

Expansion was also helped by the relatively light losses in Poland and conservation of the *Staffeln* for *'Gelb'*. The material and human losses are shown in Tables 16 and 17. A fact apparently ignored by the *Luftwaffe* was that nearly 72 per cent of the aircraft destroyed (672) were lost in accidents, a worrying indication of declining standards of training. The accidental loss rate of front-line units, aircraft destroyed or aircraft suffering more than 60 per cent damage between 1 September 1939 and

---

TABLE 16: *LUFTWAFFE* AIRCRAFT LOSSES, 1 SEPTEMBER 1939–30 APRIL 1940

Period		Missing over enemy lines	Lost over friendly territory		
			Destroyed	Damaged, repaired	Damaged, replaced
Sept–Dec 1939	EA	192	106	44	74
	Non-EA	–	326	162	212
Jan–Mar 1940	EA	30	5	8	14
	Non-EA	–	184	104	102
Apr 1940	EA	74	25	16	7
	Non-EA	–	162	79	66
Total		**296**	**638**	**413**	**475**

Note: EA = losses to enemy action; non-EA = losses not due to enemy action.
Source: IWM *Luftwaffe* Activity Book I, Reel 192, Frame 1068.

TABLE 17: *LUFTWAFFE* CASUALTIES, 1 SEPTEMBER 1939–9 MAY 1940

	Operational	Non-operational	Training
Aircraft destroyed or more than 60% damaged	514	379	567
Aircraft damaged (less than 60% damage)	189	384	501
Total	**703**	**763**	**1,068**
Dead	640	538	449
Missing/PoW	341	9	1
Injured	329	202	501
Total	**1,310**	**749**	**951**

Source: IWM Air Staff Post Hostilities Studies: Book 21, *Luftwaffe* Activity, Frames 1068, 1072.

9 May 1940 was 52 per cent of the total, although many aircrew appear to have survived. If non-operational front-line losses are added to those of the training units, then 72.5 per cent of the *Luftwaffe* aircraft destroyed or severely damaged, together with 56.5 per cent of the aircrew casualties, were due to accidents (see again Table 17). Yet there were positive aspects to these gloomy statistics. For example, although standards were declining, the training organization was able to provide a substantial increase in aircrew. Between 30 September 1939 and 30 March 1940 the number of first-line crews rose by 31 per cent, from 3,593 to 4,727, while the number of fully operational aircrew rose 19 per cent, from 3,312 to 3,941 (see Table 18). Although the expansion inevitably diluted the overall capability of the *Luftwaffe* (92 per cent of its aircrew were regarded as operationally capable in September 1939 and 83 per cent in March 1940), the greatest improvements took place in the first quarter of 1940, which saw a 25 per cent increase in aircrew numbers and a 34 per cent increase in fully qualified (operational) aircrew.

Increasing the number of aircrew at the front was a remarkable achievement for a training organization which was severely disrupted by the need to move schools at the beginning of the war and by the harsh winter which reduced flying hours. With the outbreak of hostilities in the west, a third of all flying schools were transferred to the safer skies of eastern Germany, Austria or Czechoslovakia, while simultaneously the organization expanded by 42 per cent (21 schools).[84] The increase in operational aircrew was partly due to the activities of the *Waffenschulen*, but much of the burden fell on the front-line units themselves, and a conference of *Luftflotte 3* on 23 March called for each *Kampfgruppe* to receive a complete crew from the formation replacement unit (*Ergänzungseinheiten*).[85] During the spring of 1940 there was a frenzy of training activity in the units in anticipation of *'Gelb'*, although the experience of Heinz Sellhorn, a pilot with *StG 77*, might have been exceptional, for on 1 April he flew six

Type	30 September 1939			30 December 1939			30 March 1940		
Bf 109	1,005	968	898	960	957	674	1,161	1,099	944
Bf 110	168	182	168	378	377	279	378	389	324
Bomber	1,248	1,215	1,085	1,493	1,202	959	1,726	1,644	1,236
Ju 87	381	335	322	417	382	301	420	412	378
Hs 123	39	36	36	39	42	36	39	39	39
Recon*	302	285	267	330	304	262	466	462	391
Corps	360	348	321	333	288	287	363	469	463
Naval	270	224	215	233	213	127	233	213	166
First-line total	**3,773**	**3,593**	**3,312**	**4,183**	**3,765**	**2,925**	**4,786**	**4,727**	**3,941**
Transport	225	224	186	210	190	188	463	456	398
Liaison	133	118	117	132	115	115	133	131	129
Grand total	**4,131**	**3,935**	**3,615**	**4,525**	**4,070**	**3,228**	**5,382**	**5,314**	**4,468**

Notes: The statistics are for establishments/aircrews on strength/operational aircrews. All statistics are for crews, e.g. one man in a Bf 109, two in a Bf 110 and four or more in a bomber.
Source: IWM Air Staff Post Hostilities Studies: Book 21, *Luftwaffe* Activity, Frames 1110, 1111, 1112, 1072.

bombing training sorties at Lippstadt.[86] In *III/KG 2*, raised on 1 March 1940, *Unteroffizier* Heinrich Albach's log-book showed training flights on 36 days to 9 May. The training included not only simple circuits (*Platzrunden*)—of which there were fourteen on 4 March alone—but also long-range flights, medium and low-level bombing, aerial gunnery and formation flying.[87]

During March and April the *Luftwaffe* completed its planning for *'Gelb'* and the airmen were chafing at the bit. The heavy rains which followed the thaw of early March rendered many airfields unusable, but in late March and early April they dried out. However, April brought the added complication of *Unternehmen 'Weserübung'*, which was conceived as a *coup de main* but rapidly became a prolonged campaign. Kesselring had to provide ten bomber *Gruppen* (later reinforced by *KG 54* and *LG 1*) and seven land-based transport *Gruppen* for the operation, and throughout April he was unsure when, or if, he would regain them and in what shape they would be. In the event, most of the *Gruppen* returned only at the beginning of May, bloodied but unbowed.

Meanwhile meetings and planning conferences continued to tie up the details, such as one for *Stuka* and *Schlacht Kommandeure* in Berlin on 17 February chaired by Göring. On 29 March the *Luftwaffe* Command Staff approved *Fliegerkorps VIII*'s operational plans and to support the anticipated rapid movement of units it formed a *Transportstaffel* on 17 April. On 7 May Kesselring had a short conference to work out the final details of the first-wave attack, which was to be conducted by *I/StG 2*, *II/KG 77*, a *Staffel* of *II(Schl)/LG 2* and two aircraft of *2. (F)/123* covered by *JG 27*.

The only incident of note was yet another bomb accident, on 4 May, which killed one man and wounded nine at *II (Schl)/LG 2*.[88] In *Luftflotte 3* preparations for the offensive were discussed on 23 March and it was agreed that *Oberst* Gerd von Massow (who had replaced Klein as *Jafü 3* on 5 March) would control the *Jagdstaffeln* of *Luftgau XII* shielding the main supply routes (*Rollbahnen*) west of Koblenz, which were the arteries for the Army's main thrust.[89]

In fact, despite the best efforts of the *Luftwaffe*, French reconnaissance located two of the three mechanized corps on the right flank (*XV.* and *XVI. Korps*, attached to 6. and 4. *Armee* respectively) opposite northern Belgium in mid-March but failed to discover *Panzergruppe Kleist*. A photo-reconnaissance Spitfire found it by accident on 7 May but RAF Bomber Command was reluctant to act, while the French High Command remained convinced that the main thrust would come through northern Belgium.[90] An inkling of German intentions had also come through the cracking of the *Luftwaffe*'s *Enigma* code, the French decrypting signals from 28 October and the British, separately, in January 1940. In mid-April the *Luftwaffe*'s code was changed and it was not until late May that the Allies could read it, but the signal traffic mostly related to administrative affairs. However, it did provide valuable details regarding the *Luftwaffe*'s order of battle and it gave an early warning of German intentions as the *Luftwaffe* used *Enigma* to alert units for operations.

The blow was now imminent. On 30 April Hitler set 5 May as A-Day, but the weather deteriorated and on 4 May the attack was postponed to 6 or 7 May. Good weather was essential for *Luftwaffe* support, and Hitler was willing to postpone the operation until he could get a favourable forecast. For several days the *Wekusta* toiled in their mundane but vital tasks and on 9 May the omens were finally favourable. The offensive was scheduled for the following morning, and that evening Göring went to the *Luftwaffe* staff headquarters at Kurfurst, outside Berlin, and established his own headquarters in his train '*Asia*'. (Sperrle had a similar train, code-named '*Äquator*', with seven carriages and freight wagons and 52 men.) In the west, cars began to drive up to the forward headquarters of the *Luftflotten* and *Fliegerkorps* and all awaited the final telephone call.

NOTES TO CHAPTER 7

1. Irving, *Göring*, pp.248–9.
2. *Ibid.*, pp.252–66.
3. US Monograph 151, p.41.
4. Steinhilper, p.172.
5. See Bekker, pp.42–3.
6. Diary entry 30 August, BA MA N 671/4 (hereafter Richthofen Diary).
7. See Cynk, Appendix 7.
8. Zaloga and Madej, pp.11–13.
9. For the *PLW* order of battle see Shores, *Duel for the Sky*, p.20.
10. See Bekker, pp.20–1; and *Luftflotte 4* War Diary, BA MA RL 7/331.
11. *Luftflotte 4* War Diary.
12. BA MA File RL 7/2.
13. US Monograph 151, p.184.

14. Zaloga and Madej, p.147.

15. *Ibid.*, p.116.

16. US Monograph 151, pp.183–4.

17. Losses from Cynk, pp.134, 137–9; and Shores, *Eagles*, p.24.

18. Zaloga and Madej, p.127.

19. Zaloga and Madej, p.93; after-action reports, *Einsatz im Feldzug gegen Polen*, BA MA File RL 7/2.

20. Zaloga and Madej, pp.171, 181.

21. Report by *Oberst* Wilhelm Speidel, 'Die Einsatz der Luftwaffe im Polnischen Feldzug', 1 December 1939, IWM AHB6/158, Tin 148.

22. *Fliegerdivision 2* War Diary, BA MA File RL 8/150.

23. Richthofen Diary.

24. For further details see Nielsen pp. 97–8.

25. Richthofen Diary.

26. *Fliegerdivision 1* War Diary, BA MA File RL 8/102; *Geschwader* War Diary, quoted by Schmidt, pp.285, 288, 349.

27. See Murray, *Effectiveness*, pp.108–9; and BA MA File RL 7/2.

28. Richthofen Diary.

29. *Ibid.*

30. *Ibid.*

31. *Ibid.*

32. According to Boog, p.386.

33. Though *Luftflotte 4*'s War Diary for 13 September states that 197 bombers attacked Warsaw.

34. For 13 September see Bekker, p.57; Boog, *Conduct*, specifically Boog's essay 'The Luftwaffe and Indiscriminate Bombing up to 1942' (pp.373–404) and Olaf Groehler's 'The Strategic Air War and its Impact on the German Civilian Population' (pp.279–97), quoting BA MA File RL 2/ II/51; *Fliegerdivision 1* War Diary; US Monograph 151, Frames 427–75; and Richthofen Diary, entries for 11 and 13 September. There is no mention of this attack in Gundelach.

35. US Monograph 151, Frame 454.

36. Shores, *Eagles*, p.28; Zaloga and Madej, p.148.

37. Schmidt, pp.260–302.

38. BA MA File RL 7/2; *Oberst* Weyert, *Erfahrungsberichte der Verbände der Luftflotte 1 im Polenkrieg.*

39. Figures from *Luftwaffe* personnel records: Hahn, p.196; Shores, *Eagles*, *passim*.

40. Quoted by Boog, *Conduct*, p.375.

41. BA MA File RL 2II/25: *1 Einsatz zum Schutze der Deutschen Westgrenze/Beschränkte Luftkriegführung gegen England.*

42. *Ibid.*

43. Shores, *Eagles*, pp.50–1, has the best account of air battles in the west during the '*Sitzkrieg*' and is the basis of all casualty figures. See also Martin's book for French losses.

44. For the Battle of Heligoland Bight see Shores, *Eagles*, pp.133–43, 148–50; Bekker, pp.71–9; and Winfried Bock's article.

45. Based upon Bomber Command Operations Record, PRO Air 24/200; and Advanced Air Striking Force Operations Record, PRO Air 24/21.

46. Based on Aders, pp.12–14.

47. *Lagebericht Nr 77*, IWM Ger/Misc/MCR/19, Reel 1A.

48. Kiehl, pp.56–8.

49. Dierich, *KG 51*, p.103.

50. The account of German planning of operations in the West is based on Taylor, *March*, pp.42–59, 159–70.

51. BA MA File RL 2 II/25.

52. Based on *Fliegerkorps VIII* War Diary, BA MA N671/5.

53. Balke, p.69.

54. BA MA File RL 7/2.

55. *Fliegerdivision 7* planning is based on letters from Ehrig to *Major* Werner Pissin of 12 January 1955 and from Student to Mr Vancelkenhuyzen of 15 July 1954 and on a lecture on the Koch operation by *Major* Zierach to the *Luftkriegsakademie* on 13 March 1944. Translations of these were provided to me through the kindness of Mr Alex Vanags-Baginskis. See also US Monograph 152, pp.88–139, 146.

56. *Fliegerkorps VIII* War Diary. For *Luftflotte 2* plans see US Monograph 152, pp.125–9.

57. For *Luftflotte 3* plans see Balke, pp.62–70; and US Monograph 152.

58. BA MA File RL 2 II/25.

59. *Ibid.*

60. Based on Taylor, *March*, p.61; and Bekker, pp.104–5.

61. Taylor, *March*, pp.170–4.

62. Steinhilper, pp.205–6.

63. Kiehl, p.47.

64. Some of the photographs taken on the first *AufKlGr ObdL* sorties are preserved in BA MA File RL8/3. For earlier sorties see *Lagebericht Nr 24, 31* and *32 (West)*, IWM Ger/Misc/MCR/19, Reel 1A.

65. BA MA File RL 2 II/25.

66. See Shores, p. 118; and Richey, pp.20–3.

67. *Lagebericht Nr 88.*

68. *Lagebericht Nr 77.*

69. Brookes, pp.88–90.

70. Balke, *KG 100*, pp.25–6; and Shores, *Eagles*, p.166.

71. Gunsburg, p.107; Pritchard, p.101; Wood and Dempster, p.150; PRO Air 24/507; Dispatch of Air Marshal A. S. Barratt, PRO Air 35/197; and PRO Air 2/7209: RDF System in France.

72. Brookes, pp.55–62, 65; Shores, *Eagles*, pp.182, 199, 204–5; and *Fliegerkorps VIII* War Diary.

73. Shores, *Eagles*, p.198.

74. Brookes, pp.89–90.

75. *Ibid.*, p.56.

76. *SHAA* File 2D27.

77. US Monograph 152, p.61.

78. Based on *GQG Aérienne* War Diary, *SHAA* File 1D44/1; *ZOAE* War Diary, *SHAA* File 2D27; and Shores, *Eagles*, pp.201–2.

79. PRO Air 22/54: Weekly Reports on Air Operations.

80. *SHAA* File 1D45.

81. Christienne and Lissarague, pp.335, 359.

82. PRO Air 24/507: Fighter Command Operations Record, 1 January, 8 February and 2 March 1940.

83. Figures for bomb expenditure and manufacture from Hahn, pp.198–9.

84. See Ries, *Flugzeugführerschulen*, pp.115–19, 185–7, 209–10.

85. BA MA File RL 2 II/36.

86. Log-book quoted by Smith, *Stuka Squadron*, p. 47. See also Stahl, p.51.

87. Balke, *KG 2*, p.61.

88. *Fliegerkorps VIII* War Diary.

89. BA MA File RL 2II/36: *Schutz der Deutschen Westgrenze/Feldzug gegen Frankreich.*

90. Brookes, pp.67–8; and Doughty, pp.74–6.

CHAPTER EIGHT

EAGLE OVER THE SEA

September 1939 to June 1940

T he outbreak of war created a tremendously frustrating situation for the Nazi leadership. Although it regarded Great Britain as the heart of resistance to its attempts to dominate Europe, the fact that Britain was a maritime power placed her beyond Germany's reach except by air. The German Navy could not challenge the Royal Navy, although the U-boats could impose a partial blockade upon the British Isles. To be truly effective this needed to be complemented by air attacks, but the German Navy's neglect of aviation meant that it had a small reconnaissance and minelaying force, leaving it dependent upon the *Luftwaffe* to provide bombers. The *Luftwaffe* had not ignored the problem, but its priority was to support land campaigns and at the outbreak of war the anti-shipping force consisted only of *I/KG 26* under Geisler, whose command was upgraded on 5 September to *Fliegerdivision 10* and received the newly raised *I/KG 30* equipped with the Ju 88A. Harlinghausen, *Luftflotte 2*'s Operations Officer since April 1939, assumed the same position in the *Fliegerdivision* before becoming *Fliegerkorps X*'s Chief of Staff on 1 November.

Göring dreamed of his airmen decimating the Royal Navy and proposed a mass attack on the Home Fleet's base at Scapa Flow at the outbreak of war, only to have the plan rejected by Hitler, who feared retaliatory attacks on the *Reich*. However, he encouraged attacks upon the Royal Navy at sea in a succession of *OKW* directives, and Home Fleet task forces in the North Sea were engaged on 26 September and 9 October. No damage was inflicted, and the *Luftwaffe* had the embarrassment of Goebbels' triumphant (but premature) boast to have sunk the aircraft carrier HMS *Ark Royal* when *I/KG 30* (*Hauptmann* Helmut Pohle) actually claimed only a near-miss.[1]

The *Gruppe* was undaunted and on 16 October attacked warships anchored off the Rosyth naval base in the Firth of Forth. The target was the battlecruiser HMS *Hood*, the symbol of British naval power, but she was in dock and could not be bombed. However, the cruiser HMS *Southampton* and destroyer HMS *Mohawk* were in the middle of the Firth and both were hit (the destroyer's Commander was mortally wounded) although the bombs failed to explode. Then came disaster, for, contrary to Schmid's intelligence reports, Spitfires were based nearby and shot down Pohle. Nevertheless the attack, coming immediately after a U-boat had penetrated into Scapa Flow and sunk the battleship HMS *Royal Oak*, shook British confidence and the Home Fleet sheltered in western ports until its bases had stronger defences.[2] It was a timely decision, for the following day *Fliegerdivision 10* sent seventeen bombers to Scapa Flow.

On 18 October Göring ordered that preparations be made for an attack on battleships in Rosyth Dockyard.[3] But the heavy losses of experienced Ju 88 aircrew worried the *Luftwaffe*, and on 21 October *I/KG 30* was specifically ordered (*Ia 4950/39*) not to send any operational crews to *Erprobungsgruppe 88*. An undated order sent out at about the same time stated that whenever a Ju 88 *Kette* (the word was underlined in the original document) flew over the sea, air–sea rescue floatplanes were to be alerted at List or Hörnum/Sylt.[4]

Maritime reconnaissance was largely conducted by the lumbering seaplanes of the *Küstfliegerstaffeln*, who suffered a stream of losses to the weather, mechanical failure and even British Ansons and Hudsons. The Naval Staff Operations Division's War Diary described the seaplanes on 24 December as 'in every respect inferior' to the British aircraft.[5] On 29 November five out of seventeen Do 18s (29 per cent) were lost and the Diary commented, 'This has been a black day for the *Luftstreitkräfte*.'[6] The seaplanes were easy meat for British fighters, which destroyed four out of a formation of nine He 115s from *1./KüFlGr 406* on 21 October. Four days later *FdL* units were reorganized so that the *Gruppen* (except for *KüFlGr 106*) could ease their administrative and maintenance problems (see Table 19). In addition, *KüFlGr 706* replaced its biplanes with He 111J bombers (described by the Naval Staff as 'antiquated'), but the reorganization brought no significant improvement in seaplane operations, which were further disrupted from late November when sea ice forced most bases to close for several months.

An alternative means of long-range reconnaissance was introduced on 1 October with the formation of the *Fernaufklärungsstaffel* under *Major* Edgar Petersen. This used the Focke Wulf FW 200C-0, which originated as a Japanese Navy requirement for a maritime reconnaissance version of the FW 200B airliner. The FW 200C, operating under *AufKlGr ObdL* to support U-boat operations in the Western Approaches, flew as far as Iceland and the Faeroes. Six were received in November, but one of the first missions, that on 23 November, saw the aircraft crash on take-off.[7]

Fliegerdivision 10 (renamed *Fliegerkorps X* by 27 October) also flew reconnaissance missions as part of a campaign which began on 17 October to interdict traffic off the British East Coast. Initially this targeted lone ships, but order *Ia 4986/39* of 1

TABLE 19: REORGANIZATION OF *FdL GRUPPEN*

Aircraft	New *Gruppen*	Old *Stab*
Do 18, He 59, He 115	*KüFlGr 106*	*KüFlGr 106*
Do 18	*KüFlGr 406*	*KüFlGr 306*
He 115	*KüFlGr 506*	*KüFlGr 406*
He 111J	*KüFlGr 806*	*KüFlGr 506*
He 59	*KüFlGr 906*	*KüFlGr 706*

Source: Based upon Shore, *Eagles*, pp.97–8.

TABLE 20: *LUFTWAFFE* ANTI-SHIPPING CAMPAIGN, 3 SEPTEMBER 1939–9 APRIL 1940

Month	Sorties	Losses	Sunk		
			Merchantmen	Trawlers	Warships
Sept 1939	20	–	–	–	–
Oct 1939	190	13	–	–	–
Nov 1939	70	–	–	–	–
Dec 1939	110	3	2 (714grt)	8	–
Jan 1940	115	5	9 (23,296grt)	2	–
Feb 1940	90	9	1 (629grt)	1	5 minesweepers
Mar 1940	120	6	3 (7,024grt)	4	–
1–8 Apr 1940	105	3	–	2	–
Totals	820	39	15(31,663grt)	17	5

November permitted attacks on convoys while insisting, to avoid giving offence to neutrals, that only positively identified enemy ships might be bombed. Anti-shipping attacks continued until early April but had nowhere near the anticipated effect; indeed, *KG 26* failed to sink any vessels until 17 December (see Table 20). On 21 December *Fliegerkorps X* was informed that it might operate against enemy convoy traffic to Norway and on 23 December it received permission to attack all shipping within 30 nautical miles of the British coast. (The original instruction specified 30km, but this was amended to 30nm the next day.)[8]

On 9 January a dozen ships sailing independently were bombed, and three days later *KG 26* began attacking unarmed and anchored light vessels, sinking one at the end of the month—to the displeasure of the German Navy, which feared reprisals. The *Geschwader* then concentrated upon more worthwhile targets and on 29–30 January attacked 36 independent vessels, of which four were sunk, three abandoned and thirteen damaged, although the 29 January attack upon convoy FS.83 was driven off. The following month saw further successes with eleven ships damaged, half of them on 9 February, but only the minesweeper HMS *Sphinx* and four minesweeping trawlers were sunk. The month ended in disaster when *4.* and *6./KG 26* bombed a German destroyer task force on 22 February (the Navy having failed to notify the *Luftwaffe* of the mission), sinking the *Leberecht Maas* while another destroyer, *Max Schultz*, foundered after hitting a British mine. Hitler was furious: 'I would not say anything if all the Navy were sunk in battle with the enemy, but it is inexcusable if this happens because of a failure in co-ordination.' The incident strained relations between the two Services just as they were planning for *'Weserübung'* and provided Göring with an excuse not to place his *Gruppen* under naval control.[9]

Convoys remained a difficult prospect and from 27 February the coastal sailings received permanent fighter cover which inflicted casualty rates of 27 per cent and 40 per cent upon two missions. While the attackers were frequently detected by British radar, interception was often hampered by cloud or mist. Intercepted bombers often survived because of faulty Fighter Command tactics based upon unwieldy formations, as well as poor standards of RAF gunnery, with machine guns set to give a wide pattern of fire at 400yds instead of a concentrated pattern at 250. Nevertheless, the Navy would have preferred *Zerstörer* escort for *Fliegerkorps X* and had pressed for this in vain at a conference on 8 December.

The eve of *'Weserübung'* saw the potential of anti-shipping operations demonstrated, with 57 merchantmen and 38 trawlers attacked in March 1940, seven of the former and one of the latter being seriously damaged. At the beginning of the month Harlinghausen himself sank two merchantmen (6,827grt) and severely damaged the 8,441grt passenger ship *Domala. KG 26* was confident enough on 1 March to strike into the eastern English Channel, to attack at night and to engage convoys. Attacks were made on coastal convoys FS.9 and FS.10 as well as every convoy sailing between Norway and Great Britain from 20 March until 9 April. The first Norwegian convoy (HN.20) lost one ship sunk and two damaged but strong escorts, including a radar-equipped anti-aircraft cruiser and fighter patrols, ensured that this was the only German success and the fighters claimed *Hauptmann* Otto Andreas, *Staffelkapitän* of *6./KG 26*, on 20 March. Bombs remained the preferred weapon against ships; torpedoes were used, but without success, by an *FdL* He 59 against the Polish destroyer *Blyskawica* on 7 December and by He 115s of *3./KüFlGr 506* on 12 March.[10] The torpedoes were too slow, and the Germans had to wait until 1941 when they acquired faster Italian weapons.

Starting with that on HN.20 in March 1940, attacks on convoys increasingly involved *Knickebein* (a magic raven in German folklore, but also commonly translated as 'Bent Leg') radio beam guidance, although exposure to British defences compromised its use later. An He 111 of *1./KG 26* lost on 28 October 1939 was examined by RAF intelligence officers who noted the *EBL 2* receiver for the *Lorenz* blind-landing system. In itself this was not unusual—the RAF's testing establishment at Boscombe Down had a *Lorenz* system—but these receivers were specially tuned to receive the *Knickebein* beam. In the wreckage of two *1./KG 26* aircraft were found papers referring to *Knickebein*, and after airmen prisoners were asked about it they gleefully told each other, 'They'll never find it; they'll never find it!', unaware that their cells were 'bugged'. It was a *KG 26* survivor, identified by the British as 'A231', who volunteered detailed information about the system in mid-June, and by the end of the month the British knew that the transmitters were at Cleves and Bredstedt.[11]

Stollbrock's *X-Gerät* unit, *KGr 100*, was also involved in anti-shipping operations. Its first armed reconnaissance mission was flown on 6 December and others followed until 3 April 1940, some down to the eastern Channel. They had some success, but it is astonishing that the *Luftwaffe* staff permitted the exposure of such a specialized unit to this sort of mission. It was a decision which reflected both the shortage of dedicated anti-shipping forces and the growing trend towards short-term expedience. Fortunately *X-Gerät* was not compromised, even when part of Stollbrock's aircraft was recovered by the British.

Meanwhile *KG 30* returned to its duel with the Royal Navy in December 1939, the emphasis now upon destroying installations in the Orkney and Shetland Islands, which suffered sporadic attacks in *Staffel* strength. The biggest, and most successful, of these was an evening raid on Scapa Flow on 16 March which damaged the heavy cruiser HMS *Norfolk* and the gunnery training ship HMS *Iron Duke*, a decommissioned battleship which had been the flagship of the Grand Fleet when the German High Seas Fleet was scuttled in 1918. She was saved from sinking by salvage teams seeking high-value scrap from the hulks of High Seas Fleet battleships. The strengthening of Scapa Flow's defences, including the provision of radar and three fighter squadrons, made further daylight attacks extremely hazardous, but the reason *Fliegerkorps X* ceased operations in early April was to rest units before *'Weserübung'*.

Attacks on convoys reflected a tightening of the German blockade advocated by both Raeder and Göring. On 28 November, for example, Göring sought to attack shipyards and docks, but Hitler refused permission until *'Gelb'* began. The following day *OKW Direktive Nr 9* emphasized the importance of defeating the British by crippling their economy; indeed, Hitler stated that if *'Gelb'* took the Low Countries, the *Luftwaffe* would attack enemy industry. In the meantime the *Luftwaffe* and the *Kriegsmarine* were to mine British ports, and the directive added: 'In this connection aircraft are extremely valuable at minelaying, particularly off the English West Coast ports . . . and in river estuaries.' A variety of targets were defined including shipping, storage facilities of various sorts and industrial plants 'whose loss would be of decisive significance for the military conduct of the war'. The *Luftwaffe*'s task was to maintain a watch on the ports and to develop aerial minelaying, but the earlier restrictions upon attacking centres of population and industry were not lifted.

Coeler, who relieved *Generalmajor* Hermann Bruch as *FdL West*, had always been an ardent supporter of aerial mining, but production of these weapons began only in February 1939 and although 1,250 *LMA* (550kg, 1,210lb) and 1,150 *LMB* (960kg, 2,110lb) magnetic mines were on order by September 1939 only 143 had been delivered. To lay them there were only two *Staffeln* of ancient He 59 floatplanes, *3. (M)/106* and *3. (M)/706* (later *3. (M)/906*): Jeschonnek had rejected Navy requests for modern He 111J, Do 17Z and Do 217 aircraft.

Adequate stocks of mines were not assembled until 18 November and bad weather prevented any sorties until 20 November. Then *Hauptmann* Gert Stein of *3. (M)/906* led nine aircraft from Norderney, each carrying auxiliary fuel tanks, to complete the mission off Harwich and in the Thames Estuary. Navigation problems prevented five from laying their mines, and this handicap was to remain for the elderly floatplanes. Two nights later a *Luftwaffe LMA* mine fell on sand banks, where it was recovered by the British, who developed countermeasures, including degaussing. Mines were laid not only off British ports on the East Coast but also off Dunkirk, France's primary port in the Channel.

There was a steady stream of losses because of the age of the aircraft and the difficulties of operating at night and in bad weather, while the Navy questioned the effectiveness of aerial minelaying when the task could be carried out more reliably by U-boats or surface vessels. At the end of November the *Luftwaffe*'s liaison officer with the Navy (*General der Luftwaffe beim ObdM*), *Generalmajor* Hans Ritter, protested strongly against operations which he felt were sapping the *Luftwaffe*'s strength to no good purpose. It was a view with which Göring sympathized. On 8 December Hitler attended a conference on naval matters including aerial minelaying and Coeler reviewed the situation. He sought to expand his force, which had only 29 aircraft at the beginning of the month, and to modernize it either with He 115s or with bombers. Göring promised to provide a *Kampfgruppe* and the conference decided to create a

TABLE 21: *LUFTWAFFE* **MINING OPERATIONS, 20 NOVEMBER 1939–3 MAY 1940**

Date	Sorties	Mines	Losses	Successes
Nov 1939	35	42	–	–
Dec 1939	51	44	8	–
Apr 1940	76	95	1	7 (14,564grt)
May 1940	43	81	–	–

Sources: Based upon Marchand and Huan; Rohwer and Hummelchen p.8ff; Shore, *Eagles*, pp.113, 121, 122, 128, 129, 272, 300-1, 304; and German Naval Staff Operations Division War Diary, IWM EDS 229.

reserve of mines in anticipation of a major operation against Scottish ports during the New Year. At the end of 1939 Coeler was promised *7./KG 26* and a *Staffel* of *I/KG 40* for minelaying, followed by a *Staffel* of FW 200s and ultimately three minelaying *Geschwader*. During the night of 7/8 December the first *LMBs* were laid, but the harsh winter weather prevented any further missions, which, as the Naval Staff pessimistically observed, had been less effective than U-boat and surface-warship minelaying (see Table 21).

Despite, or because of, Raeder's opposition, Göring persuaded Hitler to authorize a renewed *Luftwaffe* minelaying effort in the spring. In anticipation of this, *Fliegerdivision 9* was created under Coeler in February 1940 with *KG 4* (He 111P-4s) and *KüFlGr 106*, although *FdL* was also permitted to lay mines.[12] But Coeler's triumph was short-lived, for on 10 February *I/KG 4* was assigned to *Luftflotte 2* for training on aerial re-supply missions and its replacement, *I/KG 1*, returned to a conventional bombing mission ten days later together with *III/KG 4*, which converted to the Ju 88A. *KG 4* became a conventional anti-shipping unit and did not begin minelaying until mid-June. Coeler did receive *KGr 126* (the renamed *III/KG 26*) under Stein at the end of March, but by the beginning of the month only 566 mines were in stock, with 700 added during the next two months.

In anticipation of *'Weserübung'* it was decided to try and neutralize Scapa Flow, and on 5 April both *Fliegerkorps X* and *Fliegerdivision 9* were ordered to mine the base and bomb its facilities.[13] However, *Fliegerdivision 9* did not commence operations until the night of 17/18 April with attacks on British East Coast ports using the *LMB* mines, He 115 floatplanes and He 111 bombers. Laying mines from the He 111 posed problems since the aircraft had to fly level at a speed of no more than 300kph (186mph) at an altitude of 100m (330ft). The new minefields were far more successful (see Table 21), while losses were sharply reduced, but the diversion of *KG 4* to *'Weserübung'* and the imminence of *'Gelb'* blunted the offensive and within a month *Fliegerdivision 9* was reduced to supporting *'Gelb'*.[14]

The minelaying offensive stimulated the development of a British low-level radar system known as Chain Home Low (CHL), whose Type 2 radars had a 50-mile (80km) range but gave no indication of height. In the absence of radar-equipped night fighters, the RAF sought to destroy the threat at source through bomber and long-range fighter patrols around the seaplanes' bases. Between 12 December 1939 and 9 May 1940 the RAF flew 189 sorties in a vain attempt to catch the minelayers. During the night of 19/20 March the Hörnum/Sylt seaplane base was bombed in retaliation for the attack on Scapa Flow two nights earlier. Sylt's defences were strengthened the following afternoon when *Luftflotte 2* flew-in 60cm (24in) search-lights from *Flakbrigade VI* using transport aircraft of *Fliegerkorps VIII* and regular transport *Gruppen*. From the night of 13/14 April the RAF took a leaf from the Germans' book and conducted its own minelaying operations. By 10 May 302 sorties had been flown, 280 by Bomber Command, of which about 100 were in Scandinavian waters, for the loss of nine aircraft. These minefields proved more successful than the German ones, claiming at least twelve ships (18,355grt). German radar spotted British bombers approaching German bases but it is unclear whether or not *IV (N)/ JG 2* was ever informed, although it did succeed in destroying one of the intruders in April.

The RAF's minelaying effort was part of an Allied attempt to re-create the successful Great War blockade which had remorselessly strangled not only Germany but also her allies. After the war many German naval writers considered how the blockade might have been overcome, and one of the most influential, *Vizeadmiral* Wolfgang Wegener, suggested the solution of occupying Norway, which in the Second World War, as in the First, was neutral. Norway had played a major humanitarian role after the Great War and had temporarily adopted hundreds of starving German and Austrian children, the *Wienerbårn*, whom the Norwegians restored to health.

The embers of Polish resistance were still flickering when *Admiral* Rolf Carls, 'The Blue Czar' and commander of the Baltic-based *Marinegruppen Kommando Ost*, repeated Wegener's suggestion to Raeder. There was cautious support from the staff and senior officers, and with this Raeder raised the matter with Hitler at a routine meeting on 10 October. As Hitler was preoccupied with '*Gelb*' nothing happened, but two months later the Norwegian Fascist leader Vidkun Quisling, as a guest of the Nazi Party, warned that the Allies might occupy his country. Certainly Winston Churchill, First Lord of the Admiralty, advocated taking Narvik to reduce the flow of exports of Swedish iron ore to Germany when the Baltic was ice-bound. Later the scheme became part of an Allied plan to channel supplies to Finland after it was attacked by the Soviet Union in November.

Rumours of these plans reached Raeder through the German Naval Attaché in Oslo and the *Admiral* agreed to meet Quisling, who made the treasonable offer to assist any German occupation. Thus when Raeder met Hitler on 12 December he again proposed occupying Norway, although agreement did not come for two days, until Hitler himself had met Quisling. For security reasons *OKW* was authorized to conduct only contingency planning using a working party under Warlimont, and so the approach to both the Army and to Jeschonnek was circumspect. It produced an outline *Studie Nord* which was to be developed into a detailed plan by a tri-Service staff under a *Luftwaffe* general with a Navy Chief of Staff and an Army operations officer. Hitler's disapproval led to *Studie Nord*'s being withdrawn on 23 January and four days later Warlimont was given a special staff to produce the detailed plan code-named

'*Weserübung*'. Warlimont's staff was expanded on 5 February with representatives from the three Services, the *Luftwaffe*'s being the former Lufthansa official and *Kommodore* of *LG 1 Oberst* Robert Knauss. Impetus for the operation came eleven days later when the British destroyer HMS *Cossack* seized the German supply ship *Altmark* in Norwegian territorial waters to release captured British merchant seamen.

This acted as a catalyst and on 19 February Hitler demanded the acceleration of '*Weserübung*'. Warlimont's group now contacted the staff of *XXI.Armee Korps* (upgraded to *Gruppe XXI*) under *General* Nikolaus von Falkenhorst (who had served with Goltz in Finland in 1918) and the two conducted detailed planning at the former *RLM* premises in the Bendlerstrasse. The first *OKH* and *ObdL* knew of the operation came with the publication of an *OKW* directive on 1 March, and Göring's fury increased when he learned that the *Luftwaffe* units were to be subordinated to *Gruppe XXI*. This was a dangerous precedent, and Bodenschatz complained to Jodl that 110 *Luftwaffe* officers had been consulted without Göring's prior knowledge or permission. Göring tongue-lashed Keitel, then complained to Hitler, and within three days Jodl and Jeschonnek had worked out a compromise by which Falkenhorst would address all requests for air support to *ObdL*, which would then transmit the orders to the *Luftwaffe* units. On 4 March the *Luftwaffe* and the Army received their first formal presentation of the plan.

The occupation of Denmark ('*Weserübung Süd*') was added on 21 February, probably at Knauss's suggestion, both to extend the early warning network and to provide *Jagdgruppen* with forward bases to support the main operation. The ground operation was allocated to *Höheres Kommando XXXI* (formerly *Grenzabschnitt-kommando 1*) under Kaupisch, still a *General der Flieger* although he was transferred to the Army as a *General der Artillerie* on 17 April.[15] Small amphibious forces would take the ports and Copenhagen, but most of Kaupisch's troops would come overland. To support them small paratroop units would secure the bridges into Zealand and Aalborg West air base on Jutland.

For '*Weserübung Nord*', the assault on Norway, regimental battle groups would be landed at Narvik, Trondheim, Bergen, Stavanger, Kristiansand and Oslo to seize both the ports and nearby airfields, although Stavanger-Sola airfield would be taken by paratroops. The amphibious force would have to run the gauntlet of the Royal Navy and some troops would carried in naval task groups while others would be aboard merchantmen, which would approach their objectives covertly. Göring was extremely critical and, as Jodl commented in his diary on 5 March, 'He dominates the discussion and tries to prove that all preparations are good for nothing.' As a result, the Army and the Navy were compelled to strengthen their forces for the operation. All that remained was to decide upon the timing and whether or not it should be launched before or after '*Gelb*'. While the Russo–Finnish War raged there was a possibility of Allied intervention through Narvik, but with Finland's capitulation on 12 March everything went back into the melting pot. Having reconsidered all the options, Hitler informed Raeder and *OKW* at a conference in Berlin on 26 March that '*Weserübung*' would begin on 9 or 10 April.[16]

The problem of containing the Home Fleet when the '*Weserübung Nord*' amphibious groups set sail generated a new squabble between the *Luftwaffe* and the Navy during late March. Göring wished to bomb the Home Fleet in Scapa Flow but Raeder had little confidence in this proposal and, despite his previous criticism of the *Luftwaffe*'s minelaying capabilities, he wanted a major campaign to mine Scapa Flow and the major British estuaries. He also wished the *Luftwaffe* to take on the responsibility for protecting the troops once they had landed so that his warships could depart immediately, although Hitler's intervention was required to ensure that Göring accepted this burden.[17]

Since the prime threat to *'Weserübung'* came from the Royal Navy, the *Luftwaffe*'s air support was moulded around Geisler's anti-shipping specialists in *Fliegerkorps X*, which was reinforced by 21 *Staffeln* of fighters, *Zerstörer*, bombers and *Stukas*. Geisler's force was swollen to 48 *Staffeln*, including *KüFlGr 506* and *2. (H)/10*, with 527 combat aircraft, of which 317 were bombers (see Appendix 10). They would both shield the ships and provide the assault forces with close air support, although small forces were to cow the populations of Copenhagen and Oslo with aerial demonstrations.

In addition to its usual tasks, the *Luftwaffe* would undertake the new one of establishing, then supporting, aerial bridgeheads in Norway. Even with the transfer of *KGzbV 1* from *Fliegerdivision 7*, the *Luftwaffe*'s transport strength was inadequate and half a dozen *Gruppen* were mobilized from the *Fliegerführerschulen C* and *Blindflugschule* for this task. This brought the total transport force up to 40 *Staffeln* (533 aircraft) and in March a special command, *Lufttransportchefs Land*, was created from the headquarters of the *Lufttransportführer* and the *Kommando der Blindflugschulen*. It was given to one of Milch's few close friends, *Oberstleutnant* Karl August, *Freiherr* von Gablenz. An 'Old Eagle', Gablenz had been one of the original deputy directors of Lufthansa and was *Kommandeur der Blindflugschule der Luftwaffe*. Lufthansa provided his largest transport aircraft in the shape of twenty requisitioned four-engine airliners organized as *KGrzbV 105*. Most were Ju 90s, some retaining their livery, supplemented by FW 200s and a venerable Junkers G 38. For those places without an airfield *Lufttransportchef See* was created under *Major* Lessing with 27 floatplanes. Gablenz was to fly-in six infantry and paratroop battalions, one to secure Aalborg air base in Denmark, two to Stavanger and the remainder to Oslo, together with groundcrews and essential supplies to airfields in both Denmark and Norway. In addition *7.* and *8./KGzbV 1* were to drop paratroops in Stavanger and Denmark respectively. Paratroops flown into Oslo-Fornebu airfield were to meet the Air Attaché, *Hauptmann* Eberhard Spiller, and use transport he had assembled to rush into the city and seize King Haakon.

Gablenz's transports would carry some 2,000 men (18 per cent of the total commitment) to Norway and the material they also carried would provide interim support to the *Staffeln* at newly occupied airfields. The holds of half the 25 merchantmen bringing the remainder of the amphibious assault force contained the balance of the *Luftwaffe*'s supplies and equipment. The freighter *Rio de Janeiro* (5,261grt), for example, carried motor transport, AA guns, fuel, spares and ammunition (77 tonnes of SC 500 and SC 250 bombs) for Bergen, while the tanker *Senator* held 17,000 tonnes of high-octane petrol for Oslo. The Germans hoped to occupy Norway quickly, then transfer most *Luftwaffe Gruppen* south for *'Gelb'*, leaving a garrison of three *Kampfgruppen*, two transport *Gruppen* and a *Gruppe* each of fighters, *Zerstörer* and *Stukas* together with four reconnaissance *Staffeln* (three seaplane) protected by six *Flak* battalions. Ground support was to be provided by *Luftgaustab zbV 200* (*Generalleutnant* Wilhelm Süssmann, former *Kommodore* of *KG 55*) and for the Danish operation there was *Luftgaustab zbV 300* (*Generalleutnant* Alexander Andrae).[18]

Air opposition was expected to be weak. Intelligence estimated Norwegian air strength at some 90 aircraft when total air strength was 102, of which 74 were in front-line units: the *Haerens Flyvevåben* had 44 (28 serviceable) aircraft and the *Marinens Flyvevåben* 30 (25 serviceable). The Danish air shield was little better, with a total of 89 aircraft, a large portion of both the Danish and Norwegian Air Forces consisting of Fokker C.V corps aircraft. To protect their neutrality both countries sought modern aircraft, including fighters: the Danes had placed contracts for a dozen Macchi MC 200 fighters to supplement their Fokker D.XXIs while the Norwegians ordered 60 Curtiss Hawk 75A-6/A-8s and had a licence to build 24. But priority was given to the sword rather than the shield: the Danes were beginning to build under licence seven Fairey P.4/34 bombers (derivatives of the Battle), while the Norwegians had three Caproni Ca 310s, with a dozen Ca 312s and 36 Douglas DB-8A-5s on order. The Norwegians spent NKr4.7 million ($1,080,000) on strengthening their air forces from 1939, but by the time of the German invasion they had received only nineteen dismantled Hawks and six He 115N-I floatplanes (six He 115N-IIs, 24 Northrop N-3PB bomber floatplanes and ten FW 44 trainers were on order). On 22 February 1940 another NKr10.5 million ($2.43 million) was allocated for air defence, but this was not spent.[19]

With Norwegian air power fragmented, the Germans could anticipate no serious resistance and from 3 April their ships quietly left their berths and sailed north. The Allies, preparing their own landings, were already on the alert, especially after receiving photographs from reconnaissance Spitfires. The *Altmark* incident confirmed their suspicions that Germany was abusing Norwegian neutrality and provided a justification for a minelaying operation to disrupt the flow of Swedish iron ore from Narvik. The Home Fleet put to sea in support of this operation, scheduled for 8 April, and plans were drawn up to occupy key Norwegian ports in order to pre-empt the Germans, although there was no provision for air support.

As both fleets put to sea the two air forces began scouting for the enemy, and this led to several skirmishes. An attempted surprise attack on Scapa Flow at dusk on 8 April by 24 He 111s of *II/KG 26* was the failure which Raeder had anticipated. The attackers were detected by radar and intercepted by two Hurricane squadrons, and four bombers were either shot down or else crashed following battle damage. Two other aircraft were lost in these skirmishes, a Bf 110 to Wellington gunners and an Ar 196 which was interned after force-landing in Norwegian waters, but three British aircraft were destroyed and five were damaged. The same day a Polish submarine torpedoed and sank the *Rio de Janeiro* north-east of Kristiansand, one *Flak* battery losing 95 dead. The survivors openly stated their intentions of occupying the country and the Norwegian Government immediately ordered a partial mobilization.[20]

There was no such saving grace for Denmark, where the first hint of trouble was at 0700 when paratroops from *4./FJR 1* took Aalborg airfield in northern Jutland in a bloodless *coup de main*, the base being secured immediately afterwards by an infantry battalion flown in by *I/KGzbV 1*. The remainder of the company had hastily to organize another drop barely 36 hours before A-Day at Vordingborg, to capture the 1½-mile (2.5km) long bridge linking Zealand to Falster. This enabled troops landed by sea at Falster to march north to Copenhagen, where others under the command of *Generalmajor* Kurt Himer, Chief of Staff of *Höhere Kommando XXXI*, had disembarked during the night. The Danish King Christian and his Government were slow to respond to an ultimatum and Himer telephoned Geisler's headquarters in the Hotel Esplanade, Hamburg, to demand an aerial demonstration. Confusion over code-words led Hamburg at one point to believe that he wanted a full bombing mission. When the message was understood, 28 He 111s of *I/KG 4*, escorted by *3./ZG 1*, soon roared over Copenhagen to drop soothing leaflets, although this did not prevent a Danish machine-gunner spraying one bomber. Ground fire also damaged two Bf 110s of *Stab* and *1./ZG 1* when they raided nearby Vaerlose airfield, where four Fokker D. XXI fighters had their engines running and one was taking off. The latter was shot down by future night-fighter ace *Hauptmann* Wolfgang Falck and the others were destroyed on the ground together with ten Fokker C.Vs.

The demonstration had the desired effect and Denmark surrendered at the cost of 36 casualties. During the rest of the day *I/ZG 1* and *2./ZG 76* concentrated at Aalborg to secure air communications to southern Norway. *KGzbV 1* and *KGrzbV 101* brought in supplies and groundcrew while a *Flak* battalion was installed to protect the airfield. Also taken without incident was Esbjerg airfield, where another *Flak* battalion protected the newly arrived *II/JG 77*.21

Events did not go so smoothly in Norway and on several occasions it was the bravery and determination of *Luftwaffe* aircrew which snatched victory from disaster. This was especially so at Oslo, where Moltke's dictum that no plan survives contact with the enemy proved only too true. The plan had the intricacy of a Swiss watch. As the main force disembarked in Oslo harbour, *Oberstleutnant* Martin Drewes' *II/KGzbV 1* was to land two paratroop companies at Oslo-Fornebu airfield, where Air Attaché Spiller would meet them. Immediately afterwards *Hauptmann* Richard Wagner's *KGrzbV 103*, escorted by *Oberleutnant* Werner Hansen's *1./ZG 76*, would fly-in an infantry battalion to secure the airfield while *III/KG 26* demonstrated over the Norwegian capital.

Fog allowed the naval force to penetrate Oslo Fjord but alert defenders sank the cruiser *Blücher*, the flagship; three Ar 196 floatplanes of *5./BordFlGr 196* were lost with her. One of Drewes' transports disappeared in the fog and was believed to have crashed. Fear that the whole formation might follow suit led to permission being given to abort the operation, but not before more aircraft became lost. Geisler recalled Wagner's *Gruppe*, despite Gablenz's objections, but the *Kommandeur* believed the recall order to be an enemy ruse and ignored it. The first transport lost actually fell to Norway's seven operational fighters, Gloster Gladiators, which took off when they heard engines overhead. A mêlée then developed with Wagner and Hansen's aircraft, the latter beating off the Norwegians and destroying one fighter in the air and two on the ground as they refuelled at Fornebu. Hansen lost two Bf 110s and red lights were flashing on the instrument panels, warning of imminent fuel exhaustion. Meanwhile Wagner attempted to lead his transports on to Fornebu but was killed by ground fire.

At this moment Hansen's men themselves landed while one aircraft attacked the defenders, whose fire faded as their machine guns overheated. One Bf 110 was written off—the pilot, the future night fighter ace *Leutnant* Helmuth Lent, was uninjured—but the remainder acted as mobile machine-gun nests, taxying to the corners of the airfield to cover the transports as they landed, although the runway was so short that several crashed. They were soon joined by Drewes' lost transports and a *ZG 76* support aircraft with groundcrews and ammunition. Later in the day Hansen and Lent, in the two serviceable fighters, shot down a Sunderland probing the waters around Oslo.

Spiller ordered one transport's radio operator to report the success to *Fliegerkorps X* whose Chief of Staff, Harlinghausen, immediately ordered all transports to the airfield: 159 had arrived by the end of the day, while Gablenz flew to Fornebu to organize operations. Four hours behind schedule, the two paratroop companies returned at noon to Oslo, where Spiller packed them in buses and drove to the Norwegian Government's refuge at Hamar, 70 miles (110km) north of Oslo. There the Norwegian Inspector General of Infantry, Colonel Otto Ruge, had hastily defended the approaches to the town and in a skirmish the paratroops were driven off and Spiller was mortally wounded.[22]

The need to suppress the remaining defences around Oslo meant that the *Kampf-* and *Stukastaffeln* were extremely active, flying 125 sorties during the first day. Oslo-Kjeller airfield was attacked and two aircraft were destroyed together with four Hawks in hangars. However, the majority of Norwegian aircraft had flown north either the previous day or before Fornebu was occupied. The Norwegians had abandoned not only their bases with repair facilities, spares and ammunition but, at Oslo-Kjeller, 60 tonnes of aviation fuel—which the Germans put to good use for their transports.

The operation at Stavanger-Sola airfield was more successful, although five aircraft crashed either in take-off accidents or in collisions. The Germans arrived as nine Norwegian bombers were preparing to take off, and, despite attacks by *8./KG 4* and two Bf 110s of *3./ZG 76*, all but one Caproni managed to do so. Then 131 paratroopers under *Oberleutnant Freiherr* von Brandis dropped on to the airfield, which was quickly captured. A Caproni bomber then attacked the airfield but was driven off by an He 111 and crashed. With no other opposition there was hectic activity as fuel, ammunition and AA guns were brought in by means of 120 sorties (some involving *KGrzbV 105*), while at the end of the day *I/StG 1* arrived, supported by aircraft from *3./ZG 76* and *KüFlGr 506*, bringing the *Luftwaffe*'s strength at Stavanger to 36 aircraft.

At Stavanger, as at Trondheim, Bergen and Kristiansand, the *Luftwaffe* supported landings by bombing Norwegian coast defences. This task was performed not only by bombers but also, at Trondheim and Kristiansand, by Ar 196s of *BordFlGr 196*. Norwegian air resistance was weak, an He 111 of *7./KG 4* driving off floatplanes over Kristiansand while AA fire shielded ships elsewhere. The Norwegian air forces took wing to emergency airstrips, frozen lakes or remote fjords, where they were exposed to the elements and deprived of spares. By the end of the day their first-line strength was down to 54 aircraft (30 Fokker C-Vs, 22 floatplanes, a Gladiator and a Ca 310) and a score of trainers. The *Luftwaffe* was rushing *Staffeln* not only to former Norwegian bases such as Oslo-Fornebu and Stavanger but also to Trondheim harbour, where *KüFlGr 506* moved sixteen He 115s of *1.* and *2. Staffeln*. The *Luftwaffe* had flown some 680 sorties (400 transport) in support of the occupation and lost 23 aircraft (eleven transports), with seven more severely damaged in crash-landings.[23]

The invasion of Norway proved to be a moment of truth not only for the Norwegian people but also for the Royal Navy as fierce air–sea battles took place off the Norwegian coast. The next two months demonstrated conclusively that fleets could not operate effectively without air cover as Allied sailors paid the price for two decades of British neglect of naval air power.

Such neglect becomes even more astonishing when it is remembered that the British Admiralty in the Great War had displayed the keenest awareness of air power. This was due to the threat of German airships both to the Grand Fleet and to Great Britain herself, for whose air defence the Admiralty was responsible in 1914. At the end of the First World War the Royal Navy led the way in protecting its ships from air attack and providing them with air support. Each capital ship and cruiser had AA guns, while in the Grand Fleet and its Battlecruiser Force half the 36 capital ships, and 58 per cent of the cruisers, carried platforms so that fighters or reconnaissance aircraft could be flown off. The Grand Fleet's Flying Squadron had two floating air bases in the aircraft carriers HMS *Argus* and HMS *Furious* (the latter a converted battlecruiser), together with three seaplane carriers, while barges with flying-off platforms provided cruiser forces with air support.

This tradition of technical excellence and tactical innovation was jettisoned in the post-war world when the RAF retained control of naval air power. During the 1920s and 1930s the Admiralty cast the RAF in the role of Wicked Stepmother and itself as the Fairy Godmother as it struggled to regain control of shipborne air power, but it succeeded only in 1937. By then the newly established Fleet Air Arm (FAA) had lost too much headway, and despite the gallantry of its men it was overshadowed by the US Navy's air arm in the Second World War and only partly recovered after that. Yet the truth was that the Admiralty was the author of its own misfortune, through an ambivalent attitude towards air power between the wars.

Until 1942 all navies accepted the dominance of the gun, and naval power was assessed in terms of capital ships and cruisers. Clashes between battle lines were still regarded as the means of achieving naval supremacy, air power having a subsidiary role. Catapult-launched seaplanes were to conduct reconnaissance and to direct gunfire while aircraft carriers supported scouting forces to discover, to track and to harass the enemy battle fleet both at sea and in harbour.

Air power crusaders pointed to Mitchell's demonstrations in the early 1920s but these lacked operational credibility because the targets had been defenceless and immobile, with open watertight doors. Yet the improved performance of the aeroplane in the late 1920s could not be ignored, and by July 1931 the British Chiefs of Staff were warning that both Malta and Gibraltar might be untenable in the face of air attack.[24] The Royal Navy also recognized that there was a serious threat in narrow waters such as the Channel, the North Sea and the Mediterranean (where it was particularly exposed to attack, from numerous Italian air bases). Nevertheless, it still placed its faith in a combination of high-speed manoeuvring, armour plate and AA fire, and as late as May 1936 one commentator claimed that an aircraft attacking a capital ship stood a 94 per cent chance of being shot down.[25]

The Royal Navy was content to allow the RAF to neutralize the land-based air threat, which, it believed, posed no problem for blue-water forces, but while the admirals still believed the capital ship to be supreme in the ocean, there was a

Above: The crew of a Ju 87B of *IV(St)/LG 1* board their aircraft.
Below: A heavily bombed Polish air base. When the *Luftwaffe* conducted post-operations analysis it was discovered that most such airfields were still usable.

Left, top: The view from the nose gunner's position of an attack on Polish roads

Left, centre: Although of poor quality, this photograph shows a Ju 87B-1 whose starboard undercarriage has been shot off. The pilot then apparently jettisoned the port undercarriage to make a belly landing.

Left, bottom: *Luftwaffe* bombers swoop on a Polish road.

Right, upper: For all its hopes, the *Luftwaffe* failed to catch the Polish Air Force on the ground. While many aircraft were destroyed, they were most probably like these—trainers. (Alex Vanags-Baginskis)

Right, lower: During the *'Sitzkrieg'*, fighters and reconnaissance aircraft were most active. Corps aircraft like this Hs 126 being refuelled flew many sorties. (Alex Vanags-Baginskis)

Above: A sentry guards Ju 87Bs at a misty airfield. (Alex Vanags-Baginskis)
Below: Bf 109s at dispersal at a Western Front airfield during the autumn of 1939.
(Alex Vanags-Baginskis)

Above: A burning trawler settles by the stern during *Fliegerkorps X*'s campaign during the winter of 1939/40.
Below: A cheerful greeting from a member of *KG 26*. This *Geschwader* was the spearhead of the *Luftwaffe*'s anti-shipping effort. (Alex Vanags-Baginskis)

Left, top: A Do 18D is towed towards the slipway. These flying boats formed the backbone of *FdL West*'s reconnaissance force.

Left, centre: A Do 18 comes to grief and begins to sink. The German Naval Operations War Diary on 24 December 1939 described these and the other seaplanes as being 'in every way inferior' to their British opponents such as the Hudson and Anson. Five Do 18s were lost on 29 November alone.

Left, bottom: An He 111 on an anti-shipping mission goes down with a blazing starboard wing.

Right, top: *Aufklärungsgruppe ObdL* conducted high-altitude photographic reconnaissance during the *'Sitzkrieg'* and this extended as far west as Liverpool.

Far right, top: *Luftwaffe* aircrew examine the anti-aircraft machine guns at Oslo-Fornebu which almost prevented German paratroops from taking the airfield. One transport of *KGr zbV 103* crashed and two others crash-landed, the *Gruppe* losing three dead, including the *Kommandeur, Hauptmann* Richard Wagner.

Right, bottom: Ju 52 transports disgorge paratroops at Oslo-Fornebu on 9 April 1940 while two Norwegian Gloster Gladiators blaze in the background. (Alex Vanags-Baginskis)

Left, upper: An He 111H of *6./KG 4*
'General Wever' comes to grief at Oslo-
Fornebu. (Norwegian Armed Forces
Museum)
Left, lower: Milch makes an inspection
during his brief tour of duty as
Commander of *Luftflotte 5*.
Above: The congestion at Oslo-Fornebu
is illustrated here with a Ju 87R in the
foreground and two Ju 52s in the
background.
Right: A Ju 87R of *I/StG 1* flies low over
Norway.

Above: A Bf 109E undergoes
acceptance trials. (Greenborough
Associates)
Left: A vital feature of the
Luftwaffe's success was its reliable
and resilient signals system. Here
signallers complete the camouflage
of a radio station.
Right, top: Loading a bomb into
the cradle of a *Stuka*. The ability of
the 'Black Men' (ground crews) to
turn around aircraft quickly
allowed the *Luftwaffe* to conduct
more missions per aircraft than
the Allies in May 1940.
Right, bottom: Looking forward
from the rear seat in a Ju 87.

Left, top: Bf 110s of *ZG 76* patrol over a formation of He 111s. The Bf 110 was designed to provide fighter support to bombers thoughout their mission.

A Ju 87 swoops on targets Left, centre: somewhere in western Europe during the summer campaign.

Left, bottom: A Hurricane trails smoke and falls away after encountering a Bf 109.

Right, top: Rotterdam-Waalhaven airfield, photographed by a *Luftwaffe* reconnaissance aircraft. Right, centre: A Fairey Battle brought down by German ground fire. Right, bottom: The gutted centre of Rotterdam after the *KG 54* attacks. (Alex Vanags-Baginskis)

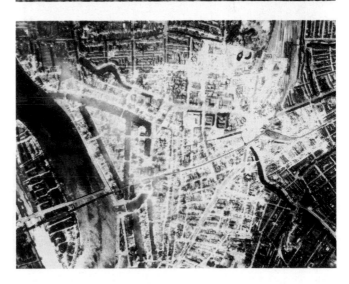

Above: Rotterdam docks were still smouldering when this photograph was taken.
Left, upper: Abandoned gliders at Fort Eben Emael. (Alex Vanags-Baginskis)
Left, lower: A *Kette* of Ju 87s is photographed from another *Kette* as they prowl the skies seeking targets.
Right, upper: Ju 88As prepare to take off from an airstrip. Similar bombers may have carried out the last bombing raid of the German campaign in the west in 1940 *after* the Armistice with the French had taken effect. (Alex Vanags-Baginskis)
Right, lower: The Hs 123s of *II(Schl)/LG 2* kept pace with the *Panzer* divisions in Poland and in western Europe. When the British broke through at Arras in May 1940 this *Gruppe* played a major part in stopping them. (Alex Vanags-Baginskis)

Left, top: Do 17Zs of *KG 2 'Holzhammer'* en route to a target in western Europe. *KG 2* suffered heavy losses, with eight bombers shot down, on the first day of the Western Campaign. (Alex Vanags-Baginskis)

Left, centre: A Do 17Z of *KG 76* dropping bombs. (Alex Vanags-Baginskis)

Left, bottom: Damaged track and burnt-out rolling stock—the typical results of a *Luftwaffe* attack on a railway station in northern France.

grudging recognition of the need to control carrier-borne aircraft to meet the Royal Navy's operational requirements—hence the Admiralty's struggle during the 1930s. Yet to many admirals the carrier, which required prodigious capital and operating expenditure, was an expensive luxury, and in the 1920s some questioned their inclusion as warships.[26]

Even for British carriers armour plate and the AA gun were the prime defence, and when the *Luftwaffe* made its first attack on HMS *Ark Royal* the carrier's air group was struck below decks and the aircraft drained of petrol to reduce the fire hazard. In the face of the new generation of fast bombers, the admirals sought refuge not in fighters to provide air cover but in converting six ships to AA cruisers. British ships generally lagged behind those of other navies in the provision both of powerful automatic weapons for the inner layer of defence and powered, dedicated (or dual-purpose) mountings for the outer layer.

There was no official requirement for high-performance aircraft, especially fighters, yet even here the RAF could not be blamed, since the specifications were drafted by the Admiralty. Robustness was essential in shipboard airframes, which were exposed to salt water and operated off heaving decks and whose landings were little more than controlled crashes. Robust airframes increased weight, which further restricted performance, while hydrodynamic effects restricted carrier speed, making it difficult to fly-off high performance-aircraft until the introduction of catapults from 1935 onwards. As carriers had only a limited hangar capacity, the Royal Navy demanded multi-role aircraft: the backbone of the FAA was the Fairey Swordfish biplane, built for the torpedo-spotter-reconnaissance (TSR) role. This extended to air defence, where the FAA relied upon the Blackburn Skua and Blackburn Roc fighter/dive bombers—the latter with a powered four-gun turret—which supplemented Gloster Sea Gladiator biplanes. Faced with these restrictions, British industry was unable to produce good naval aircraft, which would force the Royal Navy increasingly to rely on Lend-Lease American aircraft supplemented by 'navalized' landplanes which never fully met its requirements.

The Admiralty's conservatism was the result of a lack of experienced naval air officers. With promotion dependent upon traditional service at sea, or in administrative/staff roles, few young officers were willing to lose seniority by serving with carrier-borne units. Until 1937 the Royal Navy was supposed to supply 70 per cent of naval pilots, but the figure in 1930 was 65 per cent, in 1936 31 per cent and in 1939 59 per cent, forcing the RAF to make up the difference. As late as March 1939, in only one carrier had the air group staff been replaced by naval personnel. Against this background it is little wonder that the Royal Navy entered its first serious air–sea campaign weighed down with the twin millstones of technical and tactical ignorance.[27]

By contrast, naval officers had controlled US Navy aviation since before the First World War and all aircraft carrier commanders were qualified pilots. While the battleship continued to dominate US naval thought, there nevertheless developed an expertise in carrier aviation which made greater use of the flight deck for stowage and maintenance, even during landings and take-offs. British practice was to have a clear flight deck and to strike down all aircraft not immediately needed for operations, which restricted air group size to hangar capacity. This made prolonged operations difficult, for mechanical failure and accidents inevitably eroded the air group's strength—which would be further reduced in sustained operations. Consequently, even if the carriers had sufficient fuel in their bunkers, the exhaustion of their air group would reduce the effective time spent on station.

The new British carriers such as HMS *Ark Royal* with 48 aircraft had an air group twice the size of their predecessors, the *Courageous* class, but the US contemporaries of the latter (*Lexington* class) and the new-generation carriers (*Yorktown* class) each carried some 70 aircraft. Larger air groups permitted aircraft that were more specialized, of more modern construction and possessed of a higher performance than their British contemporaries (although the Grumman F4F Wildcat and Brewster F2A fighters and the Vought SB2U dive bomber were still inferior in performance to comparable *Luftwaffe* aircraft because of the constraints of carrier use). While the emphasis on offensive operations meant that, until 1942, only one-third of a US carrier's complement consisted of fighters, these had an active role in defending the carrier and engaging enemy reconnaissance aircraft.[28]

They would certainly have been useful on 9 April as the *Luftwaffe* supported the invasion with a major reconnaissance effort over the waters west of Norway, 36 sorties in the morning and thirteen in the afternoon being flown by *KüFlGr 106*, *406* and *506*. Within two hours *KüFlGr 506* was sending a stream of reports on the Home Fleet's location by means of its seaplanes, and an Ar 196 from the cruiser *Admiral Hipper* passedg British aircraft flying in the opposite direction on similar missions. Geisler had held back a strike force in anticipation of finding the British fleet and during the afternoon 110 sorties were flown by *KG 26*, *KG 30* and *StG 1*. The *Stukas* could not reach the Home Fleet but sank the Norwegian torpedo boat *Aeger* as they flew to their new base at Stavanger-Sola, though not before the Norwegian vessel had sunk the freighter *Roda*, which was carrying AA guns to Stavanger.[29] The remaining aircraft attacked the Home Fleet and sank the destroyer *Gurkha*, hit the battleship *Rodney* (the bomb failed to penetrate her armour) and inflicted slight damage or near misses on five cruisers. The ships' anti-aircraft fire shot down four of *KG 30*'s Ju 88s, including that of *III. Gruppe*'s *Kommandeur*, *Hauptmann* Siegfried Mahrenholtz; none of the crews survived.

While naval orthodoxy appeared to triumph, the prodigious use of AA ammunition (some ships' lockers were 40 per cent empty) and the lack of any capability for underway replenishment forced many ships to withdraw. Moreover, the intensity of the attack made the Royal Navy recognize that it could not operate unscathed within range of German air bases in southern Norway, and for this reason Allied strategists concentrated upon retaking Trondheim and its airfield. Consequently the *Luftwaffe* won a significant victory, despite the fact that it had had the worse of the day's encounters, losing nine aircraft to enemy action (and two to accidents) as well as an Ar 196 when the cruiser *Karlsruhe* was sunk by a submarine.

The following day saw the *Luftwaffe* consolidate its hold on southern Norway, flying some 200 transport sorties. A small airfield at Kjevik, near Kristiansand, proved to be a useful staging point for shorter-range aircraft and a base for the Bf 109s of *II/JG 77*; Trondheim's Vaernes airfield was taken, as was Oslo-Kjeller. Stavanger-Sola was reinforced by *Stab*, *2.* and *3./ZG 76* (*1.* arrived from Oslo-Fornebu on 11 April), together with *3./KG 26*, while at Oslo-Fornebu were *1./StG 1* and *8./KG 26*. The *Haerens Flyvevåben* concentrated at two bases between Trondheim and Oslo while both sides dispatched reconnaissance aircraft to determine the situation. Skirmishes occurred off the Scottish and Norwegian coasts and at dusk *KG 26* and *KG 30* made another vain attempt at a surprise attack upon Scapa Flow, losing five of the 38 aircraft taking part, including that of the *Kommandeur* of *I/KG 26*, *Oberstleutnant* Hans Ahlefeld.

Despite attacks by Blenheim IF fighters on Stavanger harbour and Sola airfield, the Germans' hold on the airfield tightened with the concentration there on 11 April of *I/ZG 76*, *Z./KG 30* and a detachment of *1. (F)/120*. At Bergen FAA dive bombers from the Orkneys sank the cruiser *Königsberg* (and another Ar 196) on the same day while on 12 April came the war's first carrier-borne strike, by Swordfish from HMS *Furious* on Trondheim. This opened a two-day campaign against German air bases and shipping involving 161 sorties. The biggest battles occurred on 12 April when *II/JG 77* shot down five of a dozen Hampdens attacking Kristiansand, but as many Bf 109s were lost. A formation of 44 Wellingtons attempted to attack the battlecruisers *Scharnhorst* and *Gneisenau* off Stavanger but were intercepted by *ZG 76* and *Z./KG 30*, which shot down five for two Ju 88s lost or severely damaged.

The three days following the German invasion cost the British nineteen aircraft but forced Bomber Command to curtail operations in and around Germany in order to concentrate on Scandinavia, including night attacks upon airfields. A number of twilight interceptions were made, but the first true night interception by the *Luftwaffe* was achieved by *Hauptmann* Falck's *I/ZG 1* at Aalborg. Falck arranged an interception procedure with the local radar commander and on 30 April came the first opportunity to practise it. Falck led three Bf 110 sorties and although these gained no victories he discussed his night fighting ideas with Udet, Milch and Kesselring, who were currently preoccupied with preparations for *'Gelb'*.[30]

Within days it was obvious that the conquest of Norway would become a prolonged struggle, and since Geisler's *Fliegerkorps* lacked administrative facilities it was decided to establish a *Luftflotte* headquarters. Göring personally selected his deputy for this assignment, and no doubt Milch was willing to seize a long awaited opportunity to 'punch his ticket' with a front-line command. Yet neither man regarded the appointment as other than temporary, and Milch ensured that he would be recalled to Berlin to participate in *'Gelb'*. Göring may have intended to prevent Milch from continuing to look over Udet's shoulder and to use him as a scapegoat if *'Weserübung'* went wrong. A superior signals network was Milch's excuse for exercising command from Hamburg, but Göring insisted upon his going to Oslo even though communications there with both Trondheim-Vaernes and Stavanger-Sola were poor.

On 12 April Milch established the new *Luftflotte*, with Förster as his Chief of Staff, but delayed his departure to Norway for four days, allegedly due to bad weather, although on 12 and 13 April the *Luftwaffe* flew 235 transport sorties to Oslo, where Milch's old friend Gablenz still controlled the air bridge.[31] Milch's arrival led to an intense period of activity as he hastily wove an infrastructure to support its operations. On 16 April he requested permission to form three repair columns with civilian workers, and ten days later permission was granted to create *Luftpark Oslo*, with *Luftflotten 1* and *4* supplying some 200 specialists and tradesmen. A signals network was also created, and gradually *Luftwaffe* operations received the support the airmen expected.[32]

This occurred as an intensification of British air attacks heralded the arrival of Allied troops in Norway, where the Germans, having secured the southern ports, had begun pushing battle groups northwards. However, the need to interdict British bridgeheads meant that most bomber support was directed against coastal targets and that the battle groups therefore received only limited assistance—some 60 bomber sorties in three missions on 11, 20 and 21 April, the majority by *KG 4*. Occasional strafing missions were flown by *ZG 76*, but for the most part *Generalleutnant* Richard Pellengahr's *196. Infanterie Division* was forced to rely upon the slender reed of *Hauptmann* Jäger's *2(H)/10* and its handful of Hs 126Bs. The operation was especially difficult because deep snow, narrow valleys and poor roads confined each German advance to a regimental-size column and eased the defenders' task. The Norwegian air forces offered some resistance, but communications were poor and stocks of supplies were very low and gradually they faded from the picture. However, the prospect of a prolonged campaign led to the decision to reinforce Geisler's *Fliegerkorps X* with *II* and *III(K)/LG 1* (Ju 88A), *I* and *II/KG 54* and the new FW 200C Condors of *1./KG 40* in mid-April—a total of 140 aircraft. In addition, *KGrzbV 108* received a number of prototype long-range seaplanes, including three Blohm und Voss Ha 139A floatplanes, two BV 138s and five Do 26 flying boats, while two four-engine BV 142 prototype airliners were assigned to *KGrzbV 105*.

In the meantime the British landed hastily assembled expeditionary forces on 14 April at Harstad, 35 miles (55km) north-west of Narvik, and Namsos, 80 miles (130km) north-east of Trondheim. Air attacks on Stavanger-Sola and Bergen covered the latter operation, inflicting minor losses and delaying the *Luftwaffe's* discovery of the landing. During the evening there was a hurried attempt to secure the Trondheim–Oslo railway when a paratroop detachment of 162 men under *Oberleutnant* Schmidt was speedily dropped at Dombås using Drewes' aircraft. The operation was a disaster: one transport was lost to ground fire, four others were written off in forced landings and one was interned in Sweden, some men were dropped too low and a Norwegian battalion near the drop zone quickly killed or captured most of Schmidt's troops (Schmidt himself was wounded). A detachment of 45 managed to block the Oslo road in the bitter cold but was surrounded and surrendered on 19 April.[33]

Earlier that day the British had landed at Åndalsnes—100 miles (160km) south-west of Trondheim, 60 miles (96km) north-west of Dombås and, like Namsos, a small fishing port with limited facilities—but the landing isolated Trondheim, defended by *138. Gebirgsjäger Regiment* (*Oberst* Wilhelm Weiss). To reinforce the 2,000-man garrison Gablenz's transports flew 600 sorties up to 25 April to bring in *Generalmajor* Kurt Woytasch of *181. Infanterie Division* with several of his battalions and their equipment, while a merchantman ran the blockade to deliver heavy weapons, including two *Flak* batteries. So serious was the situation that the Navy diverted two U-boats, *UA* and *U101*, to run in a total of 245 tonnes of aviation fuel, six tonnes of bombs and three 8.8cm AA guns during late April.[34] After 25 April the air transport effort eased, but it continued on a smaller scale until Trondheim was relieved.

The *Luftwaffe* reacted promptly when it learned of the Allied landings, *III(K)/LG 1* initiating attacks on Namsos from 15 April while a lone He 111 of *Stab KG 26* opened the aerial assault upon Åndalsnes during the afternoon of 19 April. Some 330 bomber and *Stuka* sorties were mounted against the former until 4 May and 720 against the latter until 1 May, the total losses being nine aircraft shot down or written off through being damaged by enemy action. Another two aircraft strayed into Swedish air space and were interned. At Åndalsnes *Luftwaffe* attacks sank small ships and the Norwegian torpedo boat *Trygg*, and on 26 April—ironically, the day British land-based fighters began their patrols—attacks by *LG 1* and *KG 26* set ammunition ablaze, the fire consuming the port's wooden quay.[35] But the *Luftwaffe's* greatest success was at Namsos, where the town's radio station was accidentally put out of action on 15 April when the aerial was caught by the wing of a Ju 88. Attacks intensified on 20 April after *Luftflotte 5* had received a stark order from *OKW*: 'By the *Führer's* order, the towns and railway termini of Namsos and Åndalsnes are to be destroyed without regard to the civilian population, and railways and roads effectively cut.'.[36] It was the third raid of the day, that by *II* and *III/KG 4* during the evening, which inflicted the most damage. The 24 He 111s wrecked or burned much of Namsos together with the wharves, railway station and rolling stock, most of a newly arrived French *demi-brigade's* ammunition and equipment being lost.[37] The following day the Allied commander at Namsos, Major-General Sir Adrian Carton de Wiart, informed London that there was no point in sending supplies to the port and added that unless *Luftwaffe* activity was restricted he saw little chance of conducting any decisive operations—or, indeed, any operations at all.[38]

De Wiart was a holder of the Victoria Cross, the highest order for bravery in the British Empire, and so his courage could not be questioned. Therefore, London dispatched No 263 Squadron with eighteen Gladiator fighters. It reached the frozen Lake Lesjaskog near Åndalsnes on 24 April and was joined during the next two days by four Skuas. In the meantime ineffective attempts were made to relieve the pressure by mounting attacks on German-occupied air bases. Stavanger was the prime target, 115 sorties being flown by day and night from 15 April to 3 May. The *Luftwaffe* destroyed fourteen of the attackers (another two crashed, possibly as a result of battle damage) with the aid of a *Würzburg* radar set which was operational by 24 April. The total *Luftwaffe* losses as a result of attacks upon the Stavanger area between 10 and 30 April were seventeen aircraft, including eight transports. The Oslo airhead was shielded from daylight attack by *II/JG 77* in Kristiansand, which destroyed ten enemy aircraft for the loss of five Bf 109s. RAF night attacks on Fornebu were sometimes successful and that during the night of 30 April/1 May destroyed ten Ju 52s and damaged 27. In all, the Germans lost twelve aircraft destroyed, with 31 damaged, at Fornebu up to 2 May.

RAF Bomber Command committed eighteen out of 27 squadrons to the Norway Campaign but it operated under a severe disadvantage. From their East Anglian bases the shorter-range Blenheims could barely reach Stavanger and even from Scottish bases Trondheim was beyond them. The Whitley heavy bombers could just reach Oslo from English bases and Trondheim from Scottish ones. But in a month's operations to the night of 7/8 May the squadrons flew 576 sorties and a number of minelaying sorties at the cost of 31 aircraft. However, the British effort ended with the beginning of 'Gelb'.[39]

So desperate were the British to end German air superiority that the heavy cruiser HMS *Suffolk* dashed in during the early hours of 17 April to shell Stavanger-Sola airfield and the port. The airfield had been extremely congested, with combat aircraft and transports supporting garrisons at Trondheim and Narvik, but Geisler had ordered most of the *Staffeln* to withdraw on 16 April after eleven aircraft had been damaged in accidents. By the time HMS *Suffolk* arrived there were only 37 aircraft permanently at Sola, mostly drawn from *ZG 76* and *KG 26* but including four Ju 88Cs of *Z./KG 30*. One of the latter, flown by *Oberfeldwebel* Martin Jeschke, shot down the cruiser's Walrus observation amphibian as it dropped flares to illuminate the airfield, and while these destroyed a Ju 52 this action probably saved the base. The cruiser's fire wrecked eight floatplanes in the harbour and then she began to withdraw at high speed, having stirred a hornet's nest. Avenging bombers from Stavanger-Sola, and some from as far afield as Westerland, flew 82 sorties and hit her twice, but, for once, speed, armour and AA armament—16 × 4in (101 mm) in modern, Mk XVI mountings and two quadruple 2pdr (40mm)—enabled her to reach Scapa Flow.

Nevertheless, this incident emphasized the need for carrier-borne fighter cover if the Royal Navy was to survive in Norwegian waters, and two of its largest carriers, HMS *Ark Royal* and HMS *Glorious*, were transferred to the Home Fleet from the Mediterranean. With a complement of 57 aircraft between them, the two ships took No 263 Squadron to Norway, then provided the Allied expeditionary forces with air support. On 24 April they made their first successful interceptions against two He 111s of KG 4, one carrying *Kommandeur III/KG 4*, *Major* Ernst Kusserow, whose aircraft belly-landed near German lines.

Meanwhile Lake Lesjaskog came under intense attack from 25 April, beginning with one He 111 of *Stab LG 1* which destroyed four Gladiators. During the day there were three *Gruppe*-size attacks totalling about 60 sorties by *LG 1*, some escorted by the *Zerstörer* of *ZG 76* and *Z./KG 30*, and by the end of the day nineteen British aircraft (Gladiators and Skuas) had been destroyed on the ground and the airfield made untenable. Only two of the attackers were destroyed and one was written off because of the damage received. Five surviving Gladiators moved to a new landing ground at Setnesmoen, where two were promptly lost to accident and air attack. The Gladiators, which had already flown 49 sorties without an early warning network and with the most rudimentary facilities, were now without fuel. The following day the RAF destroyed them and withdrew the men to re-form the squadron in Britain.

The destruction of the land-based fighters and the inadequacy of carrier-borne ones allowed the *Luftwaffe* to make sustained attacks upon both ports, often in *Gruppe*-size formations. To direct operations a forward command, *Fliegerführer Stavanger*, was created under the newly promoted *Kommodore* of KG 26, *Oberst* Robert Fuchs, with 2. and *Z./KG 30*, *I/ZG 76*, part of *II/JG 77*, 1. and 3. *(F)/ObdL*, 1. *(F)/120*, 1. *(F)/122* and 1.*KüFlGr 106*.[40] Dynamic German leadership, aided by 'aggressive and effective' air support,[41] especially by *KG 4*, sent the Allies reeling from 21 April when the expeditionary forces encountered German battle groups moving north: Pellengahr's from Oslo met the Åndalsnes-based force while Woytasch's troops from Trondheim engaged the Namsos-based force at Steinkjer.

The air attacks were rarely mounted in great strength and they caused few casualties (one battalion from Åndalsnes suffered only eleven), but they disrupted supplies and forced headquarters to make frequent moves. Where artillery support was weak, the Germans shielded their own troop movements by using low-flying aircraft like the *Schlasta* of the Great War to machine-gun Allied positions.[42] At Steinkjer *KüFlGr 106* helped to monitor Allied troop movements before the German attack and its seaplanes were seen to drop messages to Woytasch's infantry.[43] As the demoralized Allied troops retreated towards Namsos, the *Luftwaffe*'s aerial surveillance became so tight that, to evade it, some soldiers threw away their steel helmets in a vain attempt to look like civilian refugees.[44] Relief came only when heavy snow storms grounded the aircraft.

With both ports untenable, the Allies decided to abandon them, covering the evacuation by attacks upon Trondheim by HMS *Ark Royal* (28 April) and upon Stavanger by the RAF (30 April). These Parthian shots cost the *Luftwaffe* eight aircraft and the life of *Hauptmann* Günther Reinecke, *Kommandeur* of *I/ZG 76*, whose aircraft crashed after he had intercepted a Wellington formation. The carrier was making its second attack on Trondheim in three days, the first inflicting heavy damage on the facilities at Vaernes and destroying a dozen aircraft, seven from the newly arrived *I/StG 1*. One-third of the attackers were lost to *Flak* or fuel exhaustion, but the Germans had to draft in 800 Norwegian civilians for repair work.

Ark Royal then had to withdraw for aircraft maintenance and to rest her aircrew, leaving the defence of the two bridgeheads to intermittent patrols of RAF Blenheim IFs. On 30 April the *Luftwaffe* rendered the Norwegian airfield at Setnesmoen untenable and destroyed two Fokker C.Vs, but a token national air presence remained in central Norway until 8 May, when ten aircraft (half of them Tiger Moths) were transferred to the north. The unchallenged *Kampfgruppen* mercilessly exploited the situation with *KG 4*, *KG 30*, *LG 1* and *KGr 100* (the last moving to Trondheim-Vaernes) hammering Åndalsnes while *StG 1* concentrated upon Namsos, flying 49 sorties on 1 May to sink the destroyers *Bison* (French) and *Afridi* (British). In directing and supporting these operations, Harlinghausen played a prominent role and frequently flew personal reconnaissance missions in an He 115. The last stages of the evacuation were covered by 31 fighters from HMS *Ark Royal* and the newly returned HMS *Glorious*, despite a sustained effort by the *Luftwaffe* to sink the carriers. Some 100 sorties were flown, many by *StG 1*, but none was successful and the bloodied and battered Allies completed their withdrawal by 4 May, on which day Harlinghausen received a well deserved *Ritterkreuz*, an award soon made also to *Hauptmann* Paul-Werner Hozzel, *Kommandeur* of *I/StG 1*, who had distinguished himself in Poland.

Luftwaffe morale received a further boost the following day, 5 May, when two Ar 196s of the Aalborg-based *5./BordFlGr 196* discovered the damaged submarine HMS *Seal* trying to seek the sanctuary of Swedish waters. Led by *Leutnant zur See* Günther Mehrens, they bombed the submarine, stalling her engines, and forced the commander, Lieutenant-Commander Rupert Lonsdale, to surrender. The submarine was then towed to a German-occupied port.[45]

With the end of the threat to Trondheim, *ObdL* now pressed for the return of units for '*Gelb*'. Milch received Göring's promised summons and returned on 7 May to Germany (where he, too, received the *Ritterkreuz*), to be replaced by Stumpff, whose command of *Luftflotte 1* passed to Keller. Also transferred south were *KG 4*, *KG 30*, *KG 54*, *LG 1*, *I/ZG 1* and most of the transport *Gruppen*. As compensation, a *Staffel* of *I(J)/LG 2* arrived at Aalborg to replace *ZG 1* and was joined by *2./ZG 76*, while *5. and 6./JG 77* moved to Trondheim-Vaernes. Stumpff's command was reorganized and controlled Harlinghausen's *Fliegerführer Stavanger* with 79 fighters and reconnaissance aircraft and Fuchs' *Fliegerführer Drontheim*, the latter with 193 aircraft and the whole strike force. Geisler's *Fliegerkorps X*, with 177 aircraft including 85 transports, supported Stumpff but remained under *ObdL* until 15 May, when it was transferred to Oslo to co-ordinate operations between Harlinghausen and Fuchs (for the order of battle on 10 May, see Appendix 10).[46] There was some compensation in mid-May with the return of *6./KG 30* from *Fliegerkorps IV* followed by the remainder of *II/KG 30* during the month, but Süssmann was transferred to command the air transport forces of *Fliegerdivision 7*. With *'Gelb'* absorbing most of the *Luftwaffe*'s transport resources, air traffic to Oslo declined from a daily average of 30 sorties at the beginning of May to half a dozen by the end of the month.[47]

The threat to Trondheim was not the German High Command's only cause for concern, for until the very end of the campaign a force of some 4,500 men was isolated in Narvik. It was led by *Generalmajor* Eduard Dietl, Commander of the former Austrian *3. Gebirgsjäger Division*, who had taken the port easily with one of his mountain regiments. But prompt action by the Royal Navy saw the supporting naval task force annihilated, and although Dietl equipped the 2,500 sailors with captured weapons, he remained 400 miles (650km) from the nearest German garrison at Trondheim.[48]

With the British landing at Harstad on 14 April, the prospect grew of a blow to German military prestige as *'Gelb'* was imminent. Such a defeat would be doubly embarrassing for Hitler because Dietl was an old acquaintance, having been a participant in the Beer Hall *Putsch* of 1923. After several days' prevarication Hitler ordered Dietl on 18 April to hold on for as long as possible and placed him under *OKW* rather than Falkenhorst, although Falkenhorst was told that 'In co-operation with *Luftflotte 5*, every available means are to be used to get supplies to it as quickly as possible.'[49] However, the *Luftwaffe*'s hands were tied until Trondheim-Vaernes had been secured because few German aircraft could reach Narvik without operating at the limits of their range. The handful of FW 200 Condors of *1./KG 40* proved especially valuable, not only for maritime reconnaissance but also for bombing and even supply drops. Much of the transport effort was conducted by the long-range aircraft of *Hauptmann* Förster's *KGrzbV 108*, especially the Do 26s and Blohm und Voss flying boats which could make solo runs, although these were augmented by a *Staffel* of Ju 52/3m landplanes.

The dangers were shown in the first, abortive attempt to fly in reinforcements, heavy weapons and supplies to Dietl on 13 April. *Oberst* Wilhelm Baur de Betaz, *Kommandeur KGrzbV 102*, led a dozen Ju 52s from Oslo-Fornebu but in bad weather they flew over British warships engaged in dispatching the remnants of the German naval task force. Two aircraft were shot down, and when the remainder landed on frozen lakes around Narvik few were airworthy and most had to be abandoned. One managed to fly off a makeshift runway to Sweden, where it was interned, and Betaz's men followed suit on 19 April, but the neutral Swedes found it politic to allow them unhindered passage back to Germany. In the following weeks the abandoned transports provided target practice for Allied aircraft.[50]

The transport effort continued in spurts, due to bad weather and a lack of resources, from 15 April until 2 June, with 387 sorties flown and thirteen aircraft lost.

Two-thirds of it (265 sorties) was made in a flurry of activity organized by Gablenz before the transfer of most of his transports back to Germany. Three missions delivered two mountain infantry battalions and artillery between 20 and 27 April, then the weather intervened and by 4 May most of the transport *Gruppen* had departed or had received their transfer orders (*KGzbV 1* had returned to Germany by 8 May). *Luftflotte 5*'s reports indicate that *KGrzbV 106* remained in the theatre as late as 12 May, possibly because *KGrzbv 103* had been disbanded a week earlier. With *KGrzbV 107* fully committed to supplying Stavanger and then Trondheim, most of the effort during May was borne by Förster's *KGrzbV 108*. However, as the threat to Trondheim declined, *KGrzbV 107* began to support Narvik from 11 May, dispatching small forces, usually of half a dozen aircraft, at a time.

With few suitable sites for airfields, the Germans now relied upon seaplanes to move supplies and important items such as a 200W radio flown in by a Do 26 on 12 May. The *'Tante Jus'* brought in 584 men who were parachuted into the bridgehead after suitable drop zones had been located by air reconnaissance. Most of those dropped into the bridgehead were paratroops of *FJR 1*, who first landed on 14 May, but on 23 May a detachment of 66 mountain troops came in after a hastily organized ten-day parachuting course. The presence of Allied fighters meant that from 14 May one or two *Ketten* of Ju 88Cs from *Z./KG 30* would escort the transports, although they were never needed. Only once was an attempt made at basing *Luftwaffe* aircraft in Narvik, when four He 115s of *1./KüFlGr 106* stayed overnight on 13/14 April. But on their return journey they accidentally strayed over British warships, whose guns shot down two of their number.

Operations around Trondheim absorbed the energies of the *Kampfgruppen* for most of April, although 13 April saw 59 reconnaissance and bomber sorties flown in the Narvik area, principally by *KG 26*. During the rest of the month some 100 bomber and armed reconnaissance sorties were flown to harass enemy naval power off Narvik, but the only successes were the damaging of the carrier HMS *Furious* on 18 April by an FW 200 of *1./KG 40* and of the French destroyer *Émile Bertin* the following day by *2./KG 30*.

As the threat to Trondheim ceased, Falkenhorst pushed a relief force northwards while the *Luftwaffe*, in the shape of Fuchs' *Fliegerführer Drontheim*, also turned its attention northwards with a flurry of activity. Some 205 bomber sorties were flown by *KG 26* (supplemented by *KGr 100* and *LG 1*) in the first ten days of May, while *KGr 100* sank the Polish destroyer *Grom* and damaged the cruiser HMS *Aurora*. The intervention of the *Kampfgruppen* was timely, for the Allies were making a determined effort to seize Narvik, having previously confined their activities to the islands off the port.

French troops landed during the night of 12/13 May at Bjerkvik, about ten miles (15km) from Narvik, and slowly drove back Dietl's men towards the Swedish border. The German relief force was some 200 miles (320km) to the south, but the *Luftwaffe* again demonstrated its operational-level capability by striking enemy naval power. By the end of May some 380 bomber sorties had been flown, mostly by *KG 26*, *KG 30* and *KGr 100*, and a dozen ships had been attacked. The battleship HMS *Resolution* had to return to base after being hit on 18 May, while eight days later the radar-equipped AA cruiser HMS *Curlew* was sunk by *KGr 100* while protecting the construction of a new airfield at Skånland, 25 miles (40km) north-west of Narvik. British hopes of stopping the German relief force were dashed during the night of 14/15 May when *I/KG 26* sank the Polish liner *Chrobry* (11,500grt), killing most of the officers of a British Guards battalion and destroying the unit's equipment as well as the only British tanks in the theatre.[51] The headquarters of the French troops in Narvik was attacked on several occasions, and on 21 May one of its battalions, *6e Chasseurs Alpins*, suffered heavy losses to air attack.

The same day relief came to the Allies when HMS *Furious* flew off Gladiators of No 263 Squadron, which went first to Bardufoss, 50 miles (80km) north of Narvik; some were then transferred to Bodø, 130 miles (210km) south of Narvik, to make space for No 46 Squadron's Hurricanes. This squadron arrived at Skånland on 26 May but found the ground there too soft and joined the remainder of No 263 Squadron at Bardufoss. The British also began to build a bomber base at Banak, 200 miles (320km) north-east of Narvik, but the *Luftwaffe* were the first tenants and it became one of the Germans' primary bases in Norway. Bardufoss was completed by Norwegian civilian labour, although the workmen were summoned by civilian wireless which drew the *Luftwaffe*'s attention to the site. It lacked radar cover, and although the RAF established observer posts these had no effective radios. Out of ignorance, no attempt was made to use the female-staffed Norwegian observer posts, which continued to exist behind the German lines and could have acted like the famed Coast-Watchers of the Solomon Islands in 1942–43.[52]

Without early warning the fighters were hamstrung and the bombers reigned supreme as the Douhetists had forecast. Bodø airfield with its fighter detachment was wrecked on 27 May, mostly by long-range Ju 87Rs, which were designed for carrier use. During the evening He 111s of *KGr 100* struck the town, destroying 80 per cent of the homes and even the well-marked hospital which appeared on German maps. One of those taking part was Horst Gotz, one of the *Wienerbårn* who went to school in Bodø. At a *KG 100* reunion he told a Norwegian officer that he, and others who remembered the kindness of the Norwegians, dropped their bombs in the sea or in the mountains. Meanwhile the Narvik relief force advanced on Bodø, with some close air support from the Hs 126s, and the British evacuated the town on 31 May.[53]

The evacuation was unchallenged by the *Luftwaffe* because the latter was fully committed from 28 May in helping to repel the Allied assault on Narvik. During the first morning sea fog grounded the Bardufoss fighters as the *Stuka*s attacked a British naval task group supporting the assault, twice hitting the flagship, the AA cruiser HMS *Cairo*, and preventing her from providing gunfire support. Later attacks near-missed the cruiser HMS *Coventry* and delayed the deployment of a French battalion. When the fog cleared the RAF flew 95 sorties and shot down two of the attackers (another two fell to naval AA fire) and at the end of the day destroyed the Do 26 V1 and V3 as they unloaded. The surviving pilot of the flying boats was Hans Modrow, who became a night fighter *Experte* with 34 victories and a *Ritterkreuz* holder.

Despite the *Luftwaffe*'s efforts, Narvik fell on 28 May, but the victory came too late for the Allies. The deteriorating situation in France led the British War Cabinet to decide on 25 May to abandon northern Norway, a decision ratified six days later by the Supreme War Council. The only reason for taking Narvik now was to destroy its ore-exporting facilities and once this was achieved the Allies evacuated Norway by

8 June. Yet even here they failed, for the facilities were quickly repaired and the first shipment of ore to Germany was made seven months later.

There was no respite for the *Luftwaffe*, which maintained pressure upon Allied forces in the Narvik area. British fighters took a steady toll, and between 28 and 29 May they accounted for most of the ten aircraft shot down or induced to force-land, including an FW 200. On its last mission before transfer to *Luftflotte 2*, *KGr 100* lost its *Kommandeur*, *Hauptmann* Artur von Casimir, who was captured.[54] In exchange for *KGr 100* Stumpff received *I* and *III/KG 30* from *Fliegerkorps IV* and *II(J)/TrGr 186* from *Fliegerkorps I*, although the *Kampfgruppen* arrived piecemeal and elements continued operating along the Channel coast as late as 31 May, when a Ju 88 was lost over Dunkirk. The British fighters became so serious a menace that much consideration was given to a glider-borne assault upon Bardufoss, but events rendered this unnecessary. To prevent the Allies from exploiting Narvik, *Hauptmann* Beitzke's reformed *1./KüFlGr 706* began minelaying sorties in the fjord from 28 May using He 59Ds. Three missions were flown until 6 June, and these involved about ten sorties but had no success.

There was a brief pause at the end of May, during which Harlinghausen's reconnaissance force was reinforced by a Do 17 and sixteen Do 18 flying boats of *2.* and *3./KüFlGr 406* from *FdL West*. At the same time both sides provided tactical air support for their troops in the Narvik area, with *KG 26* and *KG 30* roving in *Ketten* and *Rotten*, seeking ground targets of opportunity as Dietl's men were pushed nearer the Swedish frontier. On 2 June the *Luftwaffe* returned in force, flying 47 bomber and *Stuka* sorties covered by eight Bf 110s of *I/ZG 76*. British fighters shot down four aircraft, including that of *Oberleutnant* Heinz Böhne, the *Staffelkapitän* of *2./StG 1*, and a *KGrzbV 107* Ju 52 which accidentally strayed into the battle, while Lent shot down a Gladiator. One of the casualties was a *Stuka* from *StG 1* whose gunner, war correspondent *Sonderführer* Brack, continued firing after the aircraft had crash-landed.[55]

The following day the Allied evacuation began, shielded by bad weather and 64 aircraft (33 fighters) from HMS *Ark Royal* and HMS *Glorious*. It was harassed by the *Luftwaffe*, which flew some 70 bomber sorties up to 9 June but was unable to prevent the withdrawal of some 24,500 troops with all their artillery and ammunition and most of their stores. The British even took the fighters, which had flown 638 sorties in the defence of Narvik, the aircraft landing on HMS *Glorious*. Tragically for the British, the carrier was sunk by the battlecruisers *Scharnhorst* and *Gneisenau* on 10 June, the day the remaining Norwegian troops under Ruge surrendered. As the Allies withdrew they were pursued by the *Luftwaffe*, which flew 70 bomber sorties and whose last mission of the campaign took place when *II/KG 26* sank two merchantmen (4,142grt) off the Faeroe Islands.

With the end of the campaign many *Gruppen* were withdrawn southwards, while *KGrzbV 107* was disbanded. *Stab KG 26* became *Fliegerführer Norwegen*, and a new *Stab* was formed under *Oberstleutnant* Karl *Freiherr* von Wechmar.[56] Gablenz's excellent handling of transport operations earned him rapid promotion and by the autumn of 1941 he had become a *Generalmajor*. At that time Milch had begun to reorganize Udet's *Generalluftzeugmeister* establishment, and Gablenz became responsible for production planning. When Udet committed suicide his last note blamed both Gablenz and Milch; nine months later, in August 1942, Gablenz was killed when his aircraft crashed during a thunderstorm.[57]

The successful conclusion of the Scandinavian campaign secured vital supplies of iron ore and provided the *Wehrmacht* with naval and air bases from which to blockade the British Isles. But the gamble came close to failure, and its ultimate success was in no small part due to the determination of relatively junior officers, especially in respect of the *Luftwaffe*, as Oslo demonstrated. The *Luftwaffe* was not debilitated by its losses of 260 aircraft (see Table 22), which included 86 transports (69 Ju 52s), and 1,130 men, of whom 341 were killed and 448 were missing (some being interned in British prison camps). By comparison the Army had suffered 3,800 casualties and the Navy 1,072. The *Luftwaffe* had accounted for 93 of the 169 British aircraft lost, including 43 shot down in air-to-air combat (sixteen and thirteen by *ZG 76* and *JG 77* respectively), 24 by *Flak* and the remainder on the ground. A few Norwegian aircraft had also been destroyed, but the majority were lost on the first day of the campaign.

The *Luftwaffe* acquitted itself well, yet neither Milch nor Stumpff made significant contributions to the operational direction of the air campaign, which remained in the hands of Harlinghausen and Fuchs (as well as Gablenz). The aggressive handling by the *Fliegerführer* of their forces made, as is widely acknowledged, a significant contribution to the German victory by rendering the Allies' maritime lines of communication insecure. In addition to warships (a cruiser, six destroyers and sloops, plus a dozen smaller vessels), the *Luftwaffe* sank 21 merchantmen (58,600grt), mostly Norwegian, ranging from the 335grt *Afjord* (4 May) to the 13,241grt *Vandyk* (9 June) as well as numerous fishing craft. The success of the *Luftwaffe*'s transport force is also worthy of comment, for it helped to tip the balance in favour of isolated garrisons such as those at Narvik and Trondheim, which survived largely thanks to the air bridge.

The *Luftwaffe* had expected to be in the forefront of the battle against the Royal Navy, yet the task proved far harder than Göring (or air power crusaders) had imagined. As the admirals recognized, moving warships were difficult to hit, and even when struck by bombs the damage could be reduced by armour plate. But without an aerial shield even the most powerful warships could not operate effectively in the

---

TABLE 22: AIRCRAFT LOSSES IN THE SCANDINAVIAN CAMPAIGN, 9 APRIL–10 JUNE 1940

Month	German			British
	Enemy action	Accidents	Total	
April	28	68	196	92
May	43	8	51	32
June	10	3	13	45 *
Totals	181	79	260	169 **

Note: * Includes 43 aircraft lost aboard HMS *Glorious*. ** 112 RAF.

shadow of land-based air power—as the Royal Navy had appreciated before the war. The Royal Navy's aggressive use of aircraft carriers had compensated for the weaknesses of British land-based air power and given a tantalizing glimpse of what might have been had the Admiralty taken its responsibilities more seriously.

Surprisingly, the *Luftwaffe*'s air superiority was rarely translated directly to the battlefield. *Hauptmann* Jäger's *2. (H)/10*, for example, was able to operate unhindered and provided valuable support to the Army, but neither *Stuka*- nor *Kampfgruppen* were used for close air support: rather, they were employed for battlefield interdiction or action at operational level, as illustrated by the relatively light Allied losses to air attack.[58] All of these characteristics were being demonstrated in Western Europe as the Norwegian campaign raged on, but on a scale unimagined by Allied leaders as *Fall 'Gelb'* was put into effect.

NOTES TO CHAPTER 8

1. Even British sailors thought that *Ark Royal* had been hit. The author's father, the late Chief Petty Officer L. C. Hooton, was a helmsman in a ship near the carrier and told of hearing a shout on the bridge, 'They've got *Ark Royal*'.
2. See Shores, *Eagles*, pp.77, 85, 88–90; and Bekker pp.65–71.
3. BA MA File RL 2 II/25.
4. *Ibid.*
5. Naval Staff Operations Division War Diary, IWM EDS 229.
6. *Ibid.*
7. *Ibid.*
8. BA MA File RL 2 II/25.
9. See Harlinghausen's report, quoted in Schmidt, *Anlage 10*; and IWM EDS 229, War Diary A, Vol. 6, pp.133–4, 140, 145, 151.
10. See Roskill, pp.137–43.
11. Based on Johnson, pp.11–22.
12. IWM EDS 229.
13. BA MA File RL 2 II/25.
14. The description of *Luftwaffe* minelaying is based upon the article by Marchand and Huan.
15. For the background to and planning of *'Weserübung'* see Taylor, *March*, pp.83–94.
16. Taylor, *March*, pp.95–6.
17. Adams, p.18; Taylor, *March*, p.97n. These events took place on 17, 26, 29 and 28 March.
18. For details of *Luftwaffe* planning see *B Nr 10055/40* of 20 March 1940 in IWM MI 14/800. See also Morzik and Hümmelchen, pp.86–9.
19. Based upon William Green's article 'Sentinel Over the Fiords' (hereafter Green, 'Sentinel'); and Shores, *Eagles*, pp.222–3. I would like to express my thanks to Mr T. H. Holm of the Norwegian Armed Forces Museum for information and photographs.
20. For air operations see Shores, *Eagles*, pp.217–26.
21. For operations in Denmark see Shores, *Eagles*, pp.226–7; Taylor, *March*, pp.112–15; Bekker, pp.79–81; and William Green's article 'Stopper in the Bottle'.
22. Adams, pp.20, 31.
23. For the occupation of Norway see Adams, pp.19–43; Air Ministry, p.60; Mason, pp.322–8; Shores, *Eagles*, pp.228–40; Taylor, *March*, pp.115–25; and Green, 'Sentinel', p.55.

24. Smith, *Air Strategy*, p.96.
25. Lt-Cdr M. H. C. Young in the *Journal of the Royal United Services Institute*, quoted by Smith, *Air Strategy*, p.95.
26. See James, p.130.
27. For the problems of British naval aviation see James, pp.119–32; and Smith, *Air Strategy*, pp.94–103.
28. The section on aircraft carriers has benefited from discussion with Mr Antony Preston and Dr Norman Friedman. The conclusions are the author's.
29 Adams, p.22; and Shores, *Eagles*, p.241. Shores on p.239 credits *8./KG 4* with the *Aeger* and fails to identify the *StG 1* victim. *Aeger* was the only Norwegian torpedo boat sunk by air attack on 9 April.
30. Aders, pp.14–15.
31. Shores, *Eagles*, pp.260–5.
32. IWM AHB 6/4.
33. Adams. p.111; Shores, *Eagles*, pp.266–7; Taylor, *March*, p.135.
34. IWM EDS 229.
35. Adams, p.132.
36. IWM EDS 229.
37. Adams, p. 97.
38. *Ibid.*, p. 173.
39. Figures based on Middlebrook and Everitt, pp.32–8.
40. For details of *Luftwaffe* operations see also *Luftflotte 5* Reports, 1–16 May, in BA MA File RL 2 II/47.
41. Taylor, *March*, p.135.
42. Adams, pp.125, 172.
43. *Ibid.*, p.98.
44. *Ibid.*, p.99.
45. Bekker, pp.90-1; Shores, *Eagles*, p.309.
46. See also BA MA File RL 2 II/47.
47. PRO Air 22/477: Air Ministry W/T Intelligence Service Daily Summary, Vol. 2.
48. Adams, pp.50–1, 54.
49. Derry, p.195; Taylor, *March*, p.139; *Luftwaffe Direktive Nr 23* of 5 May, in BA MA File RL2 II/47.
50. Details of transport operations from Morzik, pp.95–6; and Shores, *Eagles*, pp.261–340.
51. Adams, pp.67–9.
52. Adams, pp.80, 158; Derry, pp.203–5.
53. Adams, pp.81–2, 88–9; Balke, *KG 100*, p.35.
54. Balke, *KG 100*, pp.45–6 (but Shores, *Eagles*, p.340, states that *KGr 100* was transferred after 2 June).
55. Shores, *Eagles*, p.338.
56. Schmidt, p.62.
57. Irving, *Milch*, pp.133, 136, 168–9; Mitcham, pp.160–1.
58. Adams, pp.17, 172.

CHAPTER NINE

SIEG IM WESTEN

May to June 1940

A loud call of 'Everybody up! Briefing in ten minutes!' roused the slumbering airmen of *KG 2* before dawn on 10 May and was echoed in a hundred barracks across western Germany. No one needed to be told that A-Day had arrived. The alert came in the late hours of 9 May in the form of coded teletype warnings that *'Gelb'* would begin the following day, and this led to ripples of activity at air bases along the western border, with commanders contacting staff who then ordered the fuelling and arming of the aircraft. Some three or four hours later the first airmen arrived and boarded the bombers to calls from the 'Black Men' of 'Good luck' or 'Break an arm and a leg', and soon some 500 bombers were taxying to the runways. At about 0245 Berlin time *I/KG 54* led them into the air in an operation which took 45 minutes to complete and which suffered only one setback. At Roth-Kiliansdorf poor visibility caused two aircraft of *1./KG 53* to collide, and their burning wrecks blocked the runway for two hours.

Operations in support of *'Gelb'* actually began late on 9 May when *Kampfstaffeln* of *Fliegerdivision 9* flew 48 minelaying sorties off the Dutch and Belgian coasts, complemented by eleven *KüFlGr 106* sorties, to lay a total of 100 *LMB* mines, but ten aircraft were lost, the majority to accidents. These were severe misfortunes for a *Luftwaffe* which, theoretically, had only a narrow numerical superiority. On paper the *Luftwaffe* had 3,868 available aircraft (excluding 476 transports and 306 corps aircraft) but only 2,776 were serviceable, including the *Jagdstaffeln* of *Jafü 1* (*Deutsche Bucht*) and *Fernaufklärungsstaffeln* (*Heer*), of which Kesselring had 1,343 and Sperrle 1,295 (see Appendix 11). Their four enemies had some 2,600 serviceable aircraft, including seaplanes, but the *Luftwaffe* could engage them piecemeal.

The largest grouping was in northern France, where the *Armée de l'Air* (with elements of the *Aéronavale*) and the BAFF had 1,250 aircraft available which could be augmented by 200 French aircraft in *groupes* completing their re-equipment and substantial RAF forces in Britain. Yet the number of RAF aircraft actually available (i.e. serviceable) bore little relation to the establishment figures. On paper, Air Marshal Arthur Barratt's BAFF had 408 aircraft (297 serviceable), with another 22 Gladiators in squadrons currently re-equipping with the Hurricane. In Great Britain Fighter and Bomber Commands had establishments of 680 and 392 respectively, but the former had only 430 serviceable fighters (including 238 Spitfires and 154 Hurricanes) while the latter had 213 bombers, of which only 72 Blenheims were deemed suitable for daylight operations.[1]

The *Luftwaffe*'s *Staffeln* all had the benefit of modern aircraft, few having entered service before 1936, while their enemies' squadrons were burdened with obsolete or inadequate machines, but the Germans' fundamental superiority was in organization. While reconnaissance units were distributed to every level of Army command down to corps, the fighters and bombers were concentrated into *Fliegerkorps* commands supported by a dedicated signals system which allowed them to react promptly to situations throughout the operational depth. By contrast, enemy fighters and reconnaissance aircraft were divided between Army commands and Air Force-led strike forces (which also had bombers) from which the Army could claim support. This complex but fragile structure fragmented Allied air power to make effective command, communications and control impossible. The *Armée de l'Air* in particular had a labyrinthine structure headed by Vuillemin, who was responsible for strategic decisions, leaving his deputy, *Général* Marcel Têtu, as *Commandant des Forces Aériennes de Coopération du Front Nord-Est* (*FACNE*) responsible for operational-level co-ordination in support of *Général* Jacques Georges' *Théatre d'Opérations de Nord-Est* (*TONE*). But until the Germans attacked he had no contact with his three regional commands (*ZOAN*, *ZOAE* and *ZOAS*), whose forces were at the disposal of the Army Group (*Groupe des Armées*) or Army (*Armée*) commanders as well as providing strategic defence.

The British Army required air support on the battlefield such as the *Armée de l'Air* provided, but, in the Douhet tradition, the RAF remained obsessed with destroying both the enemy's means of production and his will. The RAF's plans created dissension at the highest Allied councils because they called for attacks upon enemy communications and selected industrial targets, not only to stop the German war effort but also to provide direct relief for the troops by goading the *Luftwaffe* into retaliation against the British Isles. The fear of such retaliation falling upon their heads did not make the French receptive to such concepts, but in the face of British obstinacy they agreed in April 1940 that the RAF would begin a strategic air campaign upon Germany if Hitler struck in the west.

Allied plans to meet this contingency were based on an accurate anticipation of the original '*Gelb*' plan and no allowance was made for any major revisions which the Germans might introduce. When the German offensive began, the Allied left (*1er Groupe des Armées* under *Général* Gaston Billotte, with the BEF attached) would swing into Belgium like a door hinged at Charleville and Sedan, which were held by *2e Armée* and *9e Armée*. Billotte had the largest concentration of French forces (37 divisions), including 78 per cent of the mobile divisions and, shielded by *ZOAN*, he would join the Belgians to contain the German advance on either the River Dyle or the River Escaut. The *7e Armée* at the extreme left would be catapulted into the southern Netherlands to link the Dutch and Belgian Armies, creating a continuous barrier from the Zuider Zee (Ijsselmeer) to the Swiss border.

The *Luftwaffe*'s first objective in overcoming such a barrier was to gain air superiority by wrecking enemy air bases and facilities as well as destroying aircraft on the ground. At dawn the *Kampfgruppen* attacked 47 French, fifteen Belgian and ten Dutch airfields and claimed great success. After debriefing the crews, the two *Luftflotten* claimed to have destroyed between 579 and 829 aircraft on the ground, including between 355 and 605 in hangars. In fact only 210 aircraft were destroyed, mostly in the Low Countries by *Luftflotte 2*, and Sperrle's bombastic claims of 240 to 490 aircraft were a fairy tale.

The Dutch *Luchtvaartafdeling* (*LVA*) and its naval equivalent the *Marine Luchtvaart-dienst* (*MLD*) proved to be the hardest nuts to crack: they had anticipated an attack since 2 May. Dutch expenditure on air defence had been low, Fl26,301,000 ($14,450,000) between 1933 and 1939 or 3 per cent of the total defence budget, but some modernization was taking place.[2] Re-equipment began in 1938 with orders for Fokker G.I and Koolhoven FK.58 fighters and Fokker T.V medium bombers, while Douglas DB-8A-3N light bombers had been imported for use as two-seat fighters.

The task of eliminating Dutch air power was assigned to *Oberst* Martin Fiebig's *KG 4*, and in an attempt to achieve surprise he attacked from the sea and destroyed 62 Dutch aircraft, half of them trainers at De Kooy, at the cost of eleven bombers, including his own. Fiebig, who had commanded *Bogohl 9* in 1918, spent the next five days as a prisoner and was temporarily replaced by *Oberstleutnant* Hans-Detlef Herhudt von Rhoden. The only real successes were at Amsterdam-Schipol, where the Dutch lost half their bombers (four T.Vs), and at Hague-Ypenburg, where *I/KG 4* destroyed half the 21 defending fighters to assist *KG 30* and *KG 54* in their attacks upon ports and communications.

The escorting fighters, from *JG 26* and *ZG 26*, shot down 25 aircraft (*Luftflotte 2* claimed a total of 41), but the *Luftwaffe* suffered heavily. Dutch fighters, which flew 87 of the 150 *LVA* sorties during the day, claimed most of the 21 *Luftwaffe* combat aircraft lost over the Netherlands. The *LVA* was left with only 70 aircraft by dusk, although they were well dispersed and conducted an aerial guerrilla war which destroyed another thirteen *Luftwaffe* aircraft during the next five days.[3]

Fliegerkorps IV, led by *LG 1* under *Oberst* Alfred Bülowius, had greater success against the *Aviation Militaire Belge*. Since 1936 Belgium had adopted a policy of neutrality, but most of the country's defence funds had been spent improving fortifications, delaying the modernization of the air force until late 1939. Belgium purchased fifteen former RAF Hurricanes but, as mentioned above, lost three to a Do 17 in March. Licence production arrangements were made for another 80 Hurricanes together with 32 Breguet 694 reconnaissance and Caproni Ca 335 *Maestrale* (SABCA S.47) fighter-reconnaissance aircraft, while Renard produced a number of promising fighter prototypes. Meanwhile 34 Fiat CR 42s, 40 Brewster 339s (based upon the US Navy's F2A) and twenty Koolhoven FK.56 fighters, together with 24 Ca 312 light bombers, were ordered as stop-gaps.

Only the Fiats had arrived by May 1940, and most Belgian aircraft were thus obsolete, 40 per cent of them versions of the Fairey Fox light bomber of the 1920s, including a single-seat fighter, the *Kangourou*. Nine *Kangouroux* were the only Belgian fighters to meet the *Luftwaffe* onslaught, but the Bf 109s shot down three and reduced the remainder to scrap. Bülowius's men had the good fortune to find four of the seven Belgian fighter squadrons assembled at Schaffen for exercises and promptly annihilated three of them. The survivors fled to Brusthem, to arrive immediately before another *Luftwaffe* attack which left the Belgians with only a Fiat fighter squadron operational. The destruction of two reconnaissance squadrons at Neerhespen brought the total number of Belgian first-line aircraft destroyed to 83. The only survivors, apart from the Fiats, were three Battle squadrons and six well-dispersed corps squadrons. These flew 112 sorties until 28 May but contributed nothing to the air campaign.[4]

The assault on French air bases was ruined by the inadequacies of the *Luftwaffe* reconnaissance and intelligence organizations. Of 91 airfields with first-line units in northern France only 31 were attacked (together with sixteen without first-line units), and fourteen of these housed corps squadrons (one also held an RAF fighter squadron). Thirteen of eighteen bomber bases (including *Aéronavale* dive bomber units) were untouched and significant damage at the other five occurred only at Boulogne-Alprecht, where twelve Vought dive bombers of *AB 3* were destroyed, and Roye-Amy, where *GB II/34* lost five Amiot 143s. The French lost 65 aircraft, with two *groupes de chasse* decimated, and the RAF lost none (although several were damaged),

---

TABLE 23: ATTACKS ON FRENCH TARGETS, 10 MAY–8 JUNE 1940

Date	Airfields	Railways/ stations	Factories/ camps	'Localities'	Ports
10 May	47	17	16	45	–
11 May	23	4	3	14	–
12 May	11	5	–	24	–
13 May	4	8	3	12	–
14 May	6	10	–	7	–
15 May	16	22	6	12	–
16 May	7	7	6	26	–
17 May	3	28	11	38	–
18 May	2	10	2	14	3
19 May	1	15	7	10	2
20 May	3	12	–	12	–
21 May	–	12	–	12	3
22 May	3	12	–	17	–
23 May	1	18	–	6	–
24 May	1	20	–	20	–
25 May	6	24	–	12	1
26 May	6	18	2	15	1
27 May	2	9	1	9	2
28 May	–	–	–	–	–
29 May	2	1	–	3	–
30 May	–	–	1	6	1
31 May	–	–	–	–	–
1 June	6	20	7	11	1
2 June	8	15	11	22	4
3 June	15	28	10	10	–
4 June	–	–	–	–	2
5 June	5	13	2	22	2
6 June	6	8	4	24	2
7 June	5	18	2	40	2
8 June	3	16	3	35	1

Source: *GQGAé* War Diary, *SHAA* File 2D2.

despite attacks upon four bases. Ten airfields (one RAF) in *ZOAN* were struck, mostly by *Luftflotte 2* crews, with *KG 54* of Putzier's command and *Fliegerkorps IV* distinguishing themselves at Boulogne-Alprecht, Calais-Marck and Cambrai-Niergnies, where a total of 25 aircraft were destroyed. *Luftflotte 3*'s failure to identify and to target fighter and bomber bases rendered its opening missions virtually useless, although this played no part in the day's great tragedy.

In thick cloud *III/KG 51* (*Fliegerkorps V*) made afternoon attacks upon Dijon-Longvic and Dôle-Taraux airfields. *Leutnant* Paul Seidel's *Kette* from *8./KG 51* lost the main formation and inadvertently re-crossed the Rhine. At his estimated time of arrival Seidel spotted what he believed was the target through a gap in the clouds and promptly attacked it. Unfortunately he was 200km (125 miles) off course and the bombs fell upon Freiburg, killing or wounding 158 German civilians, including 42 children. The observer network quickly identified the culprits, but Goebbels blamed the French and the raid helped release the restraints upon attacking civilian targets. Seidel himself was distraught at the tragedy and no doubt welcomed death when it claimed him and 25 of his comrades over Portsmouth on 12 August 1940.[5]

With the enemy fighter force unscathed, the second wave, totalling some 500 bombers from both *Luftflotten*, suffered badly. It had taken off from about 0415 hours to disrupt the enemy army, and Vuillemin's headquarters noted that, during the day, seventeen rail targets and sixteen factories or camps together with 45 other targets described as 'localities'—mostly communications centres, strongpoints or head-quarters—were attacked (see Table 23). The alerted Allied fighters reacted vigor-ously, flying 541 sorties to destroy most of the 39 bombers and three fighters which Sperrle lost during the day at a cost of nineteen aircraft, twelve of them French.[6] *Fliegerkorps II* alone lost 23 aircraft (nineteen Do 17Zs)—the heaviest daily losses of any *Fliegerkorps* during the campaign—and *KG 3* lost eight Dorniers to the Hawks of *GC I/5*, while other fighters dispatched a further eight of *KG 2*.

However, these battles tied down most of the Allied fighters in France just as Billotte's forces moved into Belgium, their progress monitored by *Luftwaffe* recon-naissance. Small numbers of Allied fighters were encountered by the *Luftwaffe* over the Low Countries, the majority coming from RAF Fighter Command's No 11 Group in England, which mounted 78 sorties. In the early afternoon a *Staffel* of *ZG 26* bounced six Blenheim 1F fighters of No 600 Squadron as they swept in to attack the newly occupied Rotterdam-Waalhaven airfield and destroyed five.

The seizure of Waalhaven was one of five airborne operations that day; only two were successful, and the remainder proved to be a shambles. Despite hitches, Koch's operations went well, with all three Maastricht bridges taken to open the way into the Belgian plain while *Oberleutnant* Rudolf Witzig's group landed at Fort Eben Emael (minus Witzig himself, who arrived late because his glider tow rope broke). The outnumbered Germans had to fight hard, and the fort did not fall until the following afternoon after *StG 2* and *II (Schl)/LG 2* thwarted efforts by the *7e Division d'Infanterie* to relieve its garrison. Koch's detachment lost 44 killed but took 1,940 prisoners to earn both him and Witzig the *Ritterkreuz*.[7]

In support of Kleist's advance through the Ardennes the Army extemporized two operations. *Leutnant* Wenner Hedderich flew in volunteers from *34.Infanterie Divi-sion* in 25 Fi 156 *Störche* to seize five crossings in Luxembourg which controlled routes to France. The more elaborate *Unternehmen 'Niwi'*, involving *Oberstleutnant* Eugen Garski's *III. Bataillon* of the élite *Gross Deutschland Regiment* was a farce. Crossroads at Nives and Witry were the objectives, and the *Luftwaffe* provided 98 *Störche* and three Ju 52s for re-supply missions, but the men landed at the wrong spots. They then cut telephone lines, which prevented Belgian troops in front of *XIX. Korps* from receiving orders to withdraw, with the result that the *Panzer* divisions encountered fierce resistance.[8]

The largest airborne operations were in the Netherlands and involved Sponeck's *Gruppe Nord* (based on his *22. Division*) and Student's *Gruppe Süd* (based on *7. Fliegerdivision*) with four paratroop and nine infantry battalions. Sponeck, with 9,377 men, was to decapitate the Dutch by taking The Hague, the Royal Family, the Government and the High Command; Student, with 8,100 men, was to secure the crossings across the three mighty rivers between Rotterdam and Moerdijk and be relieved by *9. Panzer Division* of *General* Rudolf Schmidt's *XXXIX. Armee Korps (Motorisiert)*. Paratroops would seize air bases and their bridgeheads would be consolidated by flying-in Sponeck's troops, although these would have to arrive in two waves three hours apart. Morzik's *KGzbV 1* was largely responsible for supporting Student while Conrad's *KGzbV 2* supported Sponeck.

Poor staff work and poor intelligence—for which Wenniger as Air Attaché must bear some responsibility—quickly wrecked the carefully crafted scheme, bringing disaster to Conrad and Sponeck. Anticipating an airborne assault, the Dutch had established an integrated air defence command under Lieutenant-General P. W. Best on 1 November 1938. Air bases received infantry garrisons, and following the *Luftwaffe*'s use of airborne troops in April the defences were strengthened, while obstacles were also placed in fields and on roads around The Hague.[9] The scale of the German assault, which followed hard on the heels of bombing attacks, surprised many Dutch airfield garrisons but they quickly recovered to conduct a defence, supported by the two-division strategic reserve, with similarities to the German response to Operation 'Market Garden' in 1944. *Gruppe Süd*, with its more experienced troops and air transport units, seized the crossings (an infantry company brought in by *Staffel Schwilden* He 59s took the bridge at Rotterdam) but were hard pressed by the Dutch. *Gruppe Nord* paid the greatest price, aggravated by the relative inexperience of most of its troops and *Staffeln*. Kesselring sent a terse signal to Göring: 'As far as we can ascertain, 22nd Airborne Division operations . . . are near-failure due to strong ground defence and enemy AA artillery.'[10] The paratroops seized the three Hague air bases (Ockenburgh, Valkenburg and Ypenburg) but the transports carrying Sponeck's troops had to pass through a hurricane of ground fire which shot down 25 aircraft.

The Dutch air bases were built on soft ground with a thin crust of turf and were used only for light, single-engine aircraft, and the surviving transports sank into the ground upon landing, rendering the airfields unusable. The follow-up aircraft tried their luck on nearby beaches with exactly the same result, and 96 were lost to this cause or to Dutch shellfire. Kesselring lost contact with Sponeck, whose radio disappeared in the confusion, and *Leutnant* Wolfgang Ludewig of *9./JG 26* was ordered to fly to Ypenburg to find the general. His Bf 109 landed on the beach and became stuck, forcing him to waste five days as an infantryman. After dusk an intermittent radio link was finally established using a portable Army set, and Kesselring ordered Sponeck to reinforce the Rotterdam garrison.[11]

The *Luftwaffe*'s air superiority over the Netherlands had little effect upon Dutch resistance and the Dutch III Corps was able to withdraw virtually unhindered in daylight to the fortifications of the Grebbe Line, where it fiercely resisted the invader. This apparent immunity was because most of *Luftflotte 2* was concentrating on Belgium, where even some of Putzier's bombers were operating, but during the afternoon approximately half of Richthofen's *Fliegerkorps VIII* was assigned to supporting Sponeck's and Student's troops. Otherwise Richthofen's men provided direct and indirect support to Reichenau's *6.Armee*, striking communications and barracks. In some *Stuka* units the crews flew six sorties while *Hauptmann* Otto Weiss's *II (Schl)/LG 2* flew eight, but losses were light at twelve aircraft, including seven *Stuka*s of *I/StG 76* to AA fire. Towards the evening there was a major attack on Antwerp to prevent the Allies from shipping-in reinforcements or French traffic from reaching the southern Netherlands through the Schelde Tunnel. *Fliegerkorps IV* provided more distant support, attacking Belgian roads and railways.

Meanwhile Kleist's troops rumbled through the Ardennes, shielded by Massow's *Jafü 3*. The French flew only eleven reconnaissance sorties during the day, losing two aircraft, and Massow's men had little difficulty holding them (and the Potez 63s of the corps units) at arm's length. Nevertheless, word filtered through to Allied headquarters, whose bombers posed a serious threat. The response was feeble, the entire force in France flying only 44 sorties on a day when the *Kampf-* and *Stukageschwader* flew at least 1,500 each. Têtu was willing to use his bombers, but Georges forbade their use against targets near towns and villages for fear of causing civilian casualties. Barratt encountered similar objections, but at midday he authorized the AASF to strike armoured columns in Luxembourg and 32 sorties were flown. However, Massow's fighters and *Flakkorps I* proved too efficient and shot down eighteen of the vulnerable bombers, which inflicted only minor damage.

By the end of the day the situation remained finely balanced. The *Luftwaffe* had lost 128 aircraft and destroyed in the air 56 British and French, whose strength remained generally undiminished and, therefore, a potential threat to the German campaign. The problem for the *Luftwaffe* was how to recover from the first day's failures and win air superiority while simultaneously supporting the Army on a battlefield covering thousands of square kilometres. Against the odds it succeeded. During the next four days it capitalized on its assets and with rapid turn-around times exploited its slim numerical superiority to fly more sorties per aircraft. Most crews managed two a day, while the *Stuka*s and *Schlachtflieger* flew about half a dozen. The unsung heroes were the 'Black Men' and the personnel of the supply organization, who worked long hours to ensure that the maximum number of aircraft were available, then snatched what food and sleep they could. Yet the strains upon machines were as great as those upon men, and serviceable strength declined: for example, *JG 27* had 85 Bf 109s serviceable on 12 May compared with 90 (not 101 as the *JG 27* history claims) on 10 May.

The bombers, which flew between 800 and 1,000 sorties daily from 11 to 15 May (compared with 1,500 on A-Day), were largely committed to attacking communications and could rarely be spared for attacks on air assets, but, while fewer airfields were attacked (as shown in Table 23), selective targeting ensured that greater damage was

TABLE 24: ALLIED AIR OPERATIONS, 10–15 MAY 1940

Date		Fighter		Bomber		Recon	
Units based in France		**Fr**	**RAF**	**Fr**	**RAF**	**Fr**	**RAF**
10 May	Sorties	363	178 (47)	12	32	11	12
	Losses	12	7 (4)	–	18	2	4
11 May	Sorties	418	179 (40)	17	11	24	7
	Losses	16	12 (2)	3	8	1	4
12 May	Sorties	508 /263	128 (38)	35	35	22	12
	Losses	13	9 (5)	10	11	4	–
13 May	Sorties	383 /260	133 (69)	36	7	9	3
	Losses	13	7 (2)	–	2	2	2
14 May	Sorties	429 /340	201 (33)	40	73	18	25
	Losses	22	15 (6)	4	41	1	2
15 May	Sorties	529 /254	191 (45)	18	11	16	5
	Losses	20	20 (11)	–	–	3	2
Totals	**Sorties**	**2,630/1,661**	**1,010 (272)**	**158**	**169**	**100**	**64**
	Losses	**96**	**70 (30)**	**17**	**80**	**13**	**14**

Date		Fighter		Bomber	
British-based units		**Defensive**	**Offensive**	**Day**	**Night**
10 May	Sorties	73	30	33	45
	Losses	–	6	3	–
11 May	Sorties	22	48	22	37
	Losses	–	6	2	3
12 May	Sorties	18	–	32	12
	Losses	–	–	17	–
13 May	Sorties	17	15	–	12
	Losses	–	6	–	–
14 May	Sorties	37	9	34	42
	Losses	–	–	7	1
15 May	Sorties	4	–	28	111
	Losses	–	–	3	1
Totals	**Sorties**	**171**	**102**	**149**	**259**
	Losses	**–**	**18**	**32**	**5**

Note: Sorties by French corps squadrons (including those by Potez 63-II) are not available and those of the RAF have been omitted. In the fighter statistics the first sortie figures for French forces are the author's calculations and the second are Martin's, while the bracketed British figures are the AASF contribution. British casualties are from Bingham, and were collated by Mr John Foreman; French casualties are from Martin. All losses are for air operations.

Sources: *GQGAé* War Diary, *SHAA* File 1D44/1; *ZOAN* War Diary, *SHAA* File 2D17; Martin, Annex 2; Paquier, *L'Aviation de Bombardement Française*; Report of AASF, PRO Air 25/255; Dispatch of Air Marshal A. S. Barratt, PRO Air 35/192; No 11 Group Operations Record, PRO Air 25/193; Franks, *Valiant Wings*; and Middlebrook and Everitt, pp.41–54.

inflicted. On 11 May there was a brief flurry of raids involving some 30 airfields, in one of which—that on Conde Vraux—*4.* and *5./KG 2* destroyed eight of the AASF's 24 Blenheims. French figures show the destruction of 68 fighters and 31 bombers on the ground between 11 and 20 May,[12] while the AASF lost at least a dozen Battles on the ground between 11 and 15 May. The pressure behind their own lines helped to prevent Anglo–French air forces from moving to Belgian airfields, although such a deployment was anticipated by the *Luftwaffe*, which continually monitored known and suspected airfield sites in Belgium.

The pressure upon Allied air forces was maintained not only by bombers but also by fighters. Although the *Jagd-* and *Zerstörergruppen* were unable to match the first day's level of activity, some 2,000 sorties, they produced a respectable daily average of some 1,500 until 15 May. Döring's *Staffeln*, under *Stäbe JG 26, JG 51* and *ZG 26*, not only escorted the bombers but increasingly conducted aggressive sweeps behind enemy lines, with *JG 51* covering the Netherlands while *JG 26* operated in northern Belgium. Between 11 and 13 May *Luftflotte 2* claimed 82 victories and *Flakkorps II* claimed 57, so that the *Luftwaffe*'s situation report of 13 May noted that enemy fighter activity in northern Belgium was weakening. In the same period Massow's fighters (*JG 2, JG 53* and *ZG 2*) covered southern Belgium and north-eastern France (with *JG 2* primarily involved over the Ardennes) to claim 66 aircraft, with another 22 by *Flakkorps I.*

The Germans remorselessly gained air superiority between 11 and 15 May as they engaged the Allied fighter force, 518 strong at the beginning of the campaign, in a battle of attrition which took place increasingly behind enemy lines. The *Luftwaffe* destroyed 147 fighters and lost 80, despite the gallantry of both British and French pilots, who were flying an average of only 620 sorties a day (some sources suggest 470). Despite substantial reinforcement, Têtu's fighter pilots rarely flew more than one sortie a day including strategic defence missions. On 13 May, for example, *ZOAN* flew 220 fighter sorties, including 29 in the Seine valley.[13] French records are contradictory but suggest a maximum of 2,267 fighter sorties in this period, although one French historian gives a figure of some 1,500 sorties.[14]

Aware of the threadbare fighter coverage, the French Government made increasingly strident demands for British help as Vuillemin stripped quiet fronts to reinforce Têtu. The newly appointed British Prime Minister, Winston Churchill, felt honour-bound to meet these demands despite protests from Dowding, who resented the erosion of the understrength Fighter Command. With the start of the German offensive the BAFF returned 22 Gladiators to operational status (their pilots were scheduled to convert to Hurricanes) and dispatched a total of 128 Hurricanes, which operated from French bases daily from 10 to 14 May. The British had an average strength of some 120 fighters (21 per cent of the Allied total) but flew 832 sorties (27 per cent of the total) between 11 and 15 May while the British-based No 11 Group flew another 72 over the Dutch and Belgian coasts.[15] Dowding's worst fears were confirmed, and by dusk on 15 May the RAF had lost 76 fighters (including accidents), 48 from the original BAFF squadrons which began the campaign with only 65 Hurricanes (see Table 24 for Allied air activity from 10 to 15 May).

Even more gratifying to the *Luftwaffe* leadership was the decimation of the Allied bomber force over Maastricht as *6.Armee* flowed through the city defended by

Flakkorps II and nine *Jagdstaffeln* under *Oberstleutnant* Max-Josef Ibel's *Stab JG 27*, part of Richthofen's *Fliegerkorps VIII*. Attacks by all four enemy air forces totalled 74 bomber sorties from 10 to 12 May and 33 were shot down, together with a few reconnaissance aircraft. Ibel's men claimed half the bombers, three each by *Hauptmann* Adolf Galland, *Oberleutnant* Gert Framm and *Oberleutnant* Walter Adolph, for the loss of five fighters and on 12 May alone flew 30 missions (350 sorties). The bombers which evaded the fighters had still to run the gauntlet of automatic fire from *Flak* batteries, which destroyed eleven out of eighteen Bre 693s on 12 May, the last day of Allied attacks. These were abandoned because the defence was simply too strong, and by 12 May had helped reduce the AASF, the Allies' primary bombing force, from 135 to 72 serviceable aircraft. The defensive successes ended the day-bombing threat to the *Reich* and allowed the *Luftwaffe* to release *Jagdgruppen* held back to defend vital areas of Germany.

Luftwaffe operations were now following a daily pattern which continued until the fall of France. Around dawn, as *Fliegerdivision 9* and *FdL* aircraft returned, the *Wekusta* set off towards the enemy coasts to obtain all-important meteorological data. Between 10 and 31 May the RAF's radio interception service detected 36 sorties by *Wekusta 26* and a similar number were flown by *Luftflotte 3*'s *Wekusta 51*. Even as they radioed back to base, the reconnaissance effort would begin with a wave of 40–50 sorties having a clear boundary between the operations of *Luftflotten* and *Heeres-aufklärungsstaffeln* and those of *AufKlGr ObdL*, the original boundary line running between Amiens and Dijon. The daily total of such sorties would be 100–150, and some were flown by the *Kampfgeschwader*, which sent out a *Kette* or two on armed reconnaissance missions to attack targets of opportunity.[16]

Attacks in support of the Army took place an average of 75km (45 miles) behind enemy lines, with bomb-lines drawn to shield friendly troops whose spearheads marked their positions with signal panels or flares or by draping the *Swastika* flag over vehicles. Nevertheless, the failure after the Polish Campaign either to redefine *Koluft* and *Flivo* responsibilities or to establish an inter-service signal network bedevilled air–ground co-ordination and there was also the problem, as Guderian's *XIX.Korps* discovered only in March 1940, that there were not even common maps! The last difficulty was resolved when staffs modified standard Army 1:25,000 scale maps, but the co-ordination of air strikes was achieved only when *Fliegerkorps* leaders flew to Army command posts. This was time-consuming, tiring (Richthofen recorded in his diary on 19 May that he was 'dead tired') and dangerous (as Richthofen discovered on 22 May when a leaking fuel tank forced him to make an emergency landing in his *Storch*). In the circumstances, the *Luftwaffe* preferred to operate deep behind the enemy lines, where it had a free hand.

Keller's *Fliegerkorps IV*, for example, struck Belgian road and rail centres both to harass Billotte's forces as they moved to the Dyle Line and to undermine their defence. Some aircraft attacked shipping supporting *7e Armée* in the southern Netherlands and around the Schelde Estuary, while Putzier's bombers struck every Belgian port. Belgian air power was by now practically non-existent, but British and French fighters (the latter including *Aéronavale* Potez 631s) were extremely active and destroyed 30 of Keller's and Putzier's aircraft between 11 and 13 May. Yet it was the Netherlands which remained the *Luftwaffe*'s ulcer, with the *LVA* flying 182 sorties for the loss of thirteen aircraft to fighters and *Flak* by 14 May (others were lost to 'friendly' AA or accidents). Putzier, reinforced by *StG 77*, attacked Dutch airfields on 11 May yet the following day the Dutch defiantly flew 69 sorties, compared with 45 the previous day.[17]

Dutch resistance, especially in the eastern fortifications, exasperated the German High Command. Halder noted in his diary on 14 May: ' The important thing now is to have *Heeresgruppe B* liquidate Holland speedily', while the previous evening the Commander of *18.Armee*, Küchler, ordered resistance at Rotterdam to be broken 'by every means'. Rotterdam held the key to Fortress Holland as a German advance through Student's bridgehead in the south could outflank the Dutch fortifications. Aerial bombardment was the only option, but low cloud on 13 May restricted *Luftwaffe* operations as *Generalleutnant* Rudolf Schmidt's *XXXIX. Korps* prepared to take Rotterdam and the following morning *OKW Direktive Nr 11* demanded the rapid crushing of Dutch resistance.

In fact surrender negotiations were already under way, and by midday the Germans recognized that the defenders were simply awaiting formal permission from The Hague. Student radioed Putzier to postpone the planned attack, but when the message reached *KG 54*'s command post the *Kommodore*, *Oberst* Walter Lackner, was approaching Rotterdam and the aircraft had reeled-in their long-range radio aerials. Haze and smoke obscured the target, and to ensure that only Dutch defences were hit Lackner brought his formation down to 700m (2,300ft). Still he failed to notice previously arranged flare signals to warn off the bombers, and 54 He 111s dropped 97 tonnes of bombs before the gaze of the appalled Schmidt and Student. As at Guernica, the bombs ignited vegetable oil and caused uncontrollable fires in which 800 Dutch civilians perished and 78,000 were rendered homeless. The toll would have been higher but the second formation saw the flares and did not bomb.

Schmidt sent a conciliatory message to the Dutch commander, who surrendered shortly afterwards. Despite the circumstances, the attack on Rotterdam was a legitimate act of war, although the consequences were undoubtedly deplored by the participants. Yet soon Putzier's bombers were dropping leaflets threatening Utrecht with a similar fate. On 15 May the Dutch, who were running short of ammunition, were encouraged to surrender. Rotterdam was seen by most of the world as a deliberate act of terror bombing, and it broke the last link of British restraint on strategic bombing.[18]

The aftermath of the Rotterdam bombing was tragic farce. As the Dutch surrendered, nervous *SS* troops arrived and, unaware of the ceasefire, began shooting wildly. Student tried to stop them and was shot in the head, his life being saved by a Dutch surgeon. A Pyrrhic victory was thus completed for the airborne forces, who lost 4,000 men (38 per cent), including 1,200 prisoners shipped to England. The transport force was decimated, with 125 Ju 52s destroyed, 53 bogged down and 47 severely damaged, and since the force drew so much of its resources from schools this adversely effected the *Luftwaffe*'s training organization.

Putzier replaced Student as commander of *Fliegerdivision 7* and his combat units were transferred to Keller. The surviving transport units, now needed to support the advance, came under Süssmann (Lackner's predecessor as *Kommodore* of *KG 54*), who flew in from Norway; he was to be killed during the Crete operation. Student and

Sponeck were both awarded the *Ritterkreuz*, but while the former went on to distinguish himself, the latter's career ended in ignominy. A corps commander in the Crimea in January 1942, he became the only senior commander to be court-martialled and sentenced to death for failing to obey Hitler's 'stand-fast' order, although the execution was delayed until 1944.

The events in Rotterdam led Bomber Command to begin strategic attacks on Germany from the night of 14/15 May. In fact, the British and French began a campaign against enemy communications in Germany west of the Rhine from the night of 10/11 May and maintained the effort for the next four nights with negligible effect. Bomber Command was in the forefront of the night bombing campaign and flew 3,300 sorties to the end of June for the loss of 48 aircraft. The *Armée de l'Air* and the *Aéronavale* continued to concentrate upon communications, but one of the latter's Centre NC 223.4s claimed to have bombed Berlin on the night of 7/8 June. The attacks achieved virtually nothing because targets were difficult to locate, and even when accurately placed the bombs were too small to do much damage. Approximately 100 people were killed (including an Englishwoman in the first attack on a German town) and 200 injured, but for the civilians, in the words of a Münster report, 'The bombing created a sensation and, for days on end, thousands of inquisitive people were attracted to the scene . . .'[19]

Although the British dispatched up to 100 bombers a night, the Germans underestimated the scale of attack and concluded that they were harassing operations; indeed, Kesselring's Chief of Staff, Speidel, stated in a post-war history that they were merely training exercises.[20] The *Flakwaffe* contained the threat but it remained an irritant and on 5 June both *Luftflotten* were ordered to take defensive measures. Kesselring used bombers in a new night-fighter area (*Gebiet*) across the mouth of the Rhine, while Sperrle dispatched two *Ketten* from *KG 2* to *Luftgau VII* (Munich) and *Luftgaue XII/XIII* (Wiesbaden/Nuremberg) for night-fighter operations using cannon-armed Do 17s, possibly Do 17Z-6 *Käuze* (Screech Owls). No victories were reported, and both *Ketten* returned to *KG 2* at the end of June.

At the front in the afternoon of 11 May *Fliegerkorps VIII* was diverted to support *II* and *III/KG 4* in halting *7e Armée*'s advance upon the Moerdijk bridgehead. During the morning it had struck Namur's fortifications but had encountered for the first time large numbers of enemy fighters, which shot down ten *Stukas* of *StG 2* despite the efforts of *JG 27* and *I/JG 51*.[21] The following day came the first clash of armour when the mechanized French *Corps de Cavallerie* met the *Panzer* divisions of *General* Erich Hoepner's *XVI.Korps* north of Namur. Richthofen's *Stukas* hamstrung enemy movement to play an important part in Hoepner's victory, at the cost of ten aircraft, including four *Stukas*, bringing its total losses in two days to 23. Simultaneously Reichenau's *6.Armee* reached the Dyle Line, and despite losing three aircraft to fighters his corps squadrons photographed enemy positions and ranged artillery for the forthcoming assault which Richthofen was to support.

However, in the late afternoon the thunderbolt struck. *Fliegerkorps VIII* was ordered to the Ardennes to support Kleist's assault across the Meuse—*the next day*. It was an extraordinary lapse of staff work for which Jeschonnek, and possibly Hoffman von Waldau, bore the responsibility, and it reflected the *Wehrmacht*'s shortage of qualified staff officers. The decision, as Speidel noted, 'presupposed a mastery of the signal communications and supply problems' which did not exist.[22] Richthofen 'chewed the carpet' and railed down the telephone until 0200 the following morning, but the order stood. The *Staffeln* could move easily enough, provided there were airfields, but Richthofen was more concerned about the lack of support facilities and especially a signal network. He was also reluctant to leave Reichenau in the lurch and a compromise was agreed whereby part of *Fliegerkorps VIII*, including *StG 2, II (Schl)/LG 2* and IV (St)/LG 1, would continue supporting *6.Armee*'s assault.

As Kleist advanced to Sedan, *Fliegerkorps I* (Grauert) and *II* (Loerzer) struck targets in southern Belgium and in northern France in a narrow sector between the *Luftflotten* boundary (Dieppe–Amiens–Valenciennes–Namur) and that of Greim's *Fliegerkorps V* (south of Verdun and Metz). Greim fought a private war with *ZOAE* and *ZOAS* over the *2e Groupe des Armées* in the Maginot Line, while, by disrupting rail communications across the Meuse, Grauert and Loerzer sought to pin down French reserves in the Line. The bombers operated at the limit of their range and Greim's *KG 51*, based at Landfeld and Lechfeld, had to stage through Mannheim and Trier on the outward journey and refuel at Frankfurt Rhein-Main or Giebelstadt on the return leg.

The activities of *KG 2* in Loerzer's command were typical. On the morning of 11 May *Stab, 1., 2., 4., 5.* and *9. Staffeln* attacked airfields at Reims-Pommery, Malmaison, Conde Vraux, and Ecury-sur-Coole while *3./KG 2* bombed troop columns and railway targets in the area Montmedy–Stenay–Mouzon. The following afternoon, while *I* and *II/KG 2* stood down to rest in anticipation of the following morning's operations, *III/KG 2* made low-level attacks upon enemy positions and rail targets around Charleville. In these operations two aircraft were lost, three suffered damage in excess of 50 per cent and another fourteen were damaged to a greater or lesser extent.[23]

Kleist's divisions made a tempting target for bombers as they drove slowly through the Ardennes, but they were shielded by Grauert's and Massow's fighters as well as by *Flakkorps I*. Yet the only bombers to attack Guderian were German, when 25 *Stuka*s from *StG 1* struck a *Panzer* division in error on 11 May. French reconnaissance aircraft occasionally got through (fighters destroyed two on 12 May), but even with photographic evidence their reports were greeted with incredulity by *9e Armée*. By the evening of 12 May Guderian was in the woods north of Sedan, ready to storm the Meuse the following day with *Generalleutnant* Georg-Hans Rheinhardt's *XLI. Korps* on his right. That evening *Generalmajor* Wolf von Stutterheim (*Kommodore* of *KG 77* and *Nahkampfführer II* under Loerzer) visited Guderian to confirm arrangements, and the *XIX. Korps* order stated that 'The attack will be conducted with the extraordinarily heavy support of the *Luftwaffe* . . . In a continuing eight-hour attack, the *Luftwaffe* will destroy the enemy's Meuse defences.'[24] This was to begin at 0800 and continue until dusk. Guderian was warned that the *Luftwaffe*'s commitments elsewhere would reduce the level of support on 14 May but that a *Jagdgeschwader* would be available to defend the bridgehead.[25] But Kleist, on his own initiative and without consulting Guderian, arranged with Sperrle a single massive air attack. Guderian learned of this only when briefing his commanding officer the night before the assault, and Kleist refused to change his mind. Meanwhile Richthofen's Chief of

Staff, Seidemann, arrived to find his worst fears confirmed, with poor signals communications and little liaison with the Army. Nevertheless, he arranged a programme of support for Rheinhardt's corps.

The *Luftwaffe*'s attack began late at 1000 hours as Guderian's forces moved forward and Loerzer adhered to the original plan, ostensibly because Sperrle's orders reached him too late for changes to be made (although this may simply have been an excuse for laziness). The first raids were made in *Gruppe* strength and concentrated upon the French main line of resistance while strafing fighter escorts hindered enemy movement and cut land-line communications. These attacks frequently shot off the radio antennas of command posts, increasing the defenders' confusion.

The French were harassed for four hours, then the main line of resistance came under increasingly heavy attack. Events reached a climax at 1500 hours as Guderian's and Rheinhardt's troops crossed the Meuse supported by 8.8cm guns, 30 per cent of whose ammunition was impact-fuzed for ground targets with another 20 per cent consisting of armour-piercing shells. In the confusion *Stuka*s hit a German motor-cycle battalion, while in a *Luftwaffe* raid upon Sedan the *XIX.Korps* military police commander was killed. *Stuka*s attacked pre-assigned bunkers but, as at Mlawa, the greatest damage was psychological rather than physical. The scream of the aircraft and the crash and concussion of the exploding bombs, with the ensuing dust and smoke, combined with the growing sense of isolation resulting from cut telephone lines to undermine the defenders' nerves. During the last two hours the French gun lines were struck, and while few batteries suffered much damage, most of the gunners fled. By the end of the day German bridgeheads were established across the Meuse and bridges were hurriedly assembled to consolidate them, protected by *102.Flak-regiment* (*Oberstleutnant* Walter von Hippel) with 36 × 8.8cm, 54 × 3.7cm and 81 × 2cm guns. During the day *Fliegerkorps II* flew 1,770 sorties in support of Guderian (including 900 bomber and 180 *Stuka* sorties) and a further 910 in support of Rheinhardt (360 bomber, 90 *Stuka*) and lost only six aircraft, half of them *Stuka*s.

During the morning of 14 May the bridges were thrown across the Meuse and with them Kleist's *Panzer* divisions tightened their grip on the west bank. The moment of decision had arrived, both on the ground and in the air, as the Allies launched all their bombers at the bridges and columns in what became a repeat of the Maastricht battle. Massow's *Jafü 3*, reinforced by *JG 26* and *JG 27*, flew 90 missions (814 sorties) and those bombers which survived were met by Hippel's *Flak*. The AASF bore the brunt, flying 81 sorties and losing 52 per cent of its aircraft, but No 2 Group shared the burden with 28 sorties and by the end of the day the RAF had lost 48 aircraft (44 per cent of those dispatched). It was the bloodiest day in RAF history, and when he learned of the casualties Barratt is reported to have burst into tears.[26] The French used nineteen bombers with an escort of 60 fighters but lost six of the bombers (37 per cent) to ground fire. While the attacks slowed movement across the bridges, none of the latter was hit—which was fortunate for the Germans, for 'Had [Guderian's] bridge been damaged even slightly, the Germans' situation would have been precarious, for most of their bridging had been used in the move through the Ardennes.'[27]

The Allied strike force had been virtually annihilated and daytime operations were reduced to unco-ordinated penny-packet attacks in which the French played a greater role as their newly re-equipped *groupes de bombardement* became operational. The

AASF took refuge in night operations, supplementing Bomber Command's efforts and those of the more elderly French bomber units, yet few of these missions proved effective. By contrast the German bombers and *Stuka*s operated throughout the day, helping the *Panzer* divisions expand their bridgeheads on 14 and 15 May and thwarting French counter-attacks, including one by an armoured division. The attacks concentrated on the support echelons (the French *2e Armée* opposite Guderian reported only two tanks destroyed by aircraft) and, by bombing villages, blocked approach routes with debris, then destroyed soft-skinned vehicles bringing urgently needed supplies.

StG 77 under *Oberst* Günter Schwartzkopff, who had earned the epithet 'The *Stuka*'s Father', played the key role, but the fluid nature of the battle inevitably led to mistakes and on 14 May *StG 77* hit *1. Panzer Division*, killing and wounding several senior officers. During the day Göring appointed Schwartzkopff Inspector of Dive Bombers but the next day, while leading his beloved *Stuka*s, he was shot down and killed by French AA in one of seven aircraft lost by the *Geschwader* in two days.[28] However, the *Stuka*s had paved the way for Kleist's armour to break through the French defences and drive westwards, paced in the north by *General* Hermann Hoth's *XV. Korps*, which crossed the Meuse at Dinant. By advancing into Belgium the Allies had hoped to slam the door in the face of the German advance: now the *Wehrmacht* had taken the door off its hinges and was pushing it towards the sea.

At this point the German High Command, frightened by its own success, suffered a crisis of nerve. The maps showed Kleist's armour thrusting like a dagger into France while *Luftwaffe* reconnaissance reported the frantic regrouping of the French Army as it sought new defensive lines where its artillery could contain an enemy they expected to be exhausted. But the *Panzer* divisions retained the initiative while the *Luftwaffe* bombed all signs of resistance, disrupting the movement of reserves and cutting communications, so the 'defensive lines' were brushed aside as easily as a spider's web. The *Panzer* commanders recognized that the enemy were retreating to the Escaut, but Hitler feared a counter-offensive from the south.

Between 16 and 18 May he tried to slip a leash on to Guderian but the cunning general was not nicknamed 'Fast Heinz' for nothing. His liberal interpretation of instructions to conduct a 'reconnaissance in force' meant that the pace of his advance went unchecked, still supported by Loerzer's *Fliegerkorps II*. Guderian had no troops to protect his southern flank, a task he left largely to Greim's *Fliegerkorps V*. Railways were a particular target, and the Ju 88s of *I* and *II/KG 51* dive-bombed them frequently. On 19 May, for example, *Luftwaffe* reconnaissance reported 33 trains halted south-west of Verdun, but the superb organization of the French rail network ensured that such delays were only temporary.

An embarrassed *ObdL* now recognized the magnitude of Richthofen's difficulties in moving *Fliegerkorps VIII* and a gradual transfer of units was permitted together with a gentle realignment of their areas of responsibility until they matched Kleist's. Newly captured airfields at Liége proved a boon, although those of *Stuka* units were sometimes shelled by Belgian-occupied forts until *StG 2* intervened. Richthofen provided daily dwindling support for Reichenau's *6. Armee* as it pushed the Allies out of the Dyle Line, then *12. Armee* benefited from the *Stukagruppen* for two days from 18 May as it widened the gap in the Allied lines south of Namur. Kesselring received compensation in the shape of Grauert's *Fliegerkorps IV* on 17 May, but most of these *Gruppen* were still east of the Rhine and Grauert simply adjusted his boundaries to support *Generaloberst* Günther Hans von Kluge's *4.Armee*.

From 18 May *KG 77* and *StG 77*, protected by sweeps from *JG 27*, reinforced the shield on Guderian's left with the former attacking columns around Montdidier while the latter struck railways running across Guderian's line of march. From the following day Guderian had Richthofen's complete support, and much of the credit for the smooth transfer must go not only to Richthofen but also his Chief of Staff, Seidemann, as well as to *Luftgaustab zbV 16* under *General der Flieger* Kurt Pflugbeil. Nevertheless, there remained difficulties, and on 18 May Richthofen noted in his diary poor deliveries of bombs and slow improvements in communications.

The concentration of *Fliegerkorps VIII* behind Guderian was timely, for having crossed the Sambre and Escaut he was poised on the Somme. With German armour only 50 miles (80km) from the Channel and nothing in front of it, Guderian was given formal permission to advance westwards on 19 May. The following day his tanks roared down the Somme valley and by dusk *2.Panzer Division* had reached the coast at Abbeville to isolate the Allied forces in Belgium. At 2142 London time on 20 May the RAF intercepted a message to *Luftwaffe* units: 'Do not attack Abbeville but Dunkirk.'[29] Just before the Germans reached the coast, most of the *Aviation Militaire Belge*, including the remaining fighters, escaped to France, leaving some twenty corps aircraft to continue the struggle.

As Guderian dashed to the Channel, Sperrle struck neighbouring airfields and the Potez factory at Albert shortly before *2.Panzer Division* took the town. The Hs 126s of *1./(H) 11 Pz*, *1./(H) 14 Pz*, *3./(H) 21 Pz* and *2./(H) 23 Pz* acted as flying eyes which called down *Fliegerkorps VIII Stukas* or *Schlachtflieger* whenever they found signs of resistance or movement. Richthofen's provision of radio-equipped forward observers who could monitor the situation from the Army's command posts and report directly to *Fliegerkorps* headquarters proved extremely useful, while less urgent traffic could plug into Kleist's land-line network. By having *Stukagruppen* on twenty-minute alert, aircraft could swoop on targets within 45 to 75 minutes of a report being received. Seidemann later commented, 'Never again during the course of the war was such a smoothly functioning system for discussing and planning joint operations achieved.'[30] *Fliegerkorps II* increasingly concentrated upon indirect support, and its *Kampfgruppen* conducted numerous night harassing attacks on the enemy rail network.

There was fragmented and weak Allied aerial resistance to Guderian. French bombers were the more active, flying some 230 sorties (70 by *Aéronavale* dive bombers) between 16 and 20 May compared with 146 British (74 by the AASF). On 19 May *KG 54* lost fifteen aircraft to fighters in the Lille–Arras area—their *Kommodore*, Lackner, was injured and taken prisoner—but *JG 26* and *JG 27* exacted revenge to shoot down 37 French (sixteen dive bombers) and 24 British aircraft, although the *Kommandeur* of *II/JG 26*, the Spanish ace *Hauptmann* Herwig Knüppel, was killed. Most of the RAF's victims fell on 17 May when *Flak* shot down one of twelve No 82 Squadron Blenheims and broke up the formation. Before the others could regain formation a Bf 109 *Staffel* pounced and destroyed all but one of the survivors. Two days later six Vought 156s from a formation of fourteen French dive bombers were shot down by Bf 109s. The German fighter effort was assisted by reinforcements from the *Reich* and when Grauert's *Fliegerkorps I* assumed responsibility for operations in northern Belgium it was reinforced by *II(J)/TrGr 186* and *I(J)/LG 2* from *Jafü Deutsche Bucht*. At about the same time *II/JG 3*, which had provided Berlin's day defence, came into action in the West, probably with *Jafü 3*.

The *Luftwaffe* did not escape unscathed, however, and between 15 and 20 May Greim, Richthofen and Kleist's *Heeresaufklärungsstaffeln* lost 55 aircraft. Nearly half (26) were lost by Greim, whose bombers often had to fly the whole length of the front from the Rhine to the sea, and in shielding them the *Zerstörer* paid a heavy price, with thirteen lost. The increase in casualties reflected the activity of Allied fighters, those of ZOAN flying more than 800 offensive sorties in northern France and Belgium. Richthofen's relatively light losses (thirteen aircraft) were partly a side-effect of Guderian's advance, which tore through the enemy air forces' ground infrastructure and forced squadrons to flee to new bases away from the armoured threat. The French and the AASF withdrew south of the Somme and the Marne, while the RAF Air Component withdrew to England from 19 May. The last had lost 203 aircraft between 10 and 20 May, including 128 (74 Hurricanes) on the ground either to the *Luftwaffe* or at the hands of RAF personnel to prevent their falling into enemy hands. Similar losses were suffered by the AASF, down to 89 aircraft (65 bombers) by 25 May, while despite substantial reinforcement ZOAN had on the previous day only 248 fighters (29 Potez 631s) and 113 bombers. Also evacuated by the British were the radars of No 14 Group, whose chain lay across Guderian's line of advance; only that near Calais fell into German hands, and it was dismissed as inferior to *Luftwaffe* equipment.

But it was fighters which the French wanted, and the failure of the British to provide them led to bitter complaints from Vuillemin after the fall of his country. Churchill's commitment to his allies remained constant, but from 16 May Fighter Command began restricting the forces it sent across the Channel. It was the day Dowding wrote his historic memorandum demanding that a limit be placed on the number of squadrons sent to France for fear of compromising the defence of Great Britain. Nevertheless, No 11 Group continued to dispatch composite squadrons daily to operate from French bases until 20 May, by which time they had flown 96 sorties. Between 16 and 31 May Vuillemin's fighters flew 4,075 sorties while the BAFF flew 451, but, as the French knew, Fighter Command's efforts over the Low Countries in the same period involved 3,072 sorties and only 820 had been flown on defensive patrols.[31]

The reorganization of Allied air power following the German advance to the Channel was reflected in a drop in the *Luftwaffe*'s losses, which totalled 94 between 21 and 25 May compared with 210 in the previous five days. However, the advance also created serious problems for the *Luftwaffe*. From 15 May single-engine strike *Gruppen* (*Jagd-*, *Stuka-* etc) supported by their *Flughafenbetriebskompanien* with their flight-line and workshop platoons, began moving into captured Belgian and French airfields as far west as Liége and Charleville, which became extremely congested, leading to a number of serious accidents. There were also severe problems supporting the *Panzer* divisions on the coast, some 125 miles (200km) away, which forced the Bf 109E and the Ju 87B with a 500kg bomb load to operate at the limits of their practical combat radius. The handful of longer-ranged Ju 87Rs with *StG 2* proved a blessing, but there were never enough and there remained considerable difficulties establishing and supplying new airfields closer to the *Panzer* divisions.

A *Fliegerkorps VIII* after-action report of 1 September 1940 noted that the constant movement between *Armee* sectors as well as difficulties with officialdom meant that units had to scrounge fuel and supplies.[32] The *Luftwaffe*'s supply organization was barely able to support units in Belgium and the Netherlands, but Kleist's advance stretched it to breaking point. With only a few roads into Kleist's corridor the problem in moving supplies was not the random Allied bomber attacks but huge traffic jams which lasted for hours at a time, while *Luftwaffe* drivers often discovered that the Army fuel depots were exhausted and sometimes used aviation fuel to keep their vehicles going. Food was less of a problem, since every unit carried five days of rations and lived either off the land or off captured supply dumps, with chocolate, wine and spirits highly prized.[33] Fuel was another matter, for only a trickle was getting through and transports were pressed into this task. Although their fuel was unsuitable for fighters they often had surrender part of it for bombers and reconnaissance aircraft. Substantial stocks were captured at enemy bases, but the Germans were reluctant to use these for fear they had been sabotaged.[34]

To overcome the problem the *Luftwaffe* had to place greater reliance on air transport than it had anticipated, but in moving fuel forward it had to balance its own needs with those of the Army. As early as 12 May *Heeresgruppe A* had demanded 59 tonnes of fuel for Guderian's spearhead, but the shortage of transport aircraft meant that only 27 tonnes was delivered, although this proved sufficient. The shortfall was due to transport losses in the Netherlands and commitments in Norway, which reduced the force in Western Europe to some 300 Ju 52s, even when extra aircraft were squeezed out of training units. If they flew only two sorties a day they were still capable of moving a nominal 1,500 tonnes, and probably more on good days. The scale of effort and the volume of traffic meant that accidents were inevitable, though total losses were only 25 aircraft destroyed and nineteen damaged.[35]

Air transport was not confined to carrying fuel: it fulfilled a variety of roles. The operations of *Luftflotte 2*'s *II/KG zbV 1* from 15 May were typical, involving the movement of airfield construction companies to Belgium and the transport of fuel, ammunition, bombs and spares for forward *Stuka-* and *Jagdgruppen* as well as tank ammunition, fuel and oil for Kleist's spearheads.[36] 'Black Men' and replacement aircrew were flown to forward units and 2,000 Army technicians were flown to Charleville to establish a vehicle repair facility (from which the *Luftwaffe* would benefit). The bonus of such operations was the swift evacuation of casualties, which boosted morale and reduced the number of fatalities. When aircraft took off they would radio ahead to prepare suitable medical facilities, a typical message intercepted by the British on 24 May being, 'Wounded on board. One shot in knee, two crushed chests, one broken knuckle, one shot in upper arm, concussion.'[37]

The rapid advance and the consequent disruption of the signals network wrecked the traditional means of obtaining replacement aircraft through *Generalmajor* Hans-Georg von Seidel's Quartermaster Staff. Increasingly the *Fliegerkorps* relied upon their *Luftflotten*, and in ensuring that both aircraft and other supplies were forthcoming, Milch, who adopted a roving commission, appears to have played an important role. By 20 May the *Luftwaffe* was beginning to feel the strain of ten days' heavy

TABLE 25: *LUFTWAFFE* PERSONNEL LOSSES, MAY 1940

Period	Killed	Missing	Wounded
10–20 May	140	1,414	333
21–30 May	290	387	Not available
Totals	**430**	**1,801**	

fighting, having lost 547 aircraft destroyed or with 60 per cent damage (excluding transports and liaison aircraft)—a figure which represented 20 per cent of its serviceable strength on 10 May. Many others had minor damage, and serviceability in many units dropped to 50 per cent. There had been some 2,600 casualties in action and a further 900 in accidents (see Table 25) including Gentzen, the Polish campaign ace, killed on 26 May. The Personnel Department noted that only *Aufklärungsstaffeln* could expect experienced replacements while the *Kampf-, Jagd-* and *Zerstörergruppen* would receive raw recruits. It requested the schools in mid-May to release 166 instrument-trained bomber crews, 40 *Stuka* crews and between 60 and 70 fighter pilots.

Typical of these replacements was Peter Stahl, who joined *KG 30* at midday on 10 May after it lost six aircraft. Aircraft were taking off and landing as he reached the command post, where crews came and went wearing oil-stained flying suits, yellow coverlets over their flying helmets, life jackets and fur-lined flying boots. Then a Ju 88 taxied up and a bloody air gunner with a neck wound emerged and was soon rushed to an ambulance. The observer was dead, leaving the pilot and flight engineer understandably shaken. 'It seems almost like a war film but this is stark reality—my first experience of war as it is,' Stahl remarked.[38]

Having isolated *1er Groupe des Armées* in Belgium, Kleist's armour turned north to compress the pocket, but this created new difficulties for Sperrle, who had to support both the mechanized divisions heading for the Belgian border and the thin defensive line with its bridgeheads along the Somme. Until German reserves reached the latter, air power was the prime means of supporting this line and possibly preventing the enemy from driving into Kleist's rear. *Fliegerkorps II* (Loerzer) and *Fliegerkorps V* (Greim) initially shared the task, but from 24 May it gradually passed to Greim's over-stretched forces. However, aggressive action actually helped the *Luftwaffe* reduce the pressure from enemy air forces upon the Kleist corridor, for it meant that between 21 and 23 May 42 per cent of *ZOAN*'s 512 fighter sorties were mounted to protect railway stations, where reinforcements were arriving to prop up the defences.[39] Rund-stedt requested an airborne operation in support of *Heeresgruppe A*'s bridgeheads on the Somme but *OKH* rejected this on 23 May after the *Luftwaffe* pointed out that it lacked spare transports, even if *Fliegerdivision 7* had recovered from its losses.

The German High Command's greatest concern was that the trapped Allies might break out to the south, and in his diary of 21 May Richthofen noted their nervousness. Yet he, too, had grounds for nervousness as the resistance of enemy fighters, many flying from England, grew stronger. Between 21 and 25 May his *Staffeln* suffered 25 per cent of the total *Luftwaffe* losses in this period and Halder's diary for 24 May noted that 'for the first time enemy air superiority has been reported by Kleist.' Major clashes with the England-based No 11 Group developed as the Germans approached Boulogne and Calais on 22 and 23 May, with Richthofen's *Staffeln* losing sixteen aircraft, including six *Stuka*s (five from *StG 77* during the evening of 22 May). Grauert's *Fliegerkorps I*, which attacked the ports from 21 May, was not immune, and *LG 1* at Bruxelles-Evère lost six bombers on 21 and 22 May until *II/JG 2* provided close escort. *JG 27* had the worst of these exchanges, losing ten Bf 109s compared with six British fighters. The two ports were defended fiercely until 26 May and were taken with the aid of *StG 2* and *StG 77*, diverted initially to Boulogne on May 23 after a personal request from Sperrle to Richthofen. Both sides continued to suffer casualties, and on 25 May *Major* Oscar Dinort, the *Kommodore* of *StG 2*, lost both his 'Kettehunde' (wing men) to Spitfires over Boulogne.

These losses took the glitter off the *Ritterkreuz* Richthofen received on 23 May and were the first signs of a major new air battle which would swirl around the Channel coast. Yet there was no sign of Allied air support the previous day when an armoured counter-attack broke through the German lines near Arras and Richthofen's airmen found themselves in the enemy's path. To provide Kleist's spearheads with tactical air support, Richthofen assigned a battle group of *I/JG 21, III/JG 54* and *II(Schl)/LG 2* to operate from bases near the front line. On 22 May this battle group was based near Arras, barely 21 miles (35km) from the front line, and during the morning, as Seidemann was meeting the *Schlachtflieger Kommandeur, Hauptmann* Otto Weiss, they were shocked to learn of the enemy break-through. Weiss's 'Ein-Zwei-Dreien' were quickly in the air, followed by the Bf 109s, to strafe and bomb British tanks, while *I/Flakregiment 33* used its 8.8cm guns to form the core of a defensive line. Prompt support from *Generalmajor* Erwin Rommel's *7.Panzer Division* stopped the enemy, who were then attacked by *StG 2* and *77*, and by the end of the day the threat was over.[40]

The Allies were actually trying to buy time while the pocket's southern defences were strengthened, but the German High Command, including the *Luftwaffe*, feared that Kleist's forces would be isolated. German pressure upon the pocket increased, but the advance was taking the tanks into marshy terrain criss-crossed with canals. This was territory where infantry was best employed, but the infantry regiments had still to arrive and the Allies were beginning to withdraw towards Dunkirk. The French port had already suffered three *Gruppe*-size raids since 16 May and Göring perceived on 23 May 'a wonderful opportunity' for the *Luftwaffe* to smash enemy resistance and allow the Army to mop up. He begged Hitler, 'My *Führer*, leave the destruction of the enemy surrounded at Dunkirk to me and my *Luftwaffe*.' With his generals increasingly fearing the destruction of the *Panzer* divisions in a battle of attrition, Hitler now had a valid excuse for formally halting Kleist's advance on 24 May.

Göring's optimism was counter-balanced by the pessimism of Kesselring and Jeschonnek, who had a more realistic view of the *Luftwaffe*'s condition. Kesselring reported that many *Gruppen* were down to fifteen serviceable aircraft, and while *Fliegerkorps IV* could operate from Dutch airfields, most of the *Kampfgruppen* were still based hundreds of miles away in Germany. They could fly only one mission a day and would still need to stage through *E-Häfen* to reach their bases. Richthofen's diary entries for 27 and 29 May suggest that Sperrle was ambivalent about the operation and had failed to make any detailed preparations (possibly because he was more concerned with the southern front), while Jeschonnek appeared deeply pessimistic. As for Göring, he spent the early part of the Dunkirk campaign scouring Dutch cities for loot, although he rushed back when he realized that things were not going to plan. By 29 May his nerves were getting the better of him, as Richthofen observed, while two days later Richthofen himself stated that the whole enterprise was a waste of time when preparations for the next offensive against the French (*Fall 'Rot'*) should have been given priority.

Despite the handicaps, *Fliegerkorps I* (Grauert), *II* (Loerzer) and *IV* (Keller) opened the campaign on 25 May with two days of heavy attacks which wrecked Dunkirk's inner harbour, forcing the Allied ships to use the limited facilities of the outer harbour and the nearby beaches. *Fliegerkorps IV* then concentrated upon anti-shipping operations and on supporting *18.Armee*'s attack upon the Belgians in the north of the pocket. The British had already begun to evacuate rear-echelon troops, but the renewed German assaults upon the Belgians and the French *1e Armée* (by Reichenau) led them to conclude that the whole of the BEF should be evacuated, and during the evening of 26 May Operation 'Dynamo' began.

It was to last until 2 June and it opened a fierce air campaign which extended well beyond the Dunkirk area. From dawn each day Air Vice-Marshal Sir Keith Park would send over alternate waves of Spitfires and Hurricanes, occasionally supplemented by Defiants and Blenheim IFs, to patrol the coast, usually at 50-minute intervals. These patrols were normally in squadron strength, but as *Luftwaffe* fighters operated in *Gruppe* strength they often overwhelmed the British, and from 28 May some station commanders tried to organize patrols similar in strength to an RAF Wing. Park's radars enabled him to concentrate his slender forces over Dunkirk by eliminating the need for many defensive standing patrols, and on occasion only three defensive sorties were flown a day, the total from 27 May to 2 June being 107. Yet there appears to have been less use of radar to direct interceptions, and certainly the radar chain proved less effective for this task over Dunkirk than it would during the Battle of Britain. (See Table 26 for main operations over Dunkirk.)

The first day of 'Dynamo' was one of sharp defeat for the *Luftwaffe*, for although the bombers dropped 320 tonnes of bombs they lost 10 per cent of the aircraft dispatched. The shocked *Luftflotten* commanders reacted quickly by increasing fighter support, both in escorts and sweeps, but the greatest handicap was the lack of an overall commander—which, no doubt, reminded Richthofen of his own experiences at Warsaw. The Dunkirk air battles involved *Luftflotten 2* and *3* and *Fliegerkorps I, II, IV* and *VIII* as well as the two *Jagdfliegerführer*, and the difficulties of co-ordination meant a succession of attacks instead of one continuous aerial assault. Consequently there were periods when few German aircraft would be over the pocket where the BEF had destroyed its heavy guns prematurely, and sometimes these would be just the times when the RAF fighters appeared.

This policy exacerbated relations between the RAF and the British Army, which thought, like all armies, of air combat in Great War terms, with opposing fighters swirling over their heads like clouds of midges. Since the late 1930s, however, air combat had consisted of high-speed battles fought over hundreds of square miles far from the front line, but the soldiers did not appreciate this and were increasingly

TABLE 26: AIR OPERATIONS IN THE DUNKIRK POCKET, 27 MAY–2 JUNE 1940

Luftwaffe

Date	Bombers	Ju 87	Hs 123	Bf 109	Bf 110	Total
27 May	225 (24)	75 (0)	–	135 (0)	110 (6)	545 (30)
28 May	50 (3)	25 (1)	–	50 (2)	50 (0)	175 (6)
29 May	175 (6)	235 (2)	20 (0)	285 (7)	45 (1)	760 (16)
30 May	45 (5)	–	–	2 (1)	25 (0)	95 (6)
31 May	195 (1)	–	–	240 (5)	20 (0)	455 (6)
1 June	160 (2)	325 (2)	–	420 (12)	110 (0)	1,015 (16)
2 June	160 (4)	145 (5)	–	440 (2)	45 (1)	790 (12)
Totals	1,010 (45)	805 (10)	20 (0)	1,595 (29)	405 (8)	3,835 (92)

RAF

Date	No 11 Group				No 2 Group	Bomber Cmnd
	Spitfire	Hurricane	Other	Total		
27 May	103 (6)	144 (14)	23 (0)	270 (20)	48 (2)	–
28 May	106 (4)	135 (8)	21 (3)	262 (15)	48 (1)	47 (1)
29 May	102 (9)	148 (10)	25 (0)	275 (19)	51 (0)	15 (0)
30 May	104 (3)	149 (3)	4 (0)	257 (6)	68 (0)	28 (1)
31 May	140 (8)	123 (7)	23 (6)	286 (21)	93 (0)	33 (2)
1 June	121 (11)	134 (5)	12 (0)	267 (16)	56 (0)	16 (0)
2 June	70 (7)	73 (2)	4 (0)	147 (9)	24 (0)	16 (0)
Totals	746 (48)	906 (49)	112 (9)	1,764 (106)	388 (3)	155 (4)

Note: Figures in parentheses are losses. In addition to these losses a number of *Luftwaffe* reconnaissance and corps aircraft as well as British Coastal command aircraft were lost in this campaign. About 110 attack sorties were flown by Coastal Command against land targets in the Dunkirk area. The *Armée de l'Air* flew half a dozen fighter sorties from England over Dunkirk without loss.
Sources: Bingham; Franks, *Dunkirk*; Huan and Marchand; Middlebrook and Everitt; No 11 Group operations record, PRO Air 25/193.

demoralized by what they interpreted as the Air Force's abandonment of them. At least one RAF fighter pilot was captured when the Army denied him permission to board a ship.

Rain, fog and low cloud restricted *Luftwaffe* operations on 28 and 30 May as well as the morning of 29 May, while the commitment of *KGr 100* and *KG 26* to Norway meant that attacks upon the bridgehead were largely confined to daylight hours. The British Official History states: '. . . the Royal Air Force had defeated the *Luftwaffe*'s intention to make evacuation impossible.'[41] Yet this was never Göring's intention, for the Germans were slow to recognize not merely the scale of the evacuation but the fact that one was taking place at all. The intention was to destroy the troops in the Allied pocket just as the *Luftwaffe* did with the Ilza and Bzura pockets in Poland, Göring informing Hitler that 'it was imperative to destroy from the air the British forces surrounded in Dunkirk,' according to Schmid, who overheard their telephone conversation. The *Luftwaffe* General Staff shared Göring's optimism, with Jeschonnek also 'absolutely convinced that the *Luftwaffe* would succeed in destroying the British Expeditionary Corps in and around Dunkirk and prevent its shipment to the British Isles,'[42] while the Chief of Operations, Hoffmann von Waldau, noted in his diary on 25 May that 'The losses . . . will be enormous.' Attacks upon shipping and port facilities, in which 89 ships (126,518grt) were sunk or damaged, were part of the process of isolating the battlefield. While shipping near the beaches was attacked, the majority of German sorties appear to have been against land targets until the beginning of June. There was no attempt to interdict traffic in the Channel or in reception ports such as Dover—indeed, the *Luftwaffe* actually transferred an anti-shipping *Gruppe* (*II/KG 30*) to Norway during 'Dynamo'.

Only on 30 May did *OKH* realize that significant numbers of the enemy were escaping, and by then the weather was restricting air operations. Nevertheless, the *Luftwaffe* believed that it had achieved its aims and the *OKW* official communiqué of 4 June noted that there were 'still quite inestimable numbers killed in action, drowned or wounded.' The following day Hoffmann von Waldau toured the battlefield with Milch and said, 'Here is the grave of British hopes in this war, and [as he tapped abandoned British helmets] these are the gravestones.' Milch, who could see few bodies, was more sceptical, but neither imagined that the Allies had evacuated 338,000 men, even if few had more than the clothes they stood in.

Although it failed in its objective, the *Luftwaffe* generally had the better of the air battle, losing about 100 aircraft (including reconnaissance and corps machines) to give an overall loss rate of about 3 per cent while Park lost 106 fighters or 6 per cent of the aircraft dispatched as well as 60 of his precious pilots. Sporadic fighting continued after 2 June but by then the *Luftwaffe* was preparing for the final offensive in France, *Unternehmen 'Rot'*.[43]

OKH planning for *'Rot'* began on 20 May and was eventually approved by Hitler eight days later. It involved two phases beginning on 5 June on the right, where Bock's *Heeresgruppe B* (44 divisions) would advance to the Seine and shatter the French left (*3e Groupe des Armées*) while *Panzergruppe Kleist* (on Bock's left flank) isolated Paris. Four days later Rundstedt's *Heeresgruppe A* (45 divisions) would strike southwards from the Aisne towards Rheims, spearheaded by *Panzergruppe Guderian*. It would shatter the French centre (*2e Groupe des Armées*) then envelop the right (*4e Groupe des Armées*) in the Maginot Line, assisted by Leeb's *Heeresgruppe C* from the *West Wall*. *OKW Direktive Nr 13* of 24 May outlined the *Luftwaffe*'s role: 'It will be the *Luftwaffe*'s task, apart from maintaining our air superiority, to give direct support to the attack, to break up any enemy reinforcements which may appear, to hamper the regrouping of enemy forces and, in particular, to protect the western flank of the attack.'

The majority of *Kampfgruppen* were assigned to Kesselring's *Luftflotte 2* (see Table 27), which retained *Fliegerkorps I* and *IV* as well as *Flakkorps II* and received *Fliegerdivision 9* because it would be covering the French coast. Sperrle, who had the remaining *Fliegerkorps* and *Flakkorps I*, also received most of the *Zerstörergruppen* because his forces would operate deeper in France. Since it would support the main armoured thrust, *Fliegerkorps VIII* was retained but was transferred from Stutterheim's *Nahkampfführer II* (to which it had been assigned since 16 May) to Massow's *Jafü 3*, where it remained until 16 June. For most of the campaign *Fliegerkorps VIII* would

Luftflotte	JGr	ZGr	KGr	Stuka/ Schlachtgr	Aufklärungs- staffeln
Luftflotte 2	8 (10)	2 (4)	21 (18)	1 (7)	6 (7)
Luftflotte 3	14 (13)	7 (5)	15 (22)	9 (3)	6 (6)
Totals	**22 (23)**	**9 (9)**	**36 (40)**	**10 (10)**	**12 (13)**

Note: Figures in parentheses are those for 10 May.

support forces which originally made up Bock's left (*Panzergruppe Kleist* and 6. and 9.*Armee*) but were increasingly committed to the centre.[44]

Most of the *Jagdgruppen* and all the *Zerstörergruppen* were assigned to the *Jafü* to co-ordinate operations in what promised to be an extremely fluid campaign (for the outline order of battle, see Appendix 11). *ObdL* feared renewed Allied daylight bombing and strengthened Germany's air defence: *Stab JG 77* returned to shield the Leuna synthetic oil works with *III/JG 52* and *I/JG 77*, *I(J)/LG 2* appears to have returned to *Jafü Deutsche Bucht* and *II(J)/TrGr 186* went to *Luftflotte 5*. Yet *ObdL* was confident enough on 3 June to inform *Oberstleunant* Otto Höhne, *Kommandeur* of *I/ KG 54* and acting *Kommodore* of *KG 54*, that the *Geschwader* was to re-equip with Ju 88As; indeed, two days later *I.* and *II.Gruppen* flew to Germany, leaving the total *Luftwaffe* strength for '*Rot*' at about 2,500 aircraft, of which Kesselring had some 1,000 (excluding seaplanes).

Despite the loss of 787 aircraft (473 fighters, 120 bombers and 194 reconnaissance/corps machines), by 5 June French first-line air strength actually rose to 2,086 (including reserves). But the disruption of the supply organization and the failure of industry to deliver vital items such as tyres and propellers meant that only 29 per cent (599 aircraft, including 340 fighters and 170 bombers) were operational.[45] To these could be added a dozen *Aéronavale* D.520 fighters, 39 Hurricanes and 82 Battles of the AASF,[46] as well as British-based squadrons for diversionary operations.

On 27 May Vuillemin reorganized his forces so that *ZOAN*, covering Paris and the *3e Groupe des Armees*, had most of his aircraft leaving the rump, with *ZOAE* to protect the remaining Army Groups. From 31 May he centralized the control of his forces, despite bitter protests from the Army whose only compensation was the knowledge that the bombers' primary role remained that of ground support. Air reconnaissance quickly showed German armour moving southwards, and to disrupt these preparations the Allies flew 454 bomber sorties (66 per cent by the French) from 28 May to 4 June. The transfer of armour is a complex business which might have been disrupted by these attacks, but, as Kesselring observed, '. . . that these movements could be carried out without a hitch in daylight is attributable only to our air superiority.'[47] The attacks achieved little, while *Flak* and the *Jagdgruppen* accounted for 24 bombers.

As French activity declined (only 65 bomber sorties were flown between 1 and 4 June), *Luftwaffe* pressure increased. During the first three days of June the *Luftwaffe* attacked some 30 airfields, 60 rail targets and 50 other targets, including ports, but the effort was concentrated in two operations.[48] The first, unnamed, by *Luftflotte 3* against communications in the Rhône valley and targets around Marseilles, was demanded by *OKH* to prevent the French bringing reinforcements from North Africa. Their concern was underlined by reports from *Aufklärungsgruppen (F) 122* and *ObdL*, who reported 300 aircraft at Marseilles' airfields and 3,600 wagons on the southern rail network, while on 1 June 48 merchantmen were discovered in the harbour with fourteen warships. The second, *Unternehmen 'Paula'*, was a *coup de main* to neutralize the *Armée de l'Air*. Although described as 'strategic' operations, both were extensions of the operational-level missions conducted by the *Luftwaffe* since the beginning of *'Gelb'* and defined by *LdV 16*'s Paragraph 150 (see above).

There are few records of the Marseilles/Rhône missions which were executed by *KG 53* (which had delivered bombs and leaflets earlier to Marseilles) and possibly *KG 55*. They involved a mass daylight attack on targets some 400 miles (650km) distant and thus near the limit of the He 111's effective operational range. The operation was one of the most dangerous mounted by the *Luftwaffe* during the Western Campaign, for the bombers had to fly along the Rhône valley, through the heart of enemy territory, and run the gauntlet of the defences.

After the low cloud and rain at the end of May, the weather on 1 June improved and during the afternoon some 60 bombers cut the rail network in seven places and sank the 20,043grt British liner *Orford*. Airfields and industrial targets around Marseilles largely escaped damage and were attacked the next day together with Lyon-Bron airfield, where fifteen aircraft were claimed as destroyed. Both missions encountered resistance, not only from the French but also from the Swiss when bombers accidentally drifted across the border and were intercepted by the *Schweizerische Flugwaffe*. The *Flugwaffe* had only 97 fighters and 131 corps aircraft when war broke out, but it defended Swiss neutrality vigorously. Its effectiveness was increased with the delivery of D-3800s (licence-built MS 406s) and Bf 109E-1/3s, some of which accounted for the six He 111s lost by *KG 53* on the first mission. The following day a *Zerstörergruppe* escort drove off *Flugwaffe* attacks and shot down one of the attackers, who claimed four victories; no German aircraft were lost. Sporadic fighting continued to 8 June, but with Italy's entry into the war on 10 June the *Luftwaffe* left the Rhône valley to the *Regia Aeronautica*.

'Paula' was, in Hoffmann von Waldau's words on 25 May, 'a blow prepared with loving care',[49] and preparations were under way as early as 23 May when the French intercepted signals traffic about the operation. The objectives were to destroy airfields and aircraft factories around Paris and also (again in Hoffmann von Waldau's words) to 'achieve a desirable influence on the morale of the capital.'[50] Reconnaissance on 1 June reported 1,150 aircraft based around Paris, but it was difficult to keep track of them owing to their constant movement. So important was the mission that both *Luftflotte 2* and *3* were involved, using most of *Fliegerkorps IV* and *Fliegerkorps II* and even permitting the He 111 *Gruppen* of *KG 54* to fly a last mission.

Cryptographic interception of *Enigma* signals traffic warned the French, allowing them both to monitor preparations and to double *ZOAN*'s fighter strength to some 120 aircraft, which were alerted on 3 June an hour before the Germans took off![51] In the early afternoon observers spotted the 640 German bombers (*LG 1* and *KG 1, 2,*

3, 4, 30 and *54*) and 460 fighters (*JG 2, 26, 27* and *53* and *ZG 2* and *76*)—other sources suggest that 1,200 aircraft participated—whose progress was monitored by shadowing Potez 631s, one of which was shot down.[52] But as Air Marshal Barratt observed, French Air Staff work tended to neglect detailed execution.[53] When the signal to take off was transmitted from the Eiffel Tower, few *groupes de chasse* heard it and only about 80 fighters got airborne, some being caught on the ground.

The *Luftwaffe* was delighted with the results of the mission, believing that it had struck a mortal blow at the *Armée de l'Air* by destroying aircraft (*KG 2* alone claimed 23 on the ground) and wrecking factories. In fact, fifteen factories reported slight damage and six out of sixteen airfields attacked were seriously damaged, but only twenty aircraft were destroyed on the ground, including five Amiot 351s. French casualties on the ground were heavy—254 dead (59 servicemen) and 652 injured (107 servicemen)—while fifteen fighters (18.75 per cent) were lost in the air, where sixteen victories were claimed. The *Luftwaffe* was lucky to lose only ten aircraft, including four bombers, but both Massow (*Jafü 3*) and Kammhuber (*Kommodore KG 51*) were shot down and captured, although the former was liberated by advancing German troops on 12 June. They were replaced by *Oberst* Werner Junck (who later returned *Jafü 3* to Massow) and *Oberst Dr* Johann-Volkmar Fisser (who was killed in August 1940) respectively.[54]

Meanwhile the *Luftwaffe* prepared for 'Rot' with the usual round of conferences, although Richthofen noted in his diary that he was able to sleep in on 2 June. There were the usual problems of finding and equipping airfields, only a limited number in Belgium and northern France being suitable for twin-engine aircraft. These remained heavily congested and, with limited air traffic control facilities, collisions were not unknown, so many *Kampfgruppen* stayed in Germany. The poor weather in late May provided the 'Black Men' with the opportunity to catch up on much-needed maintenance, and serviceability was probably around 70 per cent by the beginning of 'Rot'. The most difficult task was to re-orient the signals network from the east–west to the north–south axis, but Martini's men accomplished this on time. Martini and Milch toured the forward commands to assist, both men visiting Richthofen on 3 June.

On the eve of the offensive most *Gruppen* were rested and French records show that on 4 June only two ports were attacked. There was no rest for the *Luftwaffe*'s *Fernaufklärungsstaffeln*, which monitored not only French airfields (reporting 1,244 aircraft on 4 June, of which between 550 and 650 were single-engine) but also the entire French rail network almost to the Spanish border. The *Heeresaufklärungsgruppen* concentrated upon French tactical and operational-level defences and communications.

When 'Rot' commenced on 5 June the *Luftwaffe* rapidly established air superiority, attacking nineteen airfields to 9 June and claiming 87 aircraft destroyed on the ground. The French were not cowed and flew 1,918 sorties (1,297 fighter, 518 bomber) to 9 June but lost 130 aircraft, of which nearly 28 per cent fell to *Flak* in suicidal low-level attacks upon German columns, some Bre 693 units flying three sorties a day. A subdued AASF flew 187 sorties (155 bomber) and lost a dozen aircraft in the same period. The casualties and the disruption of the supply organization caused by the German advance led to a rapid decline in the French air effort, from a peak of 598 sorties on 6 June to 179 on 9 June.[55] In the first five days of the campaign German fighters claimed 195 victories and *Flak* claimed 26, but the *Luftwaffe* lost 101 aircraft, 37 on the first day when *KG 27* (*Fliegerkorps IV*) and *JG 27* (*Fliegerkorps VIII*) each lost five and *KG 51* (*Fliegerkorps V*) had four missing.

After the first day, however, German losses steadily declined. They totalled 199 during this campaign, and only 40 bombers fell to enemy action while supporting '*Rot*'. One reason was the activities of the *Jagd-* and *Zerstörergruppen*: on the first day of '*Rot*' alone, *JG 27* flew 265 sorties in seventeen missions. From 6 June the *Luftwaffe* lost 42 Bf 109s and nine Bf 110s (one of the former belonging to Werner Mölders, who was briefly held prisoner), but between them they destroyed 126 aircraft of the *Armée de l'Air* and many of the 39 aircraft lost by the AASF. One of the *Luftwaffe*'s most successful engagements was on 9 June when a *Jagdgruppe* escorting a Do 17 reconnaissance aircraft near Vernon encountered seventeen Caudron C 714C-1 fighters of *GC I/145* and shot down six. The campaign showed the need for experienced fighter leaders, and on 6 June *Hauptmann* Adolf Galland relieved the ineffective *Major* Ernst *Freiherr* von Berg as *Kommandeur* of *III/JG 26*. The importance of the fighters was further enhanced on 8 June by *OKW Direktive Nr 14*, which ordered Kesselring to strengthen fighter cover along the coast against RAF intrusions. These involved offensive patrols by No 11 Group, which flew 1,513 fighter sorties (29 lost) from 5 to 25 June, and raids by No 2 Group, whose bombers flew 660 sorties, including harassing attacks upon Germany from 21 June, and lost 37 Blenheims.

On 9 June Rundstedt's troops joined the offensive with *Fliegerkorps II*. A dozen of Loerzer's *Kampfgruppen*, escorted by *JG 53* and *ZG 2*, struck *4e Armée*'s reserves for two hours then spent an hour pounding its main line of resistance in missions involving 572 sorties. Many *Kampfgeschwader* were still based in Germany and had to stage through bases in France, but *KG 2* flew three missions that day. As the enemy's air power faded away, the *Luftwaffe* increasingly supported the Army by attacking defences, rail targets, communications centres and troop columns. The absence of enemy fighters allowed the *Stuka*s to display remarkable accuracy. In an attack upon *Entrepôt de l'Armée de l'Air Mantères*, near Paris, Grauert's *III/StG 51* first knocked out the electricity generating facilities then demolished fuel tanks and one of the supply halls. Other attacks disrupted the rail network—by 17 June there were seventeen troop trains trapped in Rennes alone—and thousands of refugees took to the roads, where they were exposed to air attacks, leading Vuillemin on 16 June to end day bomber raids on enemy columns.[56] On 18 June *Oberst* Wilhelm *Ritter* von Thoma, who had commanded the Condor Legion's tanks, provided Halder with 'appalling details of the effect of dive-bombing attacks upon refugee columns in the lower Seine'. The refugees were not the actual target, but they became victims either because they were intermingled with troop columns or because they were identified as troops by aircraft operating at altitudes of hundreds of feet and moving at high speed. Ironically, in 1945 Bock (who became a *Generalfeldmarschall* after '*Rot*' but was sacked in 1942) and most of his family perished in identical circumstances under the guns of RAF Typhoons.

Richthofen himself was unhappy about the offensive, commenting in his diary on 5 June that only half of the aircraft lost had been replaced and that the signals net was incomplete. Neither prevented him from providing the Army with his usual high-quality service, *StG 2* being so far forward that on 5 June its 'Black Men' could watch attacks in support of *9.Armee* from the airfield! From mid-June Richthofen ensured that close air support would be available for Kleist's mechanized forces by providing another, stronger battle group with two *Jagdgruppen* and two *Stukagruppen* as well as two *Schlachtstaffeln* which continued to be based as far forward as possible. Richthofen found out what it was like to be bombed during the night of 14/15 June when French bombs dropped within 500yds of his headquarters, which had been visited by Göring and Udet the previous day. The *Stukas* were extremely active until 18 June (*StG 77* flew three missions on 6 June and four the next day), when Richthofen began resting half the aircrew in every unit for three days until the end of the campaign. His *Gruppen* lost nineteen *Stukas* and three Hs 123s during '*Rot*', the most senior casualty being Stutterheim, *Kommodore* of *KG 77*, who had won the *Pour le Mérite* as a Guards officer and was now severely injured while making a forced landing on 18 June. The injuries caused an embolism, from which he died on 3 December that year.[57]

The French defences collapsed and on 14 June, as the *Wehrmacht* occupied Paris, *OKW Direktive Nr 15* ordered the pursuit of shattered enemy forces south of the Seine. Bock, supported by Kesselring, was to drive west and south-west into Normandy and Brittany while Rundstedt and Sperrle, having isolated the Maginot Line, were to push southwards. The *Luftwaffe* was ordered to help the Army maintain its momentum, to hinder the retreat of enemy forces and to disrupt the enemy rail network. *ObdL* assigned *Fliegerkorps II* to support *Panzergruppe Guderian* and *Fliegerkorps VIII* to *Panzergruppe Kleist*, while *Fliegerkorps V* was to support *1.Armee*'s *Unternehmen 'Tiger'*, which would complete the envelopment of the Maginot Line. The *Armée de l'Air* retreated into central France, where there were few airfields, or to training bases in southern France. Many airfields, such as that at Orléans, which was soon in German hands, had 200–300 aircraft and were vulnerable to attack. Vuillemin planned to transfer *groupes* to North Africa to continue the struggle, but the loss of Paris had taken the heart out of the French Government. The British, too, had written off France, and during mid-June the survivors of the AASF (which had lost 229 aircraft) were withdrawn.

The rhythm of air operations was disrupted from 11 June by rain and low cloud but resumed on the 14th as the skies cleared, although early morning fog and haze helped to shield the retreating enemy. The *Luftwaffe* had to adapt to increasingly fluid conditions, and from 16 June the *Fliegerkorps* largely ended their support for individual armies in favour of roving missions in the enemy's rear. The same day saw *Luftflotte 2* end tactical air support missions after attacking bridges across the Loire in conjunction with *Luftflotte 3*. Fuel shortages were the biggest impediment to the *Panzers*' advance and on 17 June *II/KG 2* found itself flying transport missions.

On that day also, the new French Government, headed by Marshal Pétain, requested an armistice, and *Kampfgruppen* operations began to wind down. This was due partly to a lack of targets and partly to the continued difficulty of finding suitable bases or facilities. On 18 June, for example, *I/KG 51* flew into Paris-Orly but was largely grounded because of a lack of fuel, while fields hastily converted into airstrips did not always prove ideal. When *II/KG 30* re-joined *Luftflotte 2* in mid-June its airfield near Le Coulot, Louvain, was so small that the Ju 88s could carry only two 500kg

(1,100lb) bombs and half-full fuel tanks, the latter being topped up at Amiens.[58] Both *Luftflotten* began to stand down their *Kampfgruppen*, although there was a brief flurry of activity on 19 June when both pounded the remnants of *3e Groupe des Armées* south of the Loire.

Luftflotte 2 was concerned not only with disrupting enemy resistance but also with preventing a mass evacuation. Night attacks upon French ports began on 9/10 June when seventeen bombers dropped 15 tonnes of bombs on Cherbourg, while *Kampfgruppen* flew up to five sorties per night in two successive attacks upon Le Havre, where a 2,949grt merchantman was sunk. Within a few nights the Atlantic ports were being bombed and on 17 June the British liner *Lancastria* (16,243grt) was sunk off St Nazaire with the loss of 5,800 lives. Attacks on ports were made both in daylight and at night until 22 June, Bordeaux, for example, being attacked by *II* and *III/LG 1* and *III/KG 1* during the night of 18/19 June. Yet the Allies were still able to evacuate nearly 192,000 men.

Throughout 'Rot', *Fliegerdivision 9*'s minelayers ranged ever further westward, the *Küstfliegergruppen* being supplemented by Kesselring's *Kampfgruppen* until 13 June. The He 115s of *KüFlGr 106* were extremely active, operating off Cherbourg (during the night of 1/2 June), off the Loire (12/13 June) and off Brest (14/15 June). However, operations at Brest were restricted by the Navy, which wished to use the port for U-boat operations.

On 21 June, under the umbrella of *II/JG 26*, Hitler and Göring presented Germany's armistice terms in the very railway carriage the Allies had used in November 1918. The terms were accepted the following day and hostilities ended on 25 June, hundreds of captured *Luftwaffe* airmen, including Mölders, Kammhuber and 118 members of *KG 54*, being given their freedom. But sporadic air fighting continued until 24 June, the last *Luftwaffe* victory being by a Bf 109 which shot down a Potez 63-II of *Groupe de Reconnaissance II/14* at Montélimar on that day. *Oberleutnant* John had the melancholy distinction of being the last *Luftwaffe* airmen to be killed in the French Campaign when his Hs 126 of *5. (H)/13* was shot down on the same day near Marchelidon by an MS 406 of *GC I/2*. By accident, rather than by design, the last *Luftwaffe* sorties appear to have been by *II/KG 30*, which attacked shipping an hour after the Armistice came into effect.[59]

The German triumph owed much to the *Luftwaffe*, which had exploited air superiority to ease the path of the *Panzer* divisions by hindering enemy troop movements and then shattering centres of resistance. Fragmentary records would indicate that, in achieving this success, each *Kampfgeschwader* flew an average of 1,500 sorties during the two campaigns while *Jagd-* and *Stukageschwader* probably flew at least twice as many (see Table 28).

The price had been high, with 1,428 aircraft or 28 per cent of the *Luftwaffe*'s initial strength lost—36 per cent if the 488 damaged aircraft are included. Enemy action accounted for 1,129 aircraft destroyed, bombers and *Zerstörer* inevitably suffering most with 25 and 24 per cent respectively of initial strength lost (90 and 438 aircraft). The human cost was also high, with 1,722 killed (1,092 aircrew), 2,897 wounded (1,395) and 2,034 missing (1,930), many of the last taken prisoner (*KG 54* had 198 missing, of whom only ten were not accounted for as prisoners of war).[60] In exchange, the *Luftwaffe* had annihilated the Belgian and Dutch Air Forces and defeated those

TABLE 28: REPRESENTATIVE DATA FOR *KAMPFGRUPPEN* PARTICIPATING IN THE WESTERN CAMPAIGNS

Gruppe	Sorties	Bombs dropped (tonnes)	Remarks
I/KG 4	567	365	Also 1,352 supply containers
II/KG 4	638	954	
III/KG 51	402	373	41 bombing missions
I/KG 54	347	409	

of France and Great Britain. *JG 26* claimed 160 victories, *JG 27* 250 and *JG 51* 103. The *Armée de l'Air* lost 574 aircraft in the air (174 to *Flak*); 460 aircrew were killed and about 120 were taken prisoner, the total of aircrew casualties including 40 per cent of officers and 20 per cent of NCOs.[61] The RAF lost 959 aircraft (including 477 fighters and 381 bombers) during the same period as well as equipment worth £1 million ($4.5 million)—equivalent to four complete aircraft parks. Worse still was the loss of 915 aircrew killed or missing (including 312 pilots), with another 184 wounded.[62]

The RAF continued to fight after the fall of France, although few doubted that the British would soon sue for peace, despite Churchill's public utterances. Until mid-June the British Isles escaped the full weight of the *Luftwaffe*, which was absorbed in supporting *'Gelb'* and *'Weserübung'*, the Germans confining themselves to reconnaissance, minelaying and harassing bombing, the last probably by reconnaissance aircraft. An air offensive against the British was forcefully advocated by Schmid in a memo of 22 November 1939 and elements of this were incorporated in *OKW Direktive Nr 9*, but the majority view was that it would have to wait until issues in Western Europe had been settled. By 24 May this was in sight and *OKW Direktive Nr 13* stated that 'The *Luftwaffe* is authorized to attack the English homeland in the fullest manner, as soon as sufficient forces are available. This attack will be opened by an annihilating reprisal for English attacks upon the Ruhr.' Göring was ordered to designate targets in accordance with the principles laid down by *OKW Direktive Nr 9*.

During the night of 2/3 June *Luftflotte 2* began harassing attacks upon British airfields by dropping 1.15 tonnes of bombs on RAF Mildenhall. Further raids followed on other airfields in East Anglia and southern England, although six northern air bases were attacked on the night of 4/5 June. They were made in *Staffel* strength by individual *Kampfgruppen* but the operations all but ceased when *'Rot'* began. When the fall of France was imminent, Göring called a planning conference at Dinant on 18 June to consider operations against Britain, and this apparently decided not only to attack every night but also to strike industrial targets. By his own account, Milch pressed for an extemporized airborne operation supported by the full weight of both *Luftflotten*, but Göring correctly dismissed the plan as 'nonsense'. The proceedings were marred by a tired Richthofen querulously complaining about his joint subordination to both *Luftflotten*; this led to an unseemly quarrel with Kesselring, who remonstrated with him for the rapid transfer to *Luftflotte 3*.

Nightly attacks upon Britain actually began on 17/18 June and continued every night until the end of the month, apart from a two-night respite because of rain from 21/22 June. Targets included English and Welsh ports as well as aircraft factories in the Bristol area, but the missions were still conducted in *Staffel* strength or smaller, largely by Kesselring's *KG 4* and *KG 27* although from 28/29 June they were joined by Sperrle's *Fliegerkorps V* (*KG 51* and *KG 55*). British records suggest that *KG 26* participated in raids on 26/27 and 27/28 June but this is not confirmed by the *Geschwader* history. The scale of these attacks was overestimated by the British (from 27 to 30 June there were some 50 bomber sorties but the defenders estimated that 150 had taken place), and even including reconnaissance and minelaying operations the total *Luftwaffe* effort over England probably amounted to fewer than 500 sorties.[63] Casualties were light, but during the night of 18/19 June *KG 4* lost six aircraft, mostly to night fighters, including that of *Major* Dietrich *Freiherr* von Massenbach, *Kommandeur* of *II/KG 4*, who fell over Newcastle-upon-Tyne. With the end of the Norwegian Campaign *KGr 100* received replacement aircrew, 70 per cent of whom came from the *Blindflugschule*, while on 14 June the *Gruppe* received a new *Kommandeur*, *Hauptmann* Kurd Aschenbrenner. A week later work began on installing new *X-Gerät* transmitters in Germany and France and on June 30 June Göring issued a directive for the conduct of the air war against Great Britain, which was to commence when the *Luftwaffe* had completed its reorganization and redisposition.

Few in the *Luftwaffe* anticipated executing the directive and this view was unanimously held by the Nazi leadership. Indeed, at a celebration of the western campaigns held in Berlin's Kroll Opera House on 19 July Hitler offered the British peace and added, 'I see no reason why the war should go on.' The occasion saw a lavish award of honours, pride of place going to Göring, for whom the rank of *Reichsmarschall* was created, together with a jumbo-size marshal's baton. Smaller batons were awarded to Kesselring, Milch and Sperrle, who leapt a rank to become *Generalfeldmarschall* while similar leaps were made by Grauert, Keller, Stumpff and Udet (to *Generaloberst*) as well as Geisler, Greim, Jeschonnek, Loerzer and Richthofen (to *General der Flieger*).

Their mettle was soon to be tested, but when the awards were made the *Luftwaffe* had reached its zenith. In twenty-two years the wheel at Compiègne had turned full circle, from German humiliation to German triumph, and this was reflected in the development of the *Luftwaffe*. Denied any form of military or paramilitary aviation, the *RWM* had secretly nurtured the seeds of an air force and turned them into seedlings, which the Nazis brought to full bloom for political and strategic ends. The hand of *General* Wever helped to hone a blade which Hitler could hold at the throats of his diplomatic enemies, and by balancing all the elements of air power Wever made the *Luftwaffe* a formidable weapon which outclassed its rivals. However, his premature death led to a subtle warping which undermined everything for which he had worked. Just how badly, Göring, Milch and Jeschonnek were to discover as their aircraft battled to control Britain's skies.

1. Based upon *ZOAN* War Diary, *SHAA* 2D17; French aircraft returns in *SHAA* 1D51/12 and 1D51/15; and figures from Air Ministry daily strength returns in PRO Air 22/32.

2. Details of Dutch expenditure have been provided by H.Th. M. Kauffman of the Dutch Air Staff Historical Section.

3. Data on Dutch air operations are based largely upon Daniel's article 'La Conquete de la Hollande'.

4. For the Belgian Air Force see Terlinden's article.

5. See Dierich, *KG 51*, pp.81–4.

6. All French losses are from Martin, Annex 2.

7. Based upon Bekker, pp.93–100; Taylor, *March*, pp.210–14; and a lecture by *Major* Zierach of the *Luftkriegsakademie* on 13 March 1944, a translation of which was kindly provided by Mr Alex Vanags-Baginskis.

8. See Doughty, pp.44–5; and Rothbrust, pp.34–8, 50–2.

9. I am indebted to Dr H. Amersfoort of the Dutch Army's Military History Section for this information.

10. Based upon a Dutch Air Staff summary of air defence operations.

11. Relations between the three key commanders were strained. Kesselring resented Student's access to Hitler, which he often used, while Student appears to have had doubts about Sponeck. For these airborne operations see Bekker, pp.100–7; Bingham, pp.50–2; Franks, *Valiant Wings*, pp.86–90; Kesselring's memoirs, p.55; Macksey, p.69; Morzik, pp.52–3, *Anlage 11*; and articles by Ausems, Boog and Daniel. A copy of a letter from *Generalleutnant* W. Ehrig to *Major* Werner Oissin of 12 January 1955 was kindly made available by Mr Alex Vanags-Baginskis.

12. *SHAA* File 1D34/12.

13. *ZOAN* War Diary, *SHAA* File 2D17.

14. The higher figures are the author's calculations based on GQGAé War Diary, *SHAA* File 1D44/1, and *ZOAN* War Diary, *SHAA* File 2D17; the lower figures are from Martin.

15. Report of AASF, PRO Air 25/255; Dispatch of Air Marshal A. S. Barratt, PRO Air 35/192; No 11 Group operations record, PRO Air 25/193; Air Support in France, PRO Air 8/287; Fighter Command Operations record, PRO AIR 24/507.

16. Based on PRO AIR 22/477: Air Ministry W/T Intelligence Service Daily Summaries, Vol. 2.

17. See Daniel's article and *Fliegerkorps VIII* War Diary, BA MA File RL 8/45.

18. For Rotterdam see Bekker, pp.107–14; Radtke, pp.29–31; and Taylor, *March*, pp.200–5. See also the article by Boog. I am indebted to Dr H. Amerfoort for additional information.

19. Quoted by Middlebrook and Everitt, p.43.

20. US Monograph 152, p.215.

21. For *Fliegerkorps VIII* operations see account in BA MA File RL 8/45; and Richthofen Diary, BA MA N 671/6.

22. US Monograph 152, p.175.

23. Balke, pp.79–87.

24. Quoted by Rothbrust, Appendix F.

25. *Ibid.*

26. Barker, p.34.

27. Doughty, p. 206. For Sedan see Balke, pp.87–90; Bekker, pp.118–19; Doughty, pp.166–216; Rothbrust pp.67–87, Appendix F; Smith, pp.51–53; Taylor, *March*, pp.219–29; *Fliegerkorps VIII* War Diary, BA MA RL8/45; and Richthofen Diary, BA MA N 671/6.

28. For Schwartzkopff see Smith, pp.20–31, 52–3.

29. PRO Air 22/477.

30. Quoted in US Monograph 152, p.182.

31. British figures from *Mission Française 'A'* to Air Ministry, *SHAA* File 1D41; French figures from Martin.

32. Kindly made available by Mr Alex Vanags-Baginskis.

33. Steinhilper, p.255.

34. *Ibid.*, pp.240–1.

35. Curiously, Morzik and Hümmelchen's history of the *Luftwaffe* transport force makes no mention of these vital operations.

36. *II/KG zbV* 1 history, p.45.

37. PRO Air 22/477.

38. Stahl, pp.42–3.

39. *ZOAN* War Diary, *SHAA* File 2D17.

40. Bekker, pp.120–2.

41. Ellis, p.246.

42. Schmid. A copy of his statement was made available to me through the courtesy of Mr Alex Vanags-Baginskis.

43. The best descriptions of the Dunkirk air operations are Huan and Marchand 's article; Bekker, pp.126–30; Franks, *Dunkirk*; and US Monograph 152, pp.321–73.

44. *Fliegerkorps VIII* after-action report of 1 September 1940.

45. Christienne and Lissarague, pp.352–3, 359–60.

46. PRO Air 35/197.

47. Kesselring, p.60.

48. GQGAé War Diary, *SHAA* File 2D2.

49. Diary, BA MA MSG 1/1 1410.

50. *Ibid.*

51. RAF Fighter Command's Operations Record for 31 May noted: 'It is learnt that the Germans contemplate making a large bombing attack on Paris.' PRO Air 24/507.

52. In March 1944 the *Luftwaffe* formed *Luftbeobachtungsstaffeln* to shadow US bomber formations.

53. Dispatches, PRO Air 35/197.

54. For the French background to *'Paula'* see Christienne and Lissarague, pp.353–4; Jackson, p.80; and Daniel's article 'Les Dix Premiers Mois d'Opérations de la Luftwaffe'.

55. GQGAé War Diary, *SHAA* File 2D2; Paquier.

56. Christienne and Lissarague, p.356.

57. For Stutterheim see Taylor, *March*, p.313n.

58. Stahl, pp.46–8.

59. Stahl, p.56.

60. Data based on Hahn, p.203; and Murray, *Strategy*, Table III.

61. Based on Christienne and Lissarague, pp.360–1, but losses amended with Martin, Annex 2.

62. Bingham, pp.148–9.

63. For these attacks see *Einsatz gegen England*, BA MA RL7/88; Air Ministry Weekly Intelligence Summary, Vol. 3, PRO Air 22/71; and Fighter Command Operations Record, Air 24/507.

APPENDICES

APPENDIX 1
GERMAN AIR FORCES, 11 NOVEMBER 1918

WESTERN FRONT

HEERESGRUPPE KRONPRINZ RUPPRECHT
Kommandeur der Flieger 4.Armee
3 *Gruppenführer der Flieger*
2 *Jagdgruppen* and 8 *Jagdstaffeln*
Bogohl 3 with 3 *Bombenstaffeln*
4 *Schlachtstaffeln*
7 *Fliegerabteilungen*
5 *Fliegerabteilungen (A)* (1 on higher establishment)
1 *Reihenbildzug* attached under *OHL* command: *RFA 500*

Kommandeur der Flieger 6.Armee
1 *Gruppenführer der Flieger*
6 *Jagdstaffeln*
Bogohl 6 with 3 *Bombenstaffeln*
1 *Fliegerabteilung*
5 *Fliegerabteilungen (A)* (2 on higher establishment)

Kommandeur der Flieger 17.Armee
4 *Gruppenführer der Flieger*
1 *Jagdgeschwader* with 4 *Jagdstaffeln*
1 *Jagdgruppe* with 4 *Jagdstaffeln*
Bogohl 5 with 3 *Bombenstaffeln*
2 *Schlachtgruppen* with 8 *Schlachtstaffeln*
1 *Fliegerabteilung*
10 *Fliegerabteilungen (A)* (7 on higher establishment)
1 *Reihenbildzug*

Kommandeur der Flieger 2.Armee
2 *Gruppenführer der Flieger*
1 *Jagdgeschwader* with four *Jagdstaffeln*
1 *Jagdgruppe* with three *Jagdstaffeln*
Bogohl 4 with three *Bombenstaffeln*
3 *Schlachtstaffeln*

3 *Fliegerabteilungen*
7 *Fliegerabteilungen (A)* (4 on higher establishment)

Kommandeur der Flieger 18.Armee
4 *Gruppenführer der Flieger*
2 *Jagdgruppen* with 8 *Jagdstaffeln*
Bogohl 1 with 3 *Bombenstaffeln*
1 *Schlachtgeschwader* with 4 *Schlachtstaffeln*
2 *Fliegerabteilungen*
11 *Fliegerabteilungen (A)* (7 on higher establishment)
1 *Reihenbildzug*

HEERESGRUPPE DEUTSCHER KRONPRINZ
Kommandeur der Flieger 7.Armee
2 *Gruppenführer der Flieger*
1 *Jagdgruppe* with 4 *Jagdstaffeln*
Bogohl 2 with 3 *Bombenstaffeln*
1 *Fliegerabteilung*
11 *Fliegerabteilungen (A)* (6 on higher establishment)
2 *Reihenbildzuge*

Kommandeur der Flieger 1.Armee
2 *Jagdstaffeln*
1 *Fliegerabteilung*
5 *Fliegerabteilungen (A)* (three on higher establishment) attached under *OHL* command: *RFA 501*

Kommandeur der Flieger 3.Armee
1 *Jagdgruppe*
8 *Jagdstaffeln*
1 *Schlachtgeschwader* with 5 *Schlachtstaffeln*
4 *Fliegerabteilungen*
7 *Fliegerabteilungen (A)* (four on higher establishment)

HEERESGRUPPE GALLWITZ
Kommandeur der Flieger 5.Armee
5 *Gruppenführer der Flieger*
Jagdgeschwader 1 with four *Jagdstaffeln*
Jagdgeschwader 3 with four *Jagdstaffeln*
1 *Jagdgruppe*
7 *Jagdstaffeln*
Bogohl 7 with three *Bombenstaffeln*
1 *Schlachtgeschwader* with 4 *Schlachtstaffeln*
1 *Schlachtgruppe* with 5 *Schlachtstaffeln*
4 *Fliegerabteilungen*
11 *Fliegerabteilungen (A)* (2 on higher
   establishment)

**Kommandeur der Flieger Armee
Abteilung 'C'**
2 *Gruppenführer der Flieger*
1 *Jagdstaffel*
1 *Schlachstaffel*
4 *Fliegerabteilungen*
8 *Fliegerabteilungen (A)* (one on higher
   establishment)

HEERESGRUPPE ALBRECHT,
HERZOG VON WÜRTTEMBERG
Kommandeur der Flieger 19.Armee
1 *Gruppenführer der Flieger*
1 *Jagdgruppe* with 4 *Jagdstaffeln*
Bogohl 8b with 3 *Bombenstaffeln*
4 *Schlachtstaffeln*
3 *Fliegerabteilungen*
5 *Fliegerabteilungen (A)*
1 *Reihenbildzug*

Armee Abteilung 'A'
1 *Jagdgruppe* with 4 *Jagdstaffeln*
Bogohl 9 with 3 *Bombenstaffeln*
4 *Fliegerabteilungen*
1 *Fliegerabteilung (A)*
1 *Reihenbildzug*

Note: *Armee Abteilung 'A'* had no
   Kommandeur der Flieger

**Kommander der Flieger Armee
Abteilung 'B'**
1 *Jagdgruppe* with 4 *Jagdstaffeln*
1 *Fliegerabteilung*
3 *Fliegerabteilungen (A)*

EASTERN FRONT

Kommandeur der Flieger 8.Armee
2 *Fliegerabteilungen*
1 *Frontstaffel*

Kommandeur der Flieger 10.Armee
1 *Fliegerabteilung (A)*

HEERESGRUPPE KIEW
4 *Fliegerabteilungen*
1 *Sonderkommando*

Deutsche Truppe in Kaukasus
1 *Fliegerabteilung*

BULGARIA

2 *Jagdstaffeln*
5 *Fliegerabteilungen*
2 *Fliegerabteilungen (A)*

MIDDLE EAST

2 *Jagdstaffeln*
4 *Fliegerabteilungen*

MARINEFLIEGER

**Kommandeur der Flieger der
Hochseestreitkräfte**
*Seeflugstationen Helgoland, Borkum,
   Norderney, List a.Sylt*
Cruiser *Stuttgart*
Seaplane carrier *Santa Elena*

**Kommandeur der Flieger beim
Festungs Gouvernement
Wilhelmshaven**
*Landflugstationen Apenrade, Bug a. Rügen,
   Flensburg, Holtenau, Kiel, Putzig,
   Warnemünde*

**Kommandeur der Flieger bei
Befehlshaber der Baltischen Gewässer**
*Seeflugstationen Libau, Papenholm, Reval,
   Windau*
Seaplane carriers *Answald, Glyndwr, Oswald*

Kommandeur der Flieger beim Marinekorps Flandern
Gruppen der Seeflieger
Seeflugstation Flandern III
Marinejagdgeschwader with 5 Jagdstaffeln

Gruppenkommandeur der Küstenflieger
3 Küstenfliegerstaffeln
2 Küstenschutzstaffeln

Gruppenkommandeur der Marine-Feldflieger
2 Marine-Feldfliegerstaffeln

Independent
Seeflugstationen Varna, Zupuldak, Constanta, Duingi

Sources: Western Front details from *Reichsarchiv* document received by Air Historical Branch, 15 January 1922 (PRO Air 1/9/15/1/22); Eastern Front details from *Darstellungen aus den Nachkriegskämpfen deutscher Truppen und Freikorps*, Band 1, pp.181–94; Bulgarian Front and Middle East Front orders of battle partly conjectural.

APPENDIX 2
AUSTRIAN AIR FORCES, 4 NOVEMBER 1918

WESTERN FRONT

3 Fliegerkompanie (F-Flik)

ITALIAN FRONT

19 Fliegerkompanie (J) (and one forming)
5 Fliegerkompanie (G)
24 Fliegerkompanie (D)
6 Fliegerkompanie (F)
14 Fliegerkompanie (S)
1 Fliegerkompanie (A)

ALBANIA

1 Fliegerkompanie (J)

ROMANIA

2 Fliegerkompanie (D)

RUSSIA

2 Fliegerkompanie (F)

AUSTRIA-HUNGARY

3 Fliegerkompanie (D)

KuK Seefliegerkommando

Seeflugstation Mitte (Santa Catarina/Pola)
Seeflugstation Trieste
Seeflugstation Kumbor

Note: A = Artillerie; D = Division; F = Fernaufklärung; G = Grossflugzeug; J = Jagd; S = Schlacht

Source: Final report of *Commission Aéronautique Interallie de Control en Autriche*, *SHAT* File 4N 105, Dossier 3

APPENDIX 3
EXPENDITURE ON AIR FORCES, 1933–1939

GERMANY

Year	Expenditure (RMm)	US equivalent ($m)	% Defence Budget
1933	76	18.09	10.0
1934	642	24.59	32.9
1935	1,036	422.57	37.4
1936	2,225	897.17	38.2
1937	3,258	1,313.71	39.4
1938	6,026	2,420.08	34.9
1939	3,942 *	1,583.13	33.1
Total	17,205	6,679.34	

* To September 1939.

FRANCE

Year	Air Budget (FFrm)	US equivalent ($m)	% Defence Budget
1933	1,490	5.82	12.7
1934	1,630	9.88	15.9
1935	2,000	13.09	19.2
1936	2,660	17.39	18.5
1937	3,890	18.17	18.7
1938	6,500	22.12	22.9
1939	38,700	97.24	33.1
Total	56,870	183.71	

GREAT BRITAIN

Year	Air Budget (£m)	US equivalent ($m)	% Defence expenditure
1933	16.78	66.45	15.55
1934	17.63	88.15	15.48
1935	27.50	135.56	19.50
1936	50.13	247.16	26.10
1937	82.29 *	404.04	31.38
1938	133.80 *	669.00	34.89
1939	105.70	489.40	36.20
Total	433.83	2,099.76	

* Figures in 1937 and 1938 include issues under the Defence Loans Act.

UNITED STATES

Year	Appropriation for USAAC ($m)	% Military appropriation
1933	25.4	8.8
1934	23.3	8.4
1935	27.4	10.7
1936	45.4	13.3
1937	59.4	15.5
1938	58.6	14.1
1939	70.6	15.7
Total	**310.1**	

Note: All years are financial years and dollar values are at contemporary exchange rates.
Sources: Figures for Germany are based on *Die Finanzierung der Aufrustung im Dritten Reich* by *Dr* Heinrich Stuebel, Europa-Archiv 12 (1951), but with amendments from *Big Business in the Third Reich* by Arthur Schweitzer, Indiana University Press (Bloomington, Indiana, 1964), p.331. Both are quoted in Homze, p.259. See also Overy, p.38. Figures for France are based on Christienne and Lissarague, p.304, as well as *France and the Coming of the Second World War* by Anthony Adamthwaite, Frank Cass and Co (London, 1977), p.164, Table 5. Figures for Great Britain are based on data in R. Higham's *Armed Forces in Peacetime* (London, 1932), quoted by Malcolm Smith in *British Air Strategy Between the Wars*, p.336; and James, p.246. Figures for the USA are from Mauer, Appendix 8. The direct appropriation represented an average of 41 per cent of actual expenditure on the USAAC between 1926 and 1934. It did not include pay, construction, supplies and other ancillaries, nor did it include material issued from the war reserve.

APPENDIX 4
PRE-WAR PRODUCTION AND PLANS

GERMANY

Production

Designation	Begins	Completion date	Aircraft
Rheinland Plan	Jan 1934	30 Sept 1935	4,021
Extended *Rheinland*	Jan 1935	30 Sept 1936	5,112
Lieferplan Nr 1	Oct 1935	1 Apr 1936	11,158
Lieferplan Nr 2	Mar 1936	31 Mar 1938	12,309
Lieferplan Nr 4	Oct 1936	31 Mar 1938	10,202
Lieferplan Nr 5	Apr 1937	1 Oct 1938	7,831
Lieferplan Nr 6	July 1937	31 Mar 1939	10,059
Lieferplan Nr 7	Jan 1938	31 Mar 1939	8,790
Lieferplan Nr 8	Apr 1938	31 Mar 1940	16,404
Lieferplan Nr 10	Jan 1939	30 June 1941	18,994
Lieferplan Nr 11	Apr 1939	31 Mar 1942	24,317

Note: *Lieferplan Nr 3* was an interim scheme. No information is available concerning *Lieferplan Nr 9*.

Order of battle plans (Staffeln)

Published	Completion	Fighter	Bomber	Recon	Naval	Total
10 Apr 1933	1 Oct 1937	6	9	16	6	37
22 May 1933	1 Oct 1935	6	27 *	12	6	51
29 Aug 1935	1 Oct 1938	45	126 **	36	23	227

* Five auxiliary bomber.   ** Six auxiliary bomber, 27 dive bomber.
Note: Reconnaissance *Staffeln* are evenly divided between long-range (*fernaufklärung*) and corps (*nahaufklärung*) units.

FRANCE

Plan	Approved	Completion date	Aircraft	Strength
I	July 1934	Dec 1939	1,365	1,010 (first line)
II	Nov 1936	Dec 1939	2,796	1,518 (first line)
				1,277 (reserves)
V	Mar 1938		4,739	2,617 (first line)

GREAT BRITAIN

Scheme	Approved	Completion date	Aircraft	
'A'	July 1934	31 Mar 1939	1,252	
'C'	May 1935	31 Mar 1937	1,512	
'F'	Feb 1936	31 Mar 1939	1,736	*
'L'	Apr 1938	31 Mar 1940	2,378	*
'M'	Nov 1938	31 Mar 1942	2,549	*

* 225 per cent reserves
Note: Aircraft strength is Home Defence, first-line.

Order of Battle Plans

Scheme	Fighter	Bomber				Recon	Total
		Light	Medium	Heavy	Torpedo		
'A'	28	25	8	8	2	13	84
'C'	35	30	18	20	2	18	123
'F'	30	–	48	20	2	24	124
'L'	38	–	26	47	–	30	141
'M'	50	–	–	85	–	28	163

APPENDIX 5
AIRCRAFT PRODUCTION, 1933–1939

GERMANY

Year	Total	(Combat)
1933	368	(197)
1934	1,968	(840)
1935	3,183	(1,823)
1936	5,112	(1,530)
1937	5,606	(2,651)
1938	5,235	(3,350)
1939 (to 31/8)	8,295	(4,733)
Total	29,767	(15,124)

FRANCE

Year	Number
1933	?
1934	197
1935	436
1936	581
1937	743
1938	1,382
1939 (to Sept))	1,092 (+ 176 from USA)

GREAT BRITAIN

Year	Total	(Combat)
1933	1,102	(?)
1934	1,108	(?)
1935	893	(499)
1936	1,830	(871)
1937	2,218	(1,294)
1938	2,828	(1,393)
1939 (year)	7,940	(3,731)
Total (so far)	15,709	(7,788)

SOVIET UNION

Year	Military aircraft
1933	2,952
1934	3,109
1935	2,529
1936	3,578
1937	4,769
1938	5,469
1939	10,382

Figures for Germany are based on the United States Strategic Bombing Survey. Those for Great Britain are based on *Statistical Digest of the War*, HMSO (London, 1951). Those for the Soviet Union are based on *Red Phoenix:The Rise of Soviet Air Power 1941–1945* by Von Hardesty, Smithsonian Institution Press, App. 9; Tyushkevich, p.191; Boyd, p.36, 98; and Groehler, p.207.

APPENDIX 6
LUFTWAFFE EXPANSION, 1935–1938

Staffeln

Date	Fighter	Bomber*	Stuka/ Schlacht	Aufklärung		Naval	Trans*	Total
				Fern	Nah			
1 Apr 1935	5	13	1 /–	5	6	4	–	34
1 Oct 1935	8	21	3 /–	5	6	4	–	47
1 Oct 1936	21(+1)	45(+1)	9 /–	15	17	4	– (+1)	111 (+3)
1 Oct 1937	43(+4)	87(+4)	18 /–	17(+1)	18	15(+1)	4	202 (+10)
1 Oct 1938	67(+3)	93(+4)	22 /18	17(+1)	22	16(+1)	8	263 (+9)

* Excludes Lufthansa auxiliary bombers/transports in *KG 172/KGzbV 172*.
Note: Figures in parentheses refer to units stationed in Spain.

APPENDIX 7
GERMAN AIRCRAFT DELIVERIES TO SPAIN, 1936–1939

1936

July/August: 15 He 51B, 21 Ju 52/3mg
September: 18 He 51B, 2 Hs 123A, 1 He 50G, 21 He 46C. 1 He 59B, 1 He 60E
October: 18 He 51B, 5 Ju 52/3mg, 2 He 70F, 1 He 59B, 1 He 60E, 1 Ju 52/3mW
November: 24 He 51B, 1 He 112V, 33 Ju 52/3mg, 1 Ju 87A, 14 He 45c, 10 He 70F, 3 He 59B, 2 He 60E, 2 Bf 108B, 3 Junkers W 34hi, 2 Ju 52/3m *See*
December: 3 Bf 109V, 25 Bü 131

1937

39 He 51C, 26 Bf 109B, 54 Ju 52, 4 Do 17E, 50 He 111B, 5 Ju 86D, 3 Hs 123, 11 Do 17F, 15 He 45, 13 He 70F, 5 He 59B, 2 He 60E, 6 Ar 66C, 25 Bü 131, 20 Bü 133, 2 Junkers W 34, 3 Kl 32

1938

17 He 51C, 19 Bf 109B, 5 Bf 109C, 28 Bf 109D, 10 Bf 109E, 3 Ar 68E, 3 He 112, 40 He 111 E/J, 12 Hs 123, 3 Ju 67A, 5 Ju 87B 8 Do 17F, 6 He 70F, 3 He 45, 6 Hs 126A 9 Ar 95A, 5 He 59B, 21 Go 145A

1939

7 Bf 109D, 35 Bf 109E, 6 He 112B, 10 He 111E, 4 Do 17P, 15 He 45, 2 He 59B, 2 He 60E, 2 He 115A

APPENDIX 8
ÖSTERREICHISCHE LUFTSTREITKRÄFTE, MARCH 1938

FLIEGERREGIMENT 1 (VIENNA)

Stabstaffel
Avro 626, De Havilland D.H.60GIII, De Havilland D.H.80A, De Havilland D.H.84A, Falke RVa, Focke Wulf FW 58, Gotha Go 145, Hopfner HS 9/36, Hopfner HS 10/35, Junkers A 35b, Junkers Ju 52

Jagdgeschwader II
Jasta 4: Fiat CR 32, Fiat CR 30★, De Havilland DH 60GIII★
Jasta 5: Fiat CR 32, Fiat CR 30★, Udet U12a★
Jasta 6: Fiat CR 32, Fiat CR 30★

Aufklärungsgeschwader I
Aufklsta 1: Fiat A 120/A 120R, Udet U12a★
Aufklsta 2: IMAM Romeo Ro 37, Hopfner HS 8/29★
Schulsta Thalerhof: DFS Habicht, Falke RVa, Focke Wulf FW 44, Gotha Go 145,

FLIEGERREGIMENT 2 (GRAZ)

Jagdgeschwader II
Jasta 1: Fiat CR 20*bis/bip*, Falke RVa★, Focke Wulf FW 44★, Focke Wulf FW 56★, Gotha Ga 145★, Phönix L2c★
Jasta 2: Fiat CR 20*bis*, Hopfner HS9/35★, Focke Wulf FW 44★, Breda Ba 28★, Gotha Go 145★
Jasta 3: Fiat CR 20*bis/bip*, Focke Wulf FW 58★, Breda Ba 28★

Bombengeschwader
Bomsta 1: Focke Wulf FW 58, Caproni Ca 133, Breda Ba 28★, Focke Wulf FW 44★
Bomsta 2: Caproni Ca 133, Focke Wulf FW 58, Junkers F 13★

Schulgeschwader
Schulsta A: Caproni Ca 100, De Havilland D.H.60GIII, Hopfner HS 9/35, Udet U12
Schulsta B: De Havilland D.H.60GIII, Focke Wulf FW 44, Gotha Go 145, Hopfner HM 13/34

★ Attached trainers

APPENDIX 9
LUFTWAFFE ORDER OF BATTLE, 1 SEPTEMBER 1939

OBERBEFEHLSHABER DER LUFTWAFFE

Stab Luft Nachrichtenabteilung 100

Köthen:	*1./Ln Abt 100*	Ju 52
	2./Ln Abt 100	He 111H

Aufklarüngsgruppe Ob.d.L

Berlin-Werder:	*8. (F)/LG 2*	Do 17 P/F
	10. (F)/LG 2	He 111, Do 18, He 115
Berlin-Gatow:	*Westa 1/Ob.d.L.*	He 111J

Fliegerdivision 7: Generalmajor Student (*Wahlstatt*)
KG zbV 1 (*Oberstleutnant* Morzik)

Liegnitz:	*Stab*	Ju 52
Schonfeld-Seifersdorf:	*I*	Ju 52
	II	Ju 52
Liegnitz u. Lüben:	*IV*	Ju 52
Aslau:	*III*	Ju 52
	KGr zbV 9	Ju 52 (attached)

Attached to *Luftgau zbV*
KG zbV 2 (*Oberst* Conrad)

Küpper-Sagan:	*Stab*	Ju 52
Sorau:	*I*	Ju 52
Freiwaldau:	*II*	Ju 52
	III	Ju 52
Breslau-Gandau:	*IV*	Ju 52

KG zbV 172 (*Major* von Gablenz)

Berlin-Tempelhof:	*Stab*	Ju 52
	II	Ju 52
	III	Ju 52

WEST

Luftflotte 2: *General der Flieger* Felmy (Brunswick)

Goslar:	*1. (F)/122*	Do 17P
Münster:	*2. (F)/122*	He 111H
Braunschweig:	*Wekusta 26*	Do 17P

Fliegerdivision 3: *Generalmajor* Putzier (Münster)
KG 54 (*Oberst* Lackner)

Fritzlar:	*Stab*	He 111P
	I	He 111P
Gütersloh:	*II*	He 111P
Jever:	*I/KG 25* (attached)	Ju 88A

Fliegerdivision 4: *General der Flieger* Keller (Brunswick)

Lubeck-Blankensee:	*I/KG 26*	He 111H

KG 27

Transferred to *Fliegerdivision 1* from midday

KG 55 (*Generalmajor* Süssmann)

Wesendorf:	*Stab*	He 111P
	II	He 111P
Dedelstorf:	*I*	He 111P

Luftgau XI (Hannover)

Nordholz:	*II/JG 77*	Bf 109E
Kiel-Holtenau:	*II(J)/TrGr 186*	Bf 109B/E

ZG 26 (*Oberst* von Doring)

Varel:	*Stab*	Bf 109D
	I	Bf 109D
Neumünster:	*III* (also designated *JGr 126*)	Bf 109D

Luftgau VI (Münster)

Koln-Ostheim:	*11. (N)/LG 2*	Bf 109D
Bonn-Hangelar:	*I/ JG 52*	Bf 109E
Werl:	*II/ZG 26*	Bf 109D

JG 26 (*Oberst* von Schleich)

Obendorf:	*Stab*	Bf 109E
	I	Bf 109E
Bonninghardt:	*II*	Bf 109E

Luftflotte 3: *General der Flieger* Sperrle (Roth bei Nürnberg)

Aufklärungsgruppe 123

Ansbach:	*1. (F)/123*	Do 17P
Würzburg:	*2. (F)/123*	Do 17P
Roth:	*Wekusta 51*	He 111J

Fliegerdivision 5: Generalmajor von Greim (Gersthofen bei Augsburg)

Biblis:	I/ZG 52 (also designated	Bf 109D
	JGr 152)	

KG 51 (Oberst Dr Fisser)

Landsberg:	Stab	He 111H
Memmingen:	I	He 111H
	III	He 111H

Fliegerdivision 6: Generalmajor Dessloch (Frankfurt/Main)

Gablingen:	II/ZG 76 (also	Bf 109D
	designated JGr 176)	
Wertheim:	III/StG 51	Ju 87B, Do 17P

KG 53 (Oberst Stahl)

Schwabisch Hall:	Stab	He 111H
	II	He 111H
Giebelstadt:	III	He 111H

Luftgau VII (Munich)

Eutingen:	I/JG 51	Bf 109E
Fürstenfeldbruck:	I/JG 71	Bf 109D
Boblingen:	10., 11./JG 72	Ar 68

Luftgau XII (Wiesbaden)

JG 53 (Oberst Klein)

Wiesbaden-Erbenheim:	Stab	Bf 109E
Kirchberg:	I	Bf 109E
Mannheim-Sandhofen:	II	Bf 109E

Luftgau XIII (Nuremberg)

Herzogenaurach:	1., 2./JG 70	Bf 109D

Fliegerführer der Luftstreitkräfte West: Generalmajor Bruch (Jever)

Wilhelmshaven:	1.BordFlGr 196	He 60

KüFlGr 106

Norderney:	1.	He 60
	2.	Do 18
	3./KüFlGr 706 (attached)	He 59
Borkum:	3.	He 59

KüFlGr 306

Hornum/Sylt:	2.	Do 18
	2./KüFlGr 506 (attached)	Do 18
	2./KüFlGr 606 (attached)	Do 18

KüFlGr 406

List/Sylt:	1.	He 115
	2.	Do 18

| | 3. | He 59 |

Koluft Heeresgruppe C

	1. (F)/22	Do 17P
AK V:	3. (H)/13	Hs 126
AK XXX:	4. (H)/12	He 45, He 46

Koluft Armee Abteilung A

| AK VI: | 2(H)/12 | Hs 126 |

Koluft 5.Armee

	2. (F)/22	Do 17P
Generalkommando der		
Grenztruppen Eifel:	1. (H)/12	Hs 126

Koluft 1.Armee

	3. (F)/22	Do 17P
AK IX:	1. (H)/13	Hs 126
AK XII:	4. (H)/22	Hs 126
Generalkommando der		
Grenztruppen Saarpfalz:	1. (H)/23	Hs 126

Koluft 7 Armee

	7. (F)/LG 2	Do 17 P/F
Generalkommando der		
Grenztruppen Oberrhein:	2. (H)/13	Hs 126

EAST

Luftflotte 1: *General der Flieger* Kesselring

Fliegerdivision 1: *Generalleutnant* Grauert (Schonfeld/Crossinsee)

Schonfeld/Crossinsee:	2. (F)/121	Do 17P/F
Malzkow:	1. (J)/LG 2	Bf 109E
Lottin:	2., 3. (J)/LG 2	Bf 109E
Lichtenau:	II/ZG 1 (initially; later to *Lehr Division*)	Bf 109E
Mühlen:	I/ZG 1	Bf 110C
Stolp-Reitz:	II/StG 2	Ju 87B, Do 17P
	IV(St)/LG 1	Ju 87B, Do 17P

| Stolp-West: | III/StG 2 | Ju 87B, Do 17P |
| | 4. (St)/TrGr 186 | Ju 87B |

KG 1 (*Generalmajor* Kessler)
Kolberg:	*Stab*	He 111H
	I	He 111E
Pinnow-Plathe:	*I/KG 152* (attached)	He 111H

KG 26 (*Generalmajor* Siburg)
Gabbert:	*Stab*	He 111H
	II	He 111H
Schonfeld/Crossinsee:	*I/KG 53*	He 111H

KG 27 (*Oberst* Behrendt) (to *Luftflotte 1* from midday)
Werneuchen:	*Stab*	He 111P
	I	He 111P
Neuhardenberg:	*II*	He 111P
Königsberg/Neumark:	*III*	He 111P

Luftwaffenlehrdivision: *Generalmajor* Foerster (Gut Wickbold/Jesau)
| Jesau: | *4. (F)/121* | Do 17P/F |
| | *I (Z)/LG 1* | Bf 110C |

LG 1 (*Oberst Dr* Knauss)
Neuhausen:	*Stab*	He 111H
Powunden:	*II(K)*	He 111H
	III(K)	He 111H

KG 2 (*Oberst* Fink)
Jesau:	*Stab*	Do 17Z
Gerdauen:	*I*	Do 17M
Schippenbeil:	*II*	Do 17Z

Luftwaffenkommando Ostpreussen: *Generalleutnant* Wimmer (Königsberg-Ballith)
Neuhausen:	*1. (F)/120*	Do 17P
Gutenfeld:	*I/JG 1* (attached to Luftgau I)	Bf 109E
	I/JG 21 (attached to Luftgau I)	Bf 109D
Elbing:	*I/StG 1*	Ju 87B, Do 17P

KG 3 (*Oberst* von Chamier-Glisczinski)
Elbing:	*Stab*	Do 17Z
Heiligenbeil:	*II*	Do 17Z
	III	Do 17Z

Luftgau III (Berlin)
JG 2 (*Oberstleutnant* von Massow)
| Doberitz: | *Stab* | Bf 109E |

	I	Bf 109E
Straussberg:	*10. (N)*	Bf 109D

Luftgau IV (Dresden)
JG 3 (Oberstleutnant Ibel)

Zerbst:	*Stab*	Bf 109E
Brandis:	*I*	Bf 109E
Sprottau:	*I/JG 20* (attached)	Bf 109E

Luftflotte 4: *General der Flieger* Lohr (Reihenbach)

Schweidnitz:	*3. (F)/123*	Do 17P
	Wekusta 76	He 111J

Fliegerdivision 2: *Generalmajor* Loerzer (Grottkau)

Woisselsdorf:	*3. (F)/122*	Do 17P
Ohlau:	*I/ZG 76*	Bf 110C
Nieder-Ellguth:	*I/StG 2*	Ju 87B, Do 17P

KG 4 (Oberst Fiebig)

Oels:	*Stab*	He 111P
	II	He 111P
Langenau:	*I*	He 111P
	III	He 111P

KG 76 (Oberst Schultheiss)

Breslau-Schongarten:	*Stab*	Do 17Z
	I	Do 17Z
Rosenborn:	*III*	Do 17Z

KG 77 (Oberst Seywald)

Grottkau:	*Stab*	Do 17E/F
	II	Do 17E
Brieg:	*I*	Do 17E
	III	Do 17E

Fliegerführer zbV: *Generalmajor* von Richthofen (Oppeln)

Schlosswalden:	*1. (F)/124*	Do 17P

LG 2 (Oberstleutnant Baier)

Nieder-Ellguth:	*Stab*	Bf 109E
Altsiedel:	*II(Schl)*	Hs 123

StG 77 (Oberst Schwartzkopff)

Neudorf:	*Stab*	Ju 87B
	II	Ju 87B, Do 17P
Ottmuth:	*I*	Ju 87B, Do 17P
Nieder-Ellguth:	*I/StG 76* (attached)	Ju 87B, Do 17P
Gross-Stein:	*I/ZG 2* (attached; also designated *JGr 102*)	Bf 109D

Luftgau VIII (Breslau)

Ottmütz:	*I/JG 76*	Bf 109E
Juliusburg-Nord:	*I/JG 77*	Bf 109E

Slovenske Vzousne Zbrane (Slovak Air Force)

	9th, 10th, 12th, 16th Squadrons	Letov S.328
	14th Squadron	Aero A.100, Letov S.328
	37th, 38th, 39th, 45th Squadrons	Avia B.534

Fliegerführer der Luftstreitkräfte Ost: Generalmajor Coeler (Dievenow)

Kiel-Holtenau:	*5./BordFlGr 196*	He 60

KüFlGr 506

Pillau:	*1.*	He 60
	3.	He 59

KüFlGr 706

Nest:	*1.*	He 60
	1./KüFlGr 306 (attached)	He 60

Koluft Heeresgruppe Nord

	2. (F)/11	Do 17P

Koluft 3.Armee

	3. (F)/10	Do 17P
AK I:	*2. (H)/10*	Hs 126
AK XXI:	*1. (H)/10*	Hs 126

Koluft 4.Armee

	3. (F)/11	Do 17P
AK II:	*3. ('H)/21*	Hs 126
AK III:	*2. (H)/21*	Hs 126
AK XIX (Mot) & *3.PzDiv:*	*9. (H)/LG 2*	Hs 126

Koluft Heeresgruppe Süd

	4. (F)/11	Do 17P
AK VII:	*4. (H)/31*	He 45, He 46

Koluft 8.Armee

	1. (H)/21	Hs 126
AK X:	4. (H)/23	He 45, He 46
AK XIII:	5. (H)/13	He 45, He 46

Koluft 10.Armee

	3. (F)/31	Do 17P
AK IV:	1. (H)/41	Hs 126
AK XIV (Mot):	3. (H)/12	Hs 126, He 46
AK XV (Mot):	1. (H)/11	Hs 126, He 46
3.lei Div:	3. (H)/41	Hs 126, He 46
AK XVI (Mot):	2. (H)/41	Hs 126
1.PzDiv:	2. (H)/23	Hs 126, He 46
4.PzDiv:	4. (H)/13	Hs 126, He 46

Koluft 14.Armee

	4. (F)/14	Do 17F
AK VIII:	1. (H)/31	Hs 126
5.PzDiv:	2. (H)/31	Hs 126, He 46
AK XVII:	3. (H)/14	Hs 126
AK XVIII:	2. (H)/14	Hs 126
2.Pz Div:	1. (H)/14	He 46, Hs 126

APPENDIX 10
LUFTWAFFE ORDER OF BATTLE IN SCANDINAVIA, 9 APRIL 1940 AND 10 MAY 1940

9 APRIL 1940

***Fliegerkorps* X: *Generalleutnant* Geisler**

Barth:	1., 2./ZG 1	Bf 110C
Flensburg:	2. (H)/10	Hs 126B
Hamburg-Fuhlsbüttel:	*Korpstransportstaffel*	Ju 52/3m
	Wetterkette	He 111H
	1. (F)/122	He 111H
Husum:	II/JG 77	Bf 109E
List/Sylt:	1., 2., 3./KüFlGr 506	He 115B/C
Lübeck-Blankensee (and Kiel-Holtenau):	1. (F)/120	Do 17P, He 111H
Kiel-Holtenau:	I/StG 1	Ju 87R
Westerland:	I/ZG 76	Bf 110C
	3./ZG 1 (attached)	Bf 110C

KG 4 (*Oberstleutnant* Fiebig)
Fassberg:	*Stab, II/KG 4*	He 111P
Lüneburg:	*III/KG 4*	
Perleberg:	*II/KG 4*	

KG 26 (*Oberstleutnant* Fuchs)
Lübeck-Blankensee:	*Stab, II, III/KG 26*	He 111H
Nordholz:	*KGr 100*	He 111H
Marx:	*I/KG 26*	He 111H

KG 30 (*Oberstleutnant* Loebel)
Westerland:	*Stab, I, II, III/KG 30*	Ju 88A
	Z./KG 30	Ju 88C

Lufttransportchef Land: Oberstleutnant von Gablenz
KG zbV 1 (*Oberstleutnant* Morzik)
Hagenow:	*IV/KG zbV 1* (also part of *II*)	Ju 52/3m
Schleswig (part at Stade):	*II/KG zbV 1*	Ju 52/3m
Übersen:	*Stab, I, III/KGzbV 1*	Ju 52/3m
Fuhlsbüttel:	*KGr zbV 107*	Ju 52/3m
Holtenau:	*KGr zbV 105*	Ju 52/3m
Neumünster:	*KGr zbV 101*	Ju 52/3m
Oldenburg:	*KGr zbV 102*	Ju 52/3m
Schleswig:	*KGr zbV 103*	Ju 52/3m
Stade:	*KGr zbV 104*	Ju 52/3m
Übersen:	*KGr zbV 106*	Ju 52/3m

Lufttransportchef See: Major Lessing
Rantum, Hörnum:	*KGrzbV 108*	He 59D, Ju 52/3m *See*

10 MAY 1940

Luftflotte 5: *General der Flieger* Stumpff (Oslo)

Oslo:	*Wekusta 5*	He 111H

Fliegerführer Stavanger: Major Harlinghausen (Stavanger)
Stavanger-Sola:	*1., 3./ZG 76*	Bf 110C/D
	4. JG 77	Bf 109E
	3. (F)/ObdL	Do 215B, He 111P
	1. (F)/122	He 111H, Ju 88A
	1./KüFlGr 106	He 115B

Fliegerführer Trondheim: Oberst Fuchs (Trondheim)
Trondheim-Vaernes:	KG 26 (*Oberst* Fuchs):	
	Stab, I, III	He 111H
	I/StG 1	Ju 87B/R
	Z./KG 30	Ju 88C

	11. (N)/JG 2	Bf 109D
	1. (F)/120	Do 17P, He 111H
Trondheim Harbour:	1., 2./KüFlGr 506	He 115C

Fliegerkorps X: Generalleutnant Geisler (Aalborg-West)

Aalborg-West:	II/KG 26	He 111H
	2./ZG 76	Bf 110C
	Korpstransportstaffel	Ju 52
Bergen (detachments at		
Aalborg and Hörnum):	KGr zbV 108	Ju 52/3m, Ju 52/3m See, He 59D, Do 24V, Do 26V, BV 138A
Copenhagen:	I/KG 40	FW 200C
Kristiansand-Kjevik:	5.,6./JG 77	Bf 109E
Oslo-Fornebu:	2. (H)/10	Hs 126
	Wetterkette X	He 111H
(Detachment at Aalborg-West)	KGrzbV 107	Ju 52/3m

APPENDIX 11
LUFTWAFFE ORDER OF BATTLE IN THE WEST, 10 MAY AND 5 JUNE 1940

10 MAY 1940

Führer der Luftstreitkräfte West: Generalmajor **Bruch** (Jever)

KüFlGr 106 (Norderney)

Borkum:	3.	He 111B, He 115C

KüFlGr 406 (Hörnum)

Hörnum:	1.	Do 18, BV 139B, Do 26
	2.	Do 18
	3.	Do 18
	2./KüFlGr 906 (attached)	Do 18
Rantum:	2./KüFlGr 106 (attached)	Do 18

KüFlGr 806 (Uetersen)

Uetersen:	2.	He 111J
	3.	He 111J
Wilhelmshaven:	1./BordFlGr 196 (attached)	Ar 196A

Fliegerdivision 9: Generalmajor **Coeler** (Jever)

Marx:	KGr 126	He 111B, He 111H

Attached from *KüFlGr 106*

Norderney:	3./KüFlGr 506	He 115C, He 115B, He 111B
	3./KüFlGr 906	He 111B

Luftflotte 2: *General der Flieger* Kesselring (Münster)

	Stab Aufkl Gr (F)/121	
	Stab Aufkl Gr (F)/122	
Münster-Loddenheide:	*2. (F)/122*	He 111H, Ju 88A
	3. (F)/122	He 111H, Ju 88A
	Wekusta 26	Do 17Z, He 111H
Goslar :	*4. (F)/122*	He 111H, Ju 88A

Fliegerkorps IV: *General der Flieger* Keller (Düsseldorf)

Münster-Handorf:	*1. (F)/121*	He 111H, Ju 88A

LG 1 (Oberst Bulowius)

Düsseldorf:	*Stab*	He 111H, Ju 88A
	I	He 111H
	II	He 111H
	III	He 111H, Ju 88A

KG 30 (Oberstleutnant Loebel)

Oldenburg:	*Stab*	Ju 88A, He 111H
	I	Ju 88A
	II	Ju 88A
Marx:	*III*	Ju 88A

KG 27 (Oberst Behrendt)

Hannover-Langenhagen:	*Stab*	He 111P, He 111D
	I	He 111P
Delmenhorst:	*II*	He 111P
Wunstorf:	*III*	He 111P

Fliegerkorps VIII: *Generalmajor* von Richthofen (Schloss Dyck/Grevenbroich)

Mönchen-Gladbach:	*2. (F)/123*	Do 17P
Duisburg:	*IV (St)/LG 1*	Ju 87B
Lauffenberg bei Neuss:	*II (Schl)/LG 2*	Hs 123

KG 77 (Generalmajor von Stutterheim)

Düsseldorf:	*Stab*	Do 17Z
	II	Do 17Z, Do 17U
	III	Do 17Z
Werl:	*I*	Do 17Z

StG 2 (Major Dinort)

Köln-Ostheim:	*Stab*	Ju 87B, Do 17M
	I	Ju 87B
	I/StG 76 (attached)	Ju 87B
Norvenich:	*III*	Ju 87B

StG 77 (*Oberst* Schwartzkoff)

Köln-Butzweilerhof:	*Stab*	Ju 87B, Do 17M
	I	Ju 87B
	II	Ju 87B

JG 27 (*Oberstleutnant* Ibel)

Mönchen-Gladbach:	*I*	Bf 109E
	II	Bf 109E
	III	Bf 109E
	I/JG 21 (attached)	Bf 109E
Gymnich:	*I/JG 1* (attached)	Bf 109E

Fliegerführer zbV 2: Generalmajor Putzier (Bremen)

Bremen:	*Auklärungsstaffel zbV*	Do 17M, He 111H

KG 4 (*Oberst* Fiebig)

Fassberg:	*Stab*	He 111P
	II	He 111P, He 111D
Gütersloh:	*I*	He 111H
Delmenhorst:	*III*	He 111P, Ju 88A

KG 54 (*Oberst* Lackner)

Quakenbrück:	*Stab*	He 111P, He 111D
	I	He 111P
Varrelbusch:	*II*	He 111P
Vechta:	*III*	He 111P

Jagdfliegerführer 2: Oberst von Doring (Dortmund)
JG 26 (*Major* Witt)

Dortmund:	*Stab*	Bf 109E
	II	Bf 109E
Essen-Mühlheim:	*III*	Bf 109E
Hopsten:	*III/JG 3* (attached)	Bf 109E

JG 51 (*Oberst* Osterkamp)

Krefeld:	*I*	Bf 109E
Bönninghardt:	*Stab*	Bf 109E
	I/JG 26 (attached)	Bf 109E
	I/JG 20 (attached)	Bf 109E
	II/JG 27 (attached)	Bf 109E, Bf 109C

ZG 26 (*Oberstleutnant* Huth)

Dortmünd:	*Stab*	Bf 110C, Bf 110D
Niedermendig:	*I*	Bf 110C, Bf 110D
Krefeld:	*III*	Bf 110C, Bf 110D
Kirchenhellen:	*I/ZG 1* (attached)	Bf 110C, Bf 110D
Gelsenkirchen-Buer:	*II/ZG 1* (attached)	Bf 110C, Bf 110D

Jagdfliegerführer Deutsche Bucht: Oberstleutnant Schumacher (Jever)
JG 1 (*Oberstleutnant* Schumacher)

Jever:	*Stab*	Bf 109E
Wangerooge:	*II(J)/TrGr 186* (attached)	Bf 109E

Also available to *Luftflotte 2*:

Wyck auf Föhr:	*I(J)/LG 2* (attached)	Bf 109E
Esbjerg:	One *Staffel* (attached)	Bf 109E
Nordholz:	*II/JG 2* (attached)	Bf 109E
Hopsten:	*10., 12. (N)/JG 2* (attached)	Bf 109D, Ar 68

Fliegerdivision 7: *Generalleutnant* Student

Parachute Operations
KGzbv 1 (*Oberstleutnant* Morzik)

I	Ju 53/3m
II	Ju 52/3m
III	Ju 52/3m
IV	Ju 52/3m

Air Landing Operations
KGzbV 2 (*Oberst* Conrad)

KGr zbV 9	Ju 52/3m
KGr zbV 11	Ju 52/3m
KGr zbV 12	Ju 52/3m
KGr zbV 172	Ju 52/3m
Staffel Schwilben	He 59

Special Operations

Auflkärungsgruppe 156	Fi 156
Sturmabteilung Koch	DFS 230

II. Flakkorps

Aufklärungsstaffel (F) des II.Flakkorps	Do 17M, Hs 126

Luftflotte 3: *General der Flieger* Sperrle (Bad Orb)

Langendiebach:	*1. (F)/123*	Do 17P, Ju 88A
	Wekusta 51	He 111H, Do 17U
Gelnhausen:	*3. (F)/123*	Do 17P, Ju 88A

Fliegerkorps I: *General der Flieger* Grauert (Cologne)

Köln-Wahn:	*5. (F)/122*	Do 17P

KG 1 (*Oberst* Exss)

Giessen:	*Stab*	He 111H
	I	He 111H
Kirtorf:	*II*	He 111H
Ettinghausen:	*III*	He 111H
Bracht:	*III/KG 28* (attached)	He 111P

KG 76 (*Oberst* Froehlich)

Nidda:	*Stab*	Do 17Z, Do 17U
	I	Do 17Z
	II	Do 17Z
	III	Do 17Z
Köln-Wahn:	*III/StG 51* (attached)	Ju 87B

JG 77 (*Oberstleutnant* von Manteuffel)

Peppenhoven:	*Stab*	Bf 109E
Odendorf:	*I*	Bf 109E
Vogelsang:	*I/JG 3* (attached)	Bf 109E

ZG 76 (*Major* Grabmann)

Köln-Wahn:	*Stab*	Bf 110C, Bf 110D
	II	Bf 110C, Bf 110D
Kaarst/Neuss:	*II/ZG 26*	Bf 110C, Bf 110D

Fliegerkorps II: *Generalleutnant* Loerzer (Frankfurt-am-Main)

Frankfurt/Main:	*3.(F)/121*	He 111H, Ju 88A

KG 2 (*Oberst* Fink)

Ansbach:	*Stab*	Do 17Z, Do 17U
	II	Do 17Z
Giebelstadt:	*I*	Do 17Z
Illesheim:	*III*	Do 17Z

KG 3 (*Oberst* von Chamier-Glisczinski)

Würzburg:	*Stab*	Do 17Z
	III	Do 17Z
Aschaffenburg:	*I*	Do 17Z
Schweinfurt:	*II*	Do 17Z

KG 53 (*Oberst* Stahl)

Roth:	*Stab*	He 111H
	I	He 111H
Oedheim:	*II*	He 111H
Schwäbisch Hall:	*III*	He 111H

StG 1 (*Oberst* Baier)

Siegburg:	*Stab*	Do 17M, Ju 87B
	II/StG 2 (attached)	Ju 87B
Hemweiler:	*I (St)/TrGr 186* (attached)	Ju 87B

Fliegerkorps V: *Generalleutnant* R. *Ritter* von Greim (Gersthofen)

Gablingen:	*4.(F)/121*	Do 17P, Ju 88A

KG 51 (*Oberst* Kammhuber)

Landsberg/Lech:	*Stab*	He 111H, Ju 88A
	III	He 111H
Lechfeld:	*I*	He 111H, Ju 88A
München-Riem:	*II*	Ju 88A

KG 55 (*Oberst* Stoeckl)

Leipheim:	*Stab*	He 111P
	II	He 111P
Neuburg/Donau:	*I*	He 111P
Gablingen:	*III*	He 111P

Also available to *Luftflotte 3*:

Neuhuasen ob Eck:	*I/ZG 52*	Bf 110C, Bf 110D
Mannheim-Sandhofen:	*V(Z)/LG 1*	Bf 110C, Bf 110D

JG 52 (*Major* von Bernegg)

Mannheim-Sandhofen:	*Stab*	Bf 109E
Lachen/Speyerdorf:	*I*	Bf 109E
Speyer:	*II*	Bf 109E

JG 54 (*Major* Mettig)

Böblingen:	*Stab*	Bf 109E
	I	Bf 109E
	II	Bf 109E

***Jagdfliegerführer 3*: *Oberst* von Massow (Wiesbaden)**

JG 2 (*Oberstleutnant* von Bulow)

Frankfurt-Rebstock:	*Stab*	Bf 109E
	I	Bf 109E
	III	Bf 109E
Ober-Olm:	*I/JG 76* (attached)	Bf 109E

JG 53 (*Oberstleutnant* von Cramon-Taubadel)

Wiesbaden-Erbenheim:	*Stab*	Bf 109E
	I	Bf 109E
	II	Bf 109E
	III	Bf 109E
Mannheim-Sandhofen:	*III/JG 52* (attached)	Bf 109E

ZG 2 (*Oberstleutnant* Vollbracht)

Darmstadt-Griesheim:	*Stab*	Bf 110C, Bf 110D
	I	Bf 110C, Bf 110D

I.Flakkorps I

Weise:	*Aufklärungsstaffel (F) des I.Flakkorps*	Do 17M

Koluft Heeresgruppe B

4.(F)/14	Do 17M, Do 17P

Koluft 18.Armee

7.(F)/LG 2	Do 17M, Do 17P

AK XXVI:	2. (H)/10	Hs 126
9. Pz Div and 1. Kav Div:	3. (H)/12Pz	Hs 126

Koluft 6.Armee

	3. (F)/11	Do 17P, Bf 110
AK I:	1. (H)/10	Hs 126
AK IV:	1(H)/41	Hs 126
AK IX:	4. (H)/22	Hs 126
AK XXVII:	1. (H)/23	Hs 126
Mot AK XVI:	2. (H)/41	Hs 126
3. Pz Div:	9. (H)/LG 2	Hs 126
4. Pz Div:	4. (H)/13	Hs 126

Koluft Heeresgruppe A

	2. (F)/11	Do 17P

Koluft 4.Armee

	4. (F)/11	Do 17M, Do 17P
AK II:	1. (H)/21	Hs 126, Do 17M
AK V:	3. (H)/13	Hs 126
AK VIII:	1. (H)/31	Hs 126
Mot AK XV:	4. (H)/31	Hs 126
5. Pz Div:	2. (H)/31Pz	Hs 126
7. Pz Div:	3. (H)/14	Hs 126

Koluft 12.Armee

	2. (F)/22	Do 17F
AK III:	2. (H)/21	Hs 126, Do 17M
AK VI:	2. (H)/12	Hs 126
AK XVIII (Geb):	2. (H)/14	Hs 126
Panzergruppe Kleist	3. (F)/10	Do 17P
AK XIX (Mot):	3. (H)/21	Hs 126
1. Pz Div:	2. (H)/23Pz	Hs 126
2. Pz Div:	1. (H)/14Pz	Hs 126
10. Pz Div:	1. (H)/11Pz	Hs 126
Mot AK XLI:		
8. Pz Div:	3. (H)/41	Hs 126

Koluft 16.Armee

	3. (F)/22	Do 17M, Do 17P, Bf 109E

AK VII:	2. (H)/13	Hs 126, Do 17M
AK XIII:	5. (H)/13	Hs 126
AK XXIII:	1. (H)/12	Hs 126

Koluft Heeresgruppe C

| | 1. (F)/22 | Do 17F, Do 17P |

Koluft 1.Armee

	Aufklärungsstaffel (F) der 1.Armee	Do 17P
AK XII:	4. (H)/12	Hs 126, Do 17M
AK XXIV:	4. (H)/21	Hs 126
AK XXX and HK XXXVII:	1. (H)/13	Hs 126, Do 17M, Bf 109D

Koluft 7.Armee

| | 3. (F)/31 | Do 17M |
| AK XXV and HK XXXIII: | 4. (H)/23 | Hs 126 |

5 JUNE 1940 (FALL 'ROT')

Luftflotte 2

2., 3., 4. (F)/122; Wekusta 26

Fliegerkorps I

KG 1	I–III; III/KG 28
KG 76	I–III; III/StG 51
KG 77	I–III
	5. (F)/122

Fliegerkorps IV

LG 1	I–III
KG 27	I–III
KG 54	I–III
	1. (F)/121

Fliegerdivision 9

KG 4	I–III
	KGr 126
	KüFlGr 106

Jafü 2
JG 26   *II, III; I and III/JG 3*
JG 51   *I/JG 26; I/JG 20; I/JG 51; II/JG 27*
ZG 76   *II; II/ZG 26*

Luftflotte 3

1., 3. (F)/123; Wekusta 51

Fliegerkorps II
KG 2   *I–III*
KG 3   *I–III*
KG 53   *I–III*
                                3. (F)/121

Fliegerkorps V
KG 51   *I–III*
KG 55   *I–III*
                                4. (F)/121

Fliegerkorps VIII
StG 1   *II/StG 2; I/TrGr 186*
StG 2   *I, III; I/StG 76*
StG 77   *I, II; IV/LG1; II/LG 2*
JG 2   *I, III; I/JG 76*
JG 27   *I; I/JG 1; 1/JG 21*

Jafü 3
JG 52   *I, II*
JG 53   *I–III*
JG 54   *I; II/JG 3; II/JG 51*
ZG 2   *I; I/ZG 52, V(Z)/LG 1*
ZG 26   *I, III; I–II/ZG 1*

Sources: Based upon US Monograph 152 and amended from unit histories.

BIBLIOGRAPHY

BOOKS

Absalon, Rudolf, (ed.), *Rangliste der Generale der Deutschen Luftwaffe dem Stand von 20. April 1945*, Podzun Verlag (Freiburg, 1984)

Adams, Jack, *The Doomed Expedition: The Campaign in Norway 1940*, Leo Cooper (London, 1989)

Adamthwaite, Anthony, *France and the Coming of the Second World War*, Frank Cass (London, 1977)

Aders, Gebhard, *History of the German Night-Fighter Force 1917–1945*, Arms & Armour Press (London, 1979)

Aders, Gebhard, and Held, Werner, *Jagdgeschwader 51 'Mölders'*, Motorbuch Verlag (Stuttgart, 1985)

Air Ministry, *The Rise and Fall of the German Air Force* (Air Ministry Pamphlet 248), (London, 1948–49). All references from reproduction by WE Inc (Old Greenwich, Conn, 1969).

Balke, Ulf, *Der Luftkrieg in Europa: Die operativen Einsätze des Kampfgeschwaders 2 im Zweiten Weltkrieg. Teil 1*, Bernard & Graefe Verlag (Koblenz, 1989)

———, *Kampfgeschwader 100 'Wiking'*, Motorbuch Verlag (Stuttgart, 1981)

Barker, A. J., *Dunkirk, the Great Escape*, J. M. Dent & Sons (London, 1977)

Bekker, Cajus, *The Luftwaffe War Diaries*, Macdonald (London), 1966

Bingham, Victor, *Blitzed: The Battle of France, May to June 1940*, Air Research Publications (New Malden, 1990)

Bley, Wulf, *Deutsche Luft Hansa AG*, Widder-Verlag (Berlin, 1932)

Bodenschatz, Karl, *Jagd in Flanderns Himmel*, Verlag Knorr und Firth (Munich, 1935)

Bond, Brian, *British Military Policy Between the Two World Wars*, Clarendon Press (Oxford, 1980)

Boog, Dr Horst, *Die Deutsche Luftwaffen Führung 1935–1945: Führungsprobleme, Spitzengliederung, Generalstabsausbildung*, Deutsche Verlags Anstalt (Stuttgart, 1982)

———, (ed.), *The Conduct of the Air War in the Second World War: An International Comparison*, Berg Publishers (Oxford, 1992)

Boyd, Alexander, *The Soviet Air Force*, Macdonald and Janes (London, 1977)

Bowyer, Michael J. F., *Air Raid! The Enemy Air Offensive against East Anglia*, Patrick Stephens (Wellingborough, 1986)

Brookes, Andrew J., *Photo-Reconnaissance: An Operational History*, Ian Allan (Shepperton, 1975)

Caldwell, Donald L., *JG 26: Top Guns of the Luftwaffe*, Orion Books (New York, 1991)

Carsten, F. L., *The Reichswehr and Politics 1918–1933*, Oxford University Press, (Oxford, 1966)

Chapman, G., *Why France Collapsed*, Cassell, London, 1968

Christienne, Charles, and Lissarrague, Pierre, (trans. Frances Kianka), *A History of French Military Aviation*, Smithsonian Institution Press (Washington, DC, 1986)

Cooper, Matthew, *The German Air Force 1933–1945: An Anatomy of Failure*, Jane's (London, 1981)

Cuneo, John R., *Winged Mars. Vol. II: The Air Weapon 1914–1916*, Military Service Publishing Co. (Harrisburg, Pa, 1947)

Cynk, Jerzy B., *History of the Polish Air Force 1918–1968*, Osprey Publishing (London, 1972)

Davies, R. E. G., *Lufthansa: An Airline and its Aircraft*, Airlife Publishing (Shrewsbury, 1991)

Deichmann, Paul, *Der Chef im Hintergrund: Ein Lebel als Soldat von der Preussischen Armee bis zur Bundeswehr*, Stalling Verlag (Hamburg, 1979)

Deist, W., *The Wehrmacht and German Rearmament*, Macmillan (1981)

Derry, T. K., *The Campaign in Norway*, HMSO (London, 1952)

Dierich, Wolfgang, (ed.), *Die Verbände der Luftwaffe 1935–1945*, Motorbuch Verlag (Stuttgart, 1976)

——, *Kampfgeschwader 51 'Edelweiss'*, Motorbuch Verlag (Stuttgart, 1973)

——, *Kampfgeschwader 'Edelweiss'*, Ian Allan (London 1975)

——, *Kampfgeschwader 55 'Greif'*, Motorbuch Verlag (Stuttgart, 1975)

Doughty, Robert Allan, *The Breaking Point: Sedan and the Fall of France*, Archon Books (Hamden, Conn, 1990)

——, *The Seeds of Disaster: The Development of French Army Doctrine 1919–1939*, Archon Books (Hamden, Conn, 1985)

Douhet, Guilio, *The Command of the Air*, Arno Press (New York, 1972)

Eberhardt, *Generalleutnant* Walter von, *Unsere Luftstreitkräfte 1914–1918*, Vaterländischen Verlag E. A.Weller (Berlin, 1930)

Ellis, Major L. F., *The War in France and Flanders 1939–1940*, HMSO (London, 1953)

Erickson, John, *The Soviet High Command*, Macmillan (London, 1962)

Faber, Harold, (ed.), *The Luftwaffe: A History*, New York Times Book Co (1977)

Farrar-Hockley, Anthony, *Student*, Ballantine Books (New York, 1973)

Fast, Niko, *Das Jagdgeschwader 52. Band I*, Bensberger Buch Verlag (1988)

Franks, Norman, *Valiant Wings: The Battle and Blenheim Squadrons over France 1940*, William Kimber (Wellingborough, 1988)

Fredette, Raymond H., *The First Battle of Britain 1917–1918*, Cassell (London, 1966)

Gabriel, Erich, *Flieger 90/71. Teil I: Militärluftfahrt und Luftabwehr in Österreich von 1890 bis 1971*, Heeresgeschichtliches Museum (Vienna, 1971)

Galland, Adolf, *The First and the Last*, Methuen (London, 1970)

Ginsburg, Jeffrey A., *Divided and Conquered:The French High Command and the Defeat of the West, 1940*, Greenwood Press (Westport, Conn, 1979)

Goulding, James, and Moyes, Philip, *RAF Bomber Command and its Aircraft 1936–1940*, Ian Allan Ltd (Shepperton, 1975)

Green, William, *Warplanes of the Third Reich*, Galahad Books (New York, 1990)

Griehl, Manfred, *Junkers Ju 88: Star of the Luftwaffe*, Arms & Armour Press (London, 1990)

Griehl, Manfred, and Dressel, Joachim, *He 177/He 277/He 274*, Motorbuch Verlag (Stuttgart 1989)

Groehler, Olaf, *Geschichte des Luftkriegs 1910 bis 1980*, Militärverlag der Deutschen Demokratischen Republik (Berlin, 1981)

Gundelach, Karl, *Kampfgeschwader 'General Wever' 4: Eine Geschichte aus Kriegstagbüchern, Dokumenten und Berichten 1939–1940*, Motorbuch Verlag (Stuttgart, 1978)

Hahn, Fritz, *Waffen und Geheimwaffen des Deutschen Heeres 1933–1945. Band 2: Panzer- und Sonderfahrzeuge, 'Wunderwaffen', Verbrauch und Verluste*, Bernard & Graefe Verlag (Coblenz 1986)

Harvey, Maurice, *The Allied Bomber War 1939–1945*, Spellmount (Tunbridge Wells, 1992)

Haubner, F., *Die Flugzeuge der Österreichen Luftstreitkrafte vor 1938 (Osterreichs Luftfahrt in Einzeldarstellung, Band 2, Band 5)*, H. Weishaupt Verlag (Graz, 1982)

Herlin, Hans, *Udet: A Man's Life*, Macdonald (London, 1960)

Hermann, Hajo, *Eagle's Wings: The Autobiography of a Luftwaffe Pilot*, Airlife Publishing (Shrewsbury, 1991)

Hildebrand, Karl Friedrich, *Die Generale der Deutschen Luftwaffe 1935–1945*, Biblio Verlag (Osnabrück, 1991)

Holley, Irving Brinton, *Buying Aircraft: Material Procurement for the Army Air Forces* (US Army in World War II Special Studies), Office of the Chief of Military History (Washington, DC, 1964)

Homze, Edward L., *Arming the Luftwaffe: The Reich Air Ministry and the German Aircraft Industry, 1919–1939*, University of Nebraska Press (Lincoln/London, 1976)

Horne, Alistair, *To Lose a Battle*, Macmillan (London, 1969)

Howson, Gerald, *Aircraft of the Spanish Civil War 1936-1939*, Putnam Aeronautical Books (London, 1990)

Imrie, Alex, *Pictorial History of the German Army Air Service*, Ian Allan (Shepperton, 1971)

Ishoven, Armand Van, *Ernst Udet:Biographie eines Grossen Fliegers*, Manfred Pawlak Verlagsgesellschaft (Herrsching, 1977)

Irving, David, *Göring: A Biography*, Macmillan (London, 1989)

———, *The Rise and Fall of the Luftwaffe: The Life of Erhard Milch*, Weidenfeld & Nicolson (London, 1973)

James, John, *The Paladins: A Social History of the RAF up to the Outbreak of World War II*, Macdonald (London, 1990)

Johnson, Brian, *The Secret War*, BBC (London, 1978)

Jones, Neville, *The Beginnings of Strategic Air Power: A Hstory of the British Bomber Force 1923–1939*, Frank Cass (London, 1987)

Kameradschaft ehemaliger Transportflieger, *Geschichte einer Transportflieger-Gruppe im II. Weltkrieg*, Druckerei Josef Grütter (Ronnenberg/Hannover, 1989)

Kappe-Hardenberg, Siegfried, *Ein Mythos wird Zerstört: Der Spanische Bürgerkrieg, Guernica und die Antideutsche Propaganda*, Kurt Vowinkel Verlag (1987)

Kesselring, Albert, *The Memoirs of Field Marshal Kesselring*, Purnell Book Services (London, 1974)

Kiehl, Heinz, *Kampfgeschwader 'Legion Condor' 53*, Motorbuch Verlag (Stuttgart, 1983)

Kilduff, Peter, *Germany's First Air Force 1914–1918*, Arms & Armour Press (London, 1991)

Klein, Burton H., *Germany's Economic Preparations for War*, Harvard University Press (Cambridge, Mass, 1959)

Koch, Horst-Adalbert, *Flak: Die Geschichte der Deutschen Flakartillerie*, Podzun Verlag (Bad Nauheim, 1965)

Longmate, N., *The Bombers: The RAF Offensive against Germany 1939–1945*, Hutchinson (London, 1983)

Lux, Albert, *Von Goering's Kriegsflugstaffeln in Goering's Zuchthausen*, Editions Sebastian Brandt (Strasbourg, 1938)

McFarland, Stephen L., and Newton, Wesley Philips, *To Command the Sky*, Smithsonian Institution Press (Washington, DC, 1991)

Macksey, Kenneth, *Kesselring: The Making of the Luftwaffe*, Batsford (London, 1978)

Maier, Klaus A., et al, *Das Deutsche Reich und der Zweite Weltkrieg. Band 2: Die Errichtung der Hegemonie auf dem Europäischen Kontinent*, Deutsche Verlags-Anstalt (Stuttgart, 1979)

Manvell, Roger and Fraenkel, Heinrich, *Göring*, NEL Books (London, 1968)

Martin, Paul, *Invisibles Vainqueurs: Exploits et Sacrifice de l'Armée de l'Air en 1939–1940*, Editions Yves Michelet (Paris, 1990)

Mason, Herbert Molloy, *The Rise of the Luftwaffe 1918–1940*, Cassell (London, 1975)

Maurer, Maurer, *Aviation in the U.S. Army 1919–1939*, Office of Air Force History, USAF (Washington DC, 1987)

Mitcham, Samuel W., *Men of the Luftwaffe*, Presidio (California, 1988)

Morrow, John H., *German Air Power in World War I*, University of Nebraska Press (Lincoln, Nebr, 1982)

Morzik, Fritz, and Hümmelchen, Gerhard, *Die Deutschen Transportflieger im Zweiten Weltkrieg*, Bernard & Graefe Verlag (Frankfurt/Main, 1966)

Murray, Williamson, *German Military Effectiveness*, The Nautical and Aviation Publishing Company of America (Baltimore, Md, 1992)

———, *Strategy for Defeat:The Luftwaffe 1933–1945*, Air University Press (Maxwell AFB, Al, 1983)

Nemeck, Vaclav, *The History of Soviet Aircraft from 1918*, Willow Books/William Collins & Sons (London, 1986)

Nowarra, Heinz, *Die Verbotenen Flugzeuge 1921–1935*, Motorbuch Verlag (Stuttgart, 1980)

———, *Fernaufklärer 1915–1945*, Motorbuch Verlag (Stuttgart, 1982)

———, *Marine Aircraft of the 1914–1918 War*, Harleyford Publications (Letchworth, 1966)

———, *Nahaufklärer 1910–1945*, Motorbuch Verlag (Stuttgart, 1981)

Overy, R. J., *Goering: The Iron Man*, Routledge & Kegan Paul (London, 1984)

Paquier, Colonel Pierre, *L'Aviation de Bombardement Française*, Éditions Berger-Levrault (Paris, 1948)

Pletschacher, Peter, *Die Königlich Bayerischen Fliegertruppen*, Motorbuch Verlag (Stuttgart, 1978)

Price, Dr Alfred, *Luftwaffe Handbook* (2nd Edn), Ian Allan (Shepperton, 1986)

———, *World War II Fighter Conflict*, Macdonald & Janes (London, 1875)

Priller, Josef, *J.G.26: Geschichte eines Jagdgeschwaders*, Motorbuch Verlag (Stuttgart, 1980)

Pritchard, David, *The Radar War*, Patrick Stephens (1989)

Proctor, Raymond L., *Hitler's Luftwaffe in the Spanish Civil War*, Greenwood Press (Westpoint, Conn, 1983)

Radtke, Siegfried, *Kampfgeschwader 54: Von der Ju 52 zur Me 262*, Schild-Verlag (Munich, 1990)

Ramsey, Winston G., (ed.), *The Blitz Then and Now*, Battle of Britain Prints International (London, 1987)

Richey, Paul, *Fighter Pilot*, Pan Books (London, 1969)

Ries, Karl, *Deutsche Flugzeugführerschulen und ihre Maschinen 1919–1945*, Motorbuch Verlag (Stuttgart, 1988)

———, *Luftwaffen-Story 1935–1939*, Verlag Dieter Hoffmann (Mainz, 1974)

———, *Die Maulwürfe: Geheimer Aufbau 1919–1935*, Verlag Dieter Hoffmann (Mainz, 1970)

Ries, Karl, and Ring, Hans, *Legion Kondor 1936–1939: Eine Illustrierte Dokumentation*, Verlag Dieter Hoffmann (Mainz, 1980)

Ring, Hans, and Gerbig, Werner, *Jagdgeschwader 27*, Motorbook Verlag (1971)

Robertson, Bruce, (ed.), *Air Aces of the 1914–1918 War*, Harleyford Publications (Letchworth, 1959)

Robinson, Douglas, H., *The Zeppelin in Combat*, G.T. Foulis & Co (London, 1962)

Rohwer, J., and Hümmelchen, G., *Chronology of the War at Sea 1939–1945. Vol. 1: 1939–1942*, Ian Allan (London, 1972)

Salas Larrazabel, Jésus, *Air War Over Spain*, Ian Allan (Shepperton, 1969–1974)

Schliephake, Hanfried, *The Birth of the Luftwaffe*, Ian Allan (Shepperton, 1971)

Schmidt, Rudi, *Achtung-Torpedos Los: Der Strategische und Operative Einsatz des Kampfgeschwaders 26*, Bernard & Graefe Verlag (Koblenz, 1991)

Schmidt-Pauli, Edgar von, *Geschichte der Freikorps 1918–1924*, Robert Lutz/Otto Schramm (Stuttgart, 1936)

Scutts, Jerry, *JG 54: Aces of the Eastern Front*, Airlife Publishing (Shrewsbury, 1992)

Seaton, Albert, *The German Army 1933–1945*, Wiedenfeld & Nicolson (London, 1982)

Shores, Christopher, *Duel for the Sky*, Blandford Books (Poole, 1985)

Shores, Christopher, et al, *Fledgling Eagles*, Grub Street (London, 1991)

Smith, J. R., and Kay, Antony, *German Aircraft of the Second World War*, Putnam & Co (London, 1972)

Smith, Malcolm, *British Air Strategy between the Wars*, Clarendon Press (Oxford, 1984)

Smith, Peter C., *Stuka Squadron: Stukagruppe 77—The Luftwaffe's Fire Brigade*, Patrick Stephens (Cambridge, 1990)

Spiedel, General Wilhelm, *The German Air Force in Poland* (USAF Historical Studies Monograph 151), US Air Force Historical Research Association Microfilm K1026-U
———, *The German Air Force in France and the Low Countries* (USAF Historical Studies Monograph 152), US Air Force Historical Research Association Microfilm K1026-U

Stahl, Peter, *The Diving Eagle: A Ju 88 Pilot's Diary (1940–1941)*, William Kimber (London, 1984)

Stutzer, Helmut, *Die Deutsche Militarflugzeuge 1919–1934*, Verlag E. S. Mittler & Sohn (Herford, Germany, 1984)

Suchenwirth, Professor Richard, *The Development of the German Air Force 1919–1939* (USAF Historical Studies Monograph 160), US Air Force Historical Division (Maxwell AFB, Al, 1968)

Taylor, Telford, *Munich:The Price of Peace*, Hodder & Stoughton (London, 1979)
———, *The March of Conquest: German Victories in Western Europe in 1940*, E. Hulton and Co (London, 1959)
———, *The Sword and the Swastika: Generals and Nazis in the Third Reich*, Simon & Schuster (New York, 1952)

Tessin, Georg, *Verbände und Truppen der Deutschen Wehrmacht und Waffen SS 1939–1945. Teil 14: Die Luftstreitkräfte*, Biblio Verlag (Osnabruck, 1980)

Thomas, Gordon, and Morgan-Witts, Max, *The Day Guernica Died*, Hodder & Stoughton (London, 1975)

Thomas, Hugh, *The Spanish Civil War* (3rd Edn), Penguin Books (London, 1986)

Titz, Zdenek, *Czechoslovakian Air Force 1918–1970*, Osprey Publications Ltd (Canterbury, 1971)

Trautloft, Hannes, et al, *Die Grünherzjäger:Bildchronik des Jagdgeschwader 54*, Podzun-Pallas Verlag (Freiburg, 1985)

Tuiider, Dr Othmar, *Die Luftwaffe in Österreich 1938–1945*, Österreichischen Bundesverlag (Vienna, 1985)

Tyushkevich, S. A., (et al), *The Soviet Armed Forces: A History of Their Organizational Development*, US Government Printing Office (Washington, 1980)

Verein zur Erhaltung der Historischen Flugwerft Oberschleissheim, *Geflogene Vergangenheit: 75 Jahre Luftfahrt in Schleissheim*, Flugzeug Publikations (Illertissen, 1988)

Völker, Karl-Heinz, *Die Deutsche Luftwaffe 1933–1939: Aufbau, Führung, Rüstung* (*Beitrage zur Militär- und Kriegsgeschichte, Band 8*), Deutsche Verlags-Anstalt (Stuttgart, 1967)
———, *Dokumente und Dokumentarfotos zur Geschichte der Deutschen Luftwaffe*, Deutsche Verlags-Anstalt (Stuttgart, 1968)
———, *Die Entwicklung der Militärischen Luftfahrt in Deutschland 1920–1933* (Part of *Beitrage zur Militär- und Kriegsgeschichte, Band 3*), Deutsche Verlags-Anstalt (Stuttgart, 1962)

Wagener, *Hauptmann, Von der Heimat Geächtet*, Belseruche Verlagsbuchhandlung (Stuttgart, 1920)

Williams, John, *The Ides of May: The Defeat of France, May–June 1940*, Constable (London, 1968)

Wood, Derek, and Dempster, Derek, *The Narrow Margin*, Arrow Books (London, 1967)

Zaloga, Steve, and Madej, Victor, *The Polish Campaign*, Hippocrene Books (New York, 1985)

Zaloga, Steven J., *Target America: The Soviet Union and the Strategic Arms Race 1945–1964*, Presidio Press (Novato, Ca, 1993)

ARTICLES AND ESSAYS

Alexandrov, Alexander, 'Junkers Planes in Russia', *Skyways: The Journal of the Airplane 1920–1940*, No 25

Andersson, Lennart, 'Secret Luftwaffe: German Military Aviation Between the Wars', *Air Enthusiast*, Vol 41

———, 'Junkers Two Seaters', *Air Enthusiast*, Vols 44 (Part 1) and 45 (Part 2)

Bartels, Dennis A., and Proctor, Raymond L., 'A Query and a Reply', *Aerospace Historian*, June 1981

Bock, Winfried, 'Die Luftschlacht über der Deutscher Bucht', *Luftfahrt International*, February and March 1983

Boog, Dr Horst, 'The Policy, Command and Direction of the Luftwaffe in World War II', *Royal Air Force Historical Society Proceedings*, No 4 (September 1988)

Botquin, Gaston, 'Pas Assez? Trop Tard? La Production Aéronautique Militaire Française 1936–1940', *Histoires Vraies de l'Avion*, Dec. 1990

Daniel, Raymond, 'La Conquet de la Hollande: Opération Secondaire?', *Icare*, No 79, Vol IX

———, 'Les Dix Premiers Mois d'Opérations de la Luftwaffe', *Icare*, No 116, Vol XIV

Destrebecq, Guy, 'The Belgian Air Force', *Air Britain Review*, Mar 1965

Echegaray, R. G., 'The Spanish Civil War 1936–1939: Nationalist and Red Fleets', *Belgian Shiplover*, No 95, 1963

Facon, Patric, 'La Visite du Général Vuillemin en Allegmagne: Recueil d'Articles et Études 1981–1983', Service Historique de l'Armée de l'Air

Green, William, 'Aerial Alpenstock', *Flying Review*, Dec 1969

———, 'Sentinel over the Fjords', *Flying Review*, Apr 1969

———, 'Stopper in the Bottle: The Royal Danish Air Force', *Air International*, Oct 1979

———, 'The Annals of the Iberian Pedros', *Air International*, June 1979

Green, William, and Swanborough, Gordon, 'Balkan Interlude: The Bulgarian Air Arm in WW II', *Air Enthusiast* Vol 39

Grosz, Peter M., and Terry, Gerard, 'The Way to the World's First All-Metal Fighter', *Air Enthusiast*, Vol 25

Huan, Claude, and Marchand, Alain, 'La Bataille Aéronavale de Dunkerque, 18. Mai–3. Juin 1940', *Revue Historique de l'Armée*, Vol 172, Sept 1988

Kössler, Karl, 'Die Erste Do 17—Wie Sie Wirklich War', *Luftfahrt International*, Vol 1/83

Krüger, Hans Eberhard, 'Jagdstaffeln über Kurland und Oberschlesien', *Jägerblatt*, Apr/May 1989

———, 'Die Reichswehr- und Polizeiflieger-Staffeln', *Jägerblatt*, Aug 1989

Marchand, Alain, and Huan, Claude, 'Achtung Minen! La Luftwaffe Mine les Ports Britanniques et Français en 1940', *La Fana de l'Aviation*, Apr 1987

Murray, Williamson, 'German Air Power and the Munich Crisis', *War and Society: A Yearbook of Military History*, Vol 2, Croom Helm (London, 1977)

Osborne, Dr Richard, 'Naval Actions of the Spanish Civil War', *Warships Supplement: Proceedings of Naval Meetings*, World Ship Society (Kendal, Cumbria, 1989)

Overy, R. J., 'The German Pre-war Aircraft Production Plans :November 1936–April 1939', *English Historical Review*, Oct 1975

Speidel, Helm, 'Reichswehr und Rote Armee', *Vierteljahrshefte für Zeitgeschichte*, Vol i, 1953

Stroud, John, 'The Birth of Air Transport', *The Putnam Aeronautical Review*, No 3 (Oct 1989)

Terlinden, Lieutenant-Colonel, 'L'Aéronautique Militaire Belge en 1940', *Icare*, No 74

Titz, Zdenek, 'The Luftwaffe's First "Ally": The Story of the Slovak Air Force', *Flying Review*, Vol 19, No 9 (May 1964)

Torroba, César O'Donnell, 'Las Pérdidas de Buques Mercantes Republicanos Causadas por Hidroaviones de la Legión Cóndor durante la Guerra Civil Española (1936–1939)', *Revista de Historia Naval*, No 43, 1993

Völker, Karl-Heinz, 'Die Geheime Luftrüstung der Reichswehr und ihre Auswirkung auf den Flugzeugbestand der Luftwaffe bis zum Beginn des Zweiten Weltkrieges', *Wehrwissenschaftliche Rundschau*, Heft 9/1962

GLOSSARY

Abteilung Waffenbeschaffungswesen (Wa B)	Weapons Procurement Department
Abteilung Waffenprüfwesen	Weapons Testing Department
Abwehr	German Military Intelligence
Aéronautique Navale (Aéronavale)	French Naval Air Force
Akademischen Fliegergruppen an Technische Hochschulen (Akaflieg)	Technical College Academic Flying Groups
Allgemeines Luftamt	General Air Directorate
Allgemeines Marineamt	General Naval Directorate (*Reichsmarine* supply and support organization)
Armee-Oberkommando	An army headquarters (usually referred to as '*Armee*')
Armée de l'Air	French Air Force
Artillerie Führer V	*Wehrkreis V* Artillery Commander
Aufklärungsgruppe (AufKlGr)	Reconnaissance Group
Aviatsiya Osobovo Naznacheniya	Aviation for Special Employment (Russian)
Bayerische Militärfliegerschule	Bavarian Military Flying School
Behelfsbomber	Auxiliary bomber
Beobachterstaffel	Observer Squadron
Beobachterlehrstaffel	Observer Demonstration Squadron
Bayerische Flugzeugwerke (BFW)	Bavarian Aircraft Works
Bayerische Motoren-Werke (BMW)	Bavarian Motor Works
Blindflugschule	Instrument Training School
Bombardement, Combate et Reconnaissance (BCR)	French multi-role aircraft requirement
Bombengeschwader der Obersten Heeresleitung (Bogohl)	First World War bomber formation of six *Staffeln*
Bordfliegergruppe (BordFlGr)	Embarked Air Group
Brigadekommandeur der Flieger	Air Brigade Commander
Bundesarchiv	German Federal Archives
Bundesministerium für Heerwesen (BMfHW)	Austrian Defence Ministry
Chef der Heeresleitung	Head of the Army Office
Chef der Luftwehr	Head of Air Defence
Chef des Ausbildungswesens	Head of Training
Comité Permanent de la Défense Nationale	Permanent National Defence Committee (French)
Deutsch-Russiche Luftverkehrs GmbH (Deruluft)	German–Russian Air Transport Ltd
Deutsche Luftfahrtverband e.V. (DLV)	German Aviation Association
Deutsche Luftfahrt GmbH	German Aviation Ltd
Deutsche Luftreederei (DLR)	German Air Transport Line
Deutschen Luftsportverband e.V. (DLV)	German Air Sport Association
Deutschösterreichische Fliegertruppe	German–Austrian Air Troops
Deutsche Verkehrsflieger-Schule (DVS)	German Air Transport School

Deutsche Versuchsanstalt für Luftfahrt (DVL)	German Air Transport Experimental Establishment
Deutsches Luftamt	German Air Directorate
Deutschösterreichische Fliegertruppe	German–Austrian Air Troops
Erkundungsflugzeug für die Divisions-nahaufklärungsstaffeln (Erkudista)	Reconnaissance Aircraft for Division Tactical Reconnaissance Squadrons
Erkundungsflugzeug für mittlere Höhen und grösste Entfernungen (Erkunigros)	Reconnaissance Aeroplane for Medium Altitudes and Long Distances
Erprobungstelle	Test Establishment
Erprobungstelle Rechlin des Reichsverbändes der deutschen Luftfahrtindustrie	Rechlin Test Establishment of the National Association of the German Aviation Industry
Fahrabteilung	Motor Transport Battalion
Fall 'Gelb'	Contingency Yellow
Fall 'Grün'	Contingency Green
Fall 'Weiss'	Contingency White
Fallschirmjägerregiment (FJR)	Paratroop Regiment
Feldflugchef	Commander of aviation units in the field
Feld Fliegerabteilung (Feld FA)	*Fliegertruppe* corps squadron
Fernaufklärungsgruppe	Long Range Reconnaissance Group
Fertigungs GmbH	Production Ltd
Fliegerabteilung (FA)	Literally 'Air Detachment'. A First World War German army corps squadron. Often with the suffix *A* (*Artillerie*) for specialist artillery and *Lb* (*Lichtbild*) for specialist photo-reconnaissance duties. Usually had an establishment of six corps aircraft but reinforced units had nine.
Fliegerabwehrkanone (Flak)	German AA guns/organization
Flieger-Ausbildungsregiment (FAR)	Air Training Regiment
Fliegerausbildungsstellen	Flight Training Facility
Flieger-Ersatzabteilung	Air Training Battalion
Fliegerführer	Air Officer Commanding (lit. 'Air Leader')
Fliegergruppenkommando	Air Group Command
Fliegerkampfgeschwader/Fliegergeschwader	*Freikorps* air formation, usually of multi-squadron strength.
Fliegerkompanie (Flik)	Air Company. A First World War Austrian squadron, usually multi-role although there were fighter (*Flik-J*) and bomber (*Flik-G*) units.
Fliegerkurierstaffeln	Air Courier Squadrons
Fliegerschule	Air Training School
Fliegerübungsstellen	Basic Training Facilities
Fliegerverbindungsoffizier (Flivo)	*Luftwaffe* Air Liaison Officer
Flottenabteilung	*Reichsmarine* Naval Staff
Fluganwärterkompanie	Air Cadet Company
Flughafenbetriebskompanie (FBK)	Airfield Support Company
Flughafenpolizeistellen	Airport Police Posts (Austrian)
Flugzeug-Beschaffungsprogramm	Aeroplane Procurement Programme
Flugzeugführerschule	Flying Training School
Freikorps	General term for Right Wing/Nationalist volunteers between 1919 and 1924.
Freiwilligen Fliegerabteilung (FFA)	Volunteer *Fliegerabteilung*

Führer der Luftstreitkräfte (FdL)	Commander of (Naval) Air Corps
Führung Abteilung	Command Department
Führungstab	Command Staff
Generalkommando	A corps-sized army task force
Generalluftzeugmeister	General Officer Commanding, Procurement and Supply
Generalstabreise	General Staff Journey
Generalstabsübung Schlesien	General Staff Manoeuvre Silesia
Geschwader	(1) In the First World War, a permanent formation of four to six *Staffeln*. Bomber units had six *Staffeln* and were designated *Bogohl*. (2) In the *Luftwaffe*, a formation of three or four *Gruppen*.
Gesellschaft zur Forderung gewerblicher Unternehmungen (GEFU)	Company for the Furtherance of Industrial Enterprises
Grenzschutz Fliegerabteilung (GFA)	Border Patrol *Fliegerabteilung*
Grenzschutz Ost	Eastern Border Patrol
Grossnachtbomber (Gronabo) Gruppe	Large Night Bomber *Gruppe*. (1) In the First World War, a permanent headquarters controlling a temporary grouping of four to six *Jasta* or *Schlasta*. (2) In the *Luftwaffe*, a permanent formation of three *Staffeln*.
Heeresaufklärungsgruppe	Army Reconnaissance Group
Heeresgruppe	Army Group. An Army command of several army headquarters (AOK).
Heereswaffenamt (HWaA)	Army Ordnance Board
Heimatjagdeinsitzer (Heitag)	Homeland Single Seat Fighter
Heimatschutzjäger	Home Defence Fighter
Höhere Fliegerkommandeure	Senior Air Commands
Höheren Kommandeure der Flakartillerie	Senior Flak Commands
Inspekteur der Flieger (Idflieg)	Inspector of Aviation
Inspekteur der Waffenschule (In 1)	Weapons Schools Inspectorate
Inspektion für Waffen und Gerät (InWG)	Weapons and Equipment Inspectorate
Istrebitel' (I)	Fighter (Russian)
Jagdfliegerführer (Jafü)	Fighter Operations Co-ordinator (lit. 'Fighter Leader')
Jagdgeschwader (JG)	(1) In the First World War, a permanent formation of four to five *Jasta*. (2) In the *Luftwaffe*, a formation of up to three *Jagdgruppen*.
Jagdgruppe (JGr)	(1) In the First World War, a permanent headquarters for a temporary grouping of four to six *Jasta*. (2) In the *Luftwaffe*, a permanent grouping of three *Staffeln*.
Jagdlehrstaffel	Fighter Demonstration Squadron
Kaiserlich und Königlich (KuK)	Imperial and Royal (prefix for units of the Austro-Hungarian Empire)
Kampfgeschwader (KG)	A *Luftwaffe* Bomber Wing of up to three *Gruppen*
Kampfgeschwader/Gruppe zur besonderen Verwendung (KG zbV)	Transport unit designation until 1943 (lit. 'Special Duties Bomber Wing/Group')
Kette	Formation of three aircraft

Kommandeur der Flieger (Kofl)	Air Commander
Kommandeur der Heeresflieger	Commander of Army Aviation
Kommandeur der Luftwaffe (Koluft)	Luftwaffe officer at Armee/Heeresgruppe headquarters responsible for Army air support.
Königlich Bayerischen Fliegertruppen	Royal Bavarian Flying Troops
Kommando der Luftstreitkräfte (KoLu)	Austrian Air Force Command
KuK Luftfahrtruppen	Austrian Army Aviation troops
KuK Seefliegerkommando	Austrian Naval Air Command
Küstfliegergruppe (KüFlGr)	Coastal Air Group
Lehr	Demonstration
Lieferplan	Production Plan
Luchtvaartafdeling (LVA)	Dutch Army Air Force
Luftdienst GmbH	Air Services Ltd
Luftfahrt GmbH	Aviation Ltd
Luftflotte	Air Fleet
Luftgau	Air District
Luftkommandoamt	Air Command Directorate
Luftkriegsakademie	Air War Academy
Luftkriegsschule (LKS)	Officer Training School (lit. 'Air War School')
Luftnachrichtenabteilung	Air Signal Battalion
Luftnachrichtenlehr- und Versuchsregiment	Air Signal Demonstration and Experimental Regiment
Luftpolizei Abteilungen	Air Police Battalion
Luftschiffer-Battaillon	Airship Battalion
Luftschutz Amt	Air Defence Directorate
Luftschutzreferat	Air Defence Desk
Luftsportlandesgruppen DLV	Provincial Air Sport Groups
Luftstreitkräfte	German Army Air Corps
Luftüberwachungs Abteilungen	Air Control Detachments
Luftverteidigungskommando	Air Defence Command
Luftverteidigungszone (Lv)	Air Defence Zone
Luftwaffendienstvorschrift 16 (LDv 16)	Air Force Service Regulations
Luftwaffengruppenkommando	Air Force Group Command
Luftwaffenkommando	Air Force Command
Luftwaffepersonalamt	Air Force Personnel Directorate
Marinekommandoamt	Naval Command Directorate
Marine Landfliegerabteilung	Naval Land-Based Fliegerabteilung
Marine Luchtvaartdienst (MLD)	Dutch Naval Air Service
Marine Luftschifferabteilung	Naval Airship Branch
Militärarchiv (MA)	Military Archives
Nachrichtendienst	Signals Service
Nachtjagd und Erkundungsflugzeug (Najuku)	Night Fighter and Reconnaissance Aeroplane
Nationalsozialistische Deutsche Arbeiter Partie (NSDAP)	National Socialist German Workers' Party
Nationalsozialistischen Fliegerkorps (NSFK)	National Socialist Air Corps
Oberbefehlshaber der Gesamten Wehrmacht	Commander-in-Chief of all the Armed Forces
Oberbefehlshaber der Luftwaffe (ObdL)	Commander-in-Chief of the Air Force
Oberkommando des Heeres (OKH)	Army High Command

Oberkommando der Kriegsmarine (OKM)	Naval High Command
Oberkommando des Wehrmachts (OKW)	Armed Forces High Command
Oberste Heeresleitung (OHL)	Senior Army Direction
Offizieranwärter	Officer cadets
Österreichische Flugpolizei	Austrian Air Police
Österreichische Luftstreitkräfte (LStrKr)	Austrian Air Force
Österreichische Luftverkehrs Aktien-gesellschaft (ÖLAG)	Austrian Air Transport Ltd
Planstudie Grün	Contingency Plan Green
Polizei-Fliegerstaffeln	Police Air Squadrons (German)
Polizei-Flugstaffeln	Police Air Squadrons (Austrian)
Polskie Lotnictwo Wojskowe	Polish Air Force
Razvedchik (R)	Reconnaissance (Russian)
Referent für Seeflugwesen	Naval Aviation Desk
Referenten zur besonderen Verwendung	Special Duty Desks
Regia Aeronautica	Italian Air Force
Reichsamt für Luftfahrt und Kraftfahrwesen	National Directorate for Air and Motor Transport
Reichs Arbeitsdienst (RAD)	National Labour Service
Reichsbahnstrecken	Cross-country training organization (lit. 'National Rail Routes')
Reichsheer	National Army (the Weimar Republic's Army)
Reichskommissar für die Luftfahrt	National Aviation Commissioner
Reichsluftamt	National Air Directorate
Reichsluftfahrtkommissariat	National Aviation Commission
Reichsluftministerium (RLM)	National Air Ministry
Reichsmarine	National Navy (the Weimar Republic's Navy)
Reichsminister für die Luftfahrt	National Minister for Aviation
Reichsverband der deutschen Luftfahrtindustrie (RDLI)	National German Aviation Manufacturers' Association
Reichsverteidigungsminister	National Defence Minister
Reichswehrministerium (RWM)	National Defence Ministry
Reichswehr	The National Forces (consisting of the *Reichsheer* and *Reichsmarine*)
Reklamestaffel	Skywriting Squadron
Riesenflugzeugabteilung (RFA)	Large Aircraft Department
Rotte	Formation of two aircraft
Roboche-Krest'yanski Krasny vozdushny flot (RKKVVF)	Workers' and Peasants' Red Air Fleet (official designation of the Soviet Air Force until 1924)
Schlachtstaffel (Schlasta)	Close-support squadron of six light attack (CL) aircraft in the First World War
Schwarm	Formation of two *Rotten*
Seeflugzeug-Erprobungsstelle (SES)	Seaplane Experimental Station
Seeflugzeug-Versuchsabteilung GmbH	Experimental Seaplane (Severa) Department Ltd
Seetransportabteilung	Marine Transport Division
Service Historique de l'Armée de l'Air (SHAA)	French Air Force Historical Service
Service Historique de l'Armée de Terre (SHAT)	French Army Historical Service

Sicherheitsabteilung	Security Battalion
Sicherheitspolizei (Sipo)	Security Police
Skorostnoi Bombardirovshchik (SB)	Fast Bomber (Russian)
Sonderfall	Special Contingency
Sondergruppe R	Special Section R (Russia)
Sonderstab England	Special Staff England
Sportflug GmbH	Light Aviation Ltd
Staatssekretär der Luftfahrt	State Secretary for Aviation
Stabsoffizier der Flieger (Stofl)	Aviation Staff Officer
Stationskommandos Nordsee/Ostsee	North Sea/Baltic Naval Stations
Sturzkampfflugzeug (Stuka)	Dive bomber
Stukageschwader/gruppe	Dive Bomber Wing/Group
Trägergruppe (TrGr)	Aircraft Carrier Group
Truppenamt	Cover name for the General Staff in the *Reichsheer* (lit. 'Troop Directorate')
Truppenamt Heeresabteilung (T1)	*Truppenamt* Operations Department
Truppenamt Heeres-Organisationsabteilung (T2)	*Truppenamt* Organization Department
Truppenamt Heeres-Statistische Abteilung (T3)	*Truppenamt* Intelligence Department (to 1929)
Truppenamt Abteilung Fremde Heer (T3)	*Truppenamt* Intelligence Department (from 1929)
Truppenamt Heeres-Ausbildungsabteilung (T4)	*Truppenamt* Training Policy Department
Truppen-Fliegerstaffel	Troop Air Squadron. A *Reichsheer* air squadron of six to ten aircraft.
Tsentralnii Aero-Gidrodinamicheskii Institut (TsAGI)	Central Aero and Hydrodynamic Institute (Russian)
Tyazhyoly Bombardirovshchik (TB)	Heavy bomber (Russian)
Unternehmen	Operation
Volkswehr	Austrian People's Militia
Voenno-Vozdushnye Sily Raboche-Krestyanskaya Krasnaya Armiya (VVS-RKKA)	Military Aviation Forces of the Workers' and Peasants' Red Army. Official designation of the Soviet Air Force from 1924.
Waffenamt Prüfwesen 8 (Wa Prw 8)	Ordnance Board's Testing Section 8
Waffenschule	Operational Training Schools
Wehrmacht	Armed Forces
Wetterkundungsstaffel (Wekusta)	Meteorological Squadron
Wissenschaftliche Versuchs- und Prufanstalt für Luftfahrzeuge (Wivupal)	Scientific Experimental and Testing Establishment for Aviation
Waffenschulen	Operational Training Schools
'Wasserkante'	'Seaboard'
Wehrkreis	Military District
Wiener Sicherheitswache	Vienna Security Guard
Zerstörergeschwader/gruppe	Heavy fighter Wing/Group (lit. 'Destroyer Wing/Group')
Zentrale Moskau (ZMo)	Moscow Centre
Zone d'Operations Aérienne Est (ZOAE)	Eastern Air Zone (French)
Zone d'Operations Aérienne Nord (ZOAN)	Northern Air Zone (French)
Zone d'Operations Aérienne Sud (ZOAS)	Southern Air Zone (French)
Zur besonderen Verwendung (zbV)	For special employment

INDEX

General Index

Aero-engines:
As 10, 56
BMW II, 48
BMW III, 48
BMW IV, 48
BMW VI (M-17), 47, 48, 56, 65, 78, 79
BMW 801, 153
Bristol Jupiter (Sh22/SAM 22B), 56, 78, 103
Daimler-Benz DB 600, 77, 107
Daimler-Benz DB 601, 109, 156
Daimler-Benz DB 602, 66
Daimler-Benz DB 606, 156, 157
Junkers Jumo 205, 107, 155
Napier Lion, 46, 56
Rolls-Royce Buzzard, 78
Rolls-Royce Kestrel V, 78
Rolls-Royce Kestrel VI, 78
Siemens Jupiter, 65
Aeronautical Guarantee Committee, 34, 39, 40, 52
Aerosport GmbH, 57
Aéronavale, 239, 242, 248, 250, 254, 262
Airborne formations:
Fallschirmjäger Regiment (FJR) 1, 233; *1./FJR 1*, 194; *4./FJR 1*, 221; *III/FJR 1*, 185
III/Regiment General Göring, 161; *Gruppe Nord*, 244; *Gruppe Süd*, 244
Aircraft:
AEG G.V, 30
Aero: A 100, 133; A 300/304, 165
Aichi D1A, 107
Albatros: B.II, 39; C.XV, 32, 36; D.III, 36; D.Va, 20; L 35, 58; L 65/II, 63; L 68a, 47; L 70, 63; L 75a, 69; L 76, 47,66; L 77, 47, 65; L 78, 47,65,66; L 82,69; L 84, 78
Amiot: 143, 82, 242; 351, 264
ANF Les Mureaux 113/115/117, 81, 201
Arado: Ar 66, 158; Ar 67, 78; Ar 68, 153, 166, 200; Ar 80, 78; Ar 96, 158, 170; Ar 196, 221-3, 226, 231; SC I/II, 58; SD I, 65, 78; SD II, 78; SD III, 78; SD IV/Ar 64, 77-8 ; SD V/Ar 65, 77-8, 108; SSD 1, 65
Armstrong Whitworth Whitley, 164, 229
Avia: B.135, 135; B.534, 170
Aviatik (Berg) D.I, 25
Avro: Anson, 213 ; Lancaster, 157; Manchester (Type 679), 157
BFW: M20, 104; M 22, 65
Blackburn: Roc, 225; Skua, 225, 230
Bloch: MB 131, 197; MB 135, 84; MB 151/152, 85; MB 162, 84; MB 174, 197; MB 200/210, 82
Blohm und Voss: BV 138, 228; Ha 139A, 228; BV 142, 228

Boeing: B-17, 84-5, 89, 109, 116; Model 247 (Y1B-9), 83; Model 299 (Y1B-17), 84; P-26, 84; XB-40/YB-40, 83
Boulton Paul Defiant, 259
Brandenburger C.I, 80
Breda Ba 65, 160
Breguet: Bre 14, 81; Bre 19, 124; Bre 27, 81; Bre 482, 84; Bre 693/694, 241, 248, 264
Brewster 339/F2A Buffalo, 226, 241
Bristol: Blenheim, 83, 164, 197, 204, 227, 229, 231, 239, 243, 247, 254, 259; F2B Fighter ('Brisfit'), 82
Bücker Bü 131, 158
CAO 700, 84
Caproni: Ca 133, 112,; Ca 310/312, 221, 223, 241; Ca 335 Maestrale/SABCA S.47, 241
Caspar: C 36, 65; S1, 33; U1, 33; U2, 33
Caudron: C.714, 265; R.11, 82
Centre NC 223.4, 70, 109, 250
Consolidated: Model 32/B-24 Liberator, 165; P-30 (PB-2)/A-11, 82
Curtiss: F8C Helldiver, 107; F11C Goshawk/Hawk II, 107; Hawk 75 (P-36), 85, 116, 165, 201, 202, 221, 223, 243; Hawk 81 (P-40/Warhawk/Tomahawk), 165
De Havilland: Mosquito, 84; Moth, 69, 112, 231
Dewoitine: D.371, 84; D.500/501/510, 84; D.520, 85, 262
DFS 230, 170
DFW C.V/VI, 36
Dietrich: DP I/IIa, 39; DP IX, 58
Dornier: Cs II *Delphin I*, 33; Do A *Libelle*, 83; Do B *Merkur*, 47, 63; C I/Do 10, 78; Do F/Do 11, 78, 104-5; Do H *Falke*, 33; Do J *Wal*, 65, 79; 120; Do N (Kawasaki Type 87), 55; Do P, 73, 78; Do R *Superwal*, 65, 78; Do X, 65,77; Do 11C, 103, 105; Do 13, 105-6; Do 17, 105, 107, 128, 129, 133, 135, 140, 142-3, 154-5, 159-61, 166, 197-8, 200, 216, 235, 241, 243, 250, 265; Do 18, 213, 235; Do 19, 108-9; Do 22, 78; Do 23, 105-6; Do 26, 228, 232, 234; Do 215, 200; Do 217, 156, 216; *Komet III*, 58
Douglas: B-18, 116; DB-7, 165; DB-8A-5, 221, 241
Fairey: Battle, 90, 105, 164, 178, 191, 194, 197-8, 221, 247, 262; Fox, 241; *Kangourou*, 241; P.4/34, 221; Swordfish, 225, 227
Farman F.221/222, 73, 84
Feiseler Fi 156 *Storch*, 170, 184, 194, 243-4, 248

Fiat: CR 20, 111; CR 30, 111; CR 32, 112, 123, 162; CR 42, 241
Fiat–Ansaldo A 120, 80, 111
Focke Wulf: FW 44 *Stieglitz*, 112, 158, 221; FW 56 Stösser, 110 ; FW 57, 109; FW 58, 158; FW 189, 144, 170; FW 190, 153; FW 200, 125, 168, 204, 213, 217, 228, 232, 235; S 39, 78; S 40, 78; W 7, 79
Fokker: C.V, 221-3, 231; D.VII, 17, 20-1, 23, 36, 39, 47; D.VIII, 20; D.XI, 40; D.XIII, 40, 47, 63, 78-9, 222; D.XXI, 84, 221; G.I, 241; T.V, 241
Friedrichshafen: FF.49, 29, 57; G.IIIa, 30
Gloster Gladiator/Sea Gladiator, 84, 204, 222-3, 225, 229-30, 234-5, 239, 247
Gotha Go 145, 158
Grumman F4F Wildcat, 226
Halberstadt: C.V, 20, 32; CL.II/IV, 20
Handley Page: HP 56/Halifax, 84, 109, 157; Hampden, 227; Harrow, 164; Heyford, 164
Hannover CL.V, 20
Hansa-Brandenburg C.I, 25
Hawker: Demon, 82; Fury, 83, 85; Hardy, 82; Hart, 82, 90, 114; Hurricane, 85, 164, 178, 198, 200-2, 204, 221, 234, 239, 241, 247, 255, 259, 262; Typhoon, 265
Heinkel: HD 15, 65; HD 16, 66,79; HD 17, 46; HD 20, 47; HD 21, 47, 57; HD 24, 58; HD 25 (Aichi Type 2), 55; HD 30, 79; HD 32, 39; HD 33, 63; HD 35, 39; HD 37 (I-7), 55, 65; HD 38/He 38, 65, 77-9; HD 41, 65, 78; HD/He 42, 58; HD 46/He 46, 77-8; HE 1, 33, 60; HE 5, 65-6, 79; HE 7, 65; HE 9, 65-6, 79; HE 10, 66; HE 31, 65; He 45, 65, 77-8, 128, 130, 135, 137, 166; He 50, 107, 123; He 51, 77-9, 108, 110, 113, 121, 123-4, 127, 130, 132-5, 138-9, 161; He 59, 77, 79, 138-9, 215-16, 235, 244; He 60, 124; He 63, 77; He 70, 124-5, 146; He 72, 156; He 111, 77, 83, 104-5, 107, 128-9, 133-6, 142-3, 148, 154-5, 159, 166-7, 197, 199, 213, 215-17, 221-3, 228-30, 234, 249, 263; He 100, 165; He 112, 78, 84, 86, 127, 132, 159, 166; He 115, 139, 213, 215-17, 221, 223, 231, 233, 267; He 118, 78, 152; He 177, 84, 100, 150, 153, 156-7, 168-9; He 179, 157; He 277, 157
Henschel: Hs 123, 108, 123-12, 129, 131, 137, 166, 181, 258, 266; Hs 124, 109; Hs 126, 142, 159, 170, 228, 254, 267; Hs 127, 120; Hs

129, 144
Hopfner HS 8-29, 80
Ilyushin: DB-3, 116; Il-2, 144
Junkers: A 20, 46, 58; A 35/K 53, 58,
  60, 66, 80; A 48, 78–9, 113; CL.I,
  19–20, 23, 37; D.I, 20,23, 37; F 13,
  32, 34, 37, 46, 58, 60, 158; G 24,
  47, 55, 58, 60, 62–3, 66; G 31, 55;
  G 38, 55, 220; J 20 (Yu 20), 38, 48;
  J 21 (Yu 21), 38, 48; J 22, 38; K 30
  (R–42), 47–8; K 37 (Mitsubishi
  Type 93), 55; K 47, 78, 107; Ju
  52ce, 79; Ju 52/3m, 62, 103, 105–6,
  115, 120–5, 129, 133, 158, 160–1,
  187–8, 194, 200, 230, 232–3, 236,
  244, 256; Ju 86, 105, 107, 128,
  154–5, 159, 200; Ju 87, 78, 108,
  127, 131, 137, 140, 142–3, 155,
  159, 177, 179, 183, 185, 188, 193,
  226, 234–5, 250, 252–3, 255, 258,
  265–6; Ju 88, 83–4, 110, 113, 143,
  153, 155–6, 169, 200, 205, 213,
  217, 226–7, 229–30, 233, 235, 253,
  257, 262; Ju 89, 108; Ju 90, 109,
  168, 200; W 33 (PS-4), 48, 54, 58,
  79, 158; W 34/K 43, 58, 78, 158
Klemm: Kl 25, 104; Kl 35, 158
Koolhoven: FK 56, 241; FK 58, 241
Kotscherigin DI-6, 82
Lloyd C.III, 25
Loire et Olivier: LeO 206, 73; LeO
  451, 190, 197
Lockheed Hudson, 165, 213
Löhner C.I, 25
LVG: B.III, 57; C.IV, 32; C.V/VI, 20,
  36
LWS *Mewa*, 178
Macchi MC 200, 221
Martin: 123 (XB-907), 83; 139 (B-10/
  B-12), 83, 89, 106 ; 167 Maryland,
  165
Messerschmitt: Bf 108, 195; Bf 109,
  77–8, 84, 86, 110, 127–9, 133–6,
  138–9, 142–3, 152, 154–5, 159,
  161, 164, 166–7, 169–70, 172,
  176–7, 179–80, 191, 200–2, 205,
  226, 244–5, 254–5, 258, 263, 265,
  267; Bf 110, 109, 143, 155, 165,
  167, 170, 176–7, 179, 190, 200,
  202, 205, 221–3, 227, 235, 265; Bf
  162, 110; Me 210, 110, 153, 169;
  Me 262, 128
Morane: MS 225, 84; MS 406/D-
  3800, 85, 164, 201–2, 263, 267
Nieuport 17, 23
Northrop N-3PB, 221
Öfag-Albatros D.III, 25, 41
Petlyakov TB-7 (Pe-8), 84
Phönix: C.I, 41; D.I/II/III, 25, 41;
  L2c, 80
Polikarpov: I-15 '*Chato*', 84, 125–6,
  129, 143; I-152 '*Super Chato*', 84,
  135; I-16 '*Mosca*'/'*Rata*', 85, 116,
  125–6, 134, 137–8
Potez: 39, 81; 63/631/633/637/639,
  109, 190, 197, 245, 248, 264, 267
PZL: P.7, 84, 178; P.11, 84, 178–9;
  P.23 *Karas*, 178, 180–1; P.37 *Los*,
  178, 181; P.46 *Sum*, 178; P.48
  Lampart, 178; P.50 *Jastrzab*, 178
Rohrbach: E 20, 32–4; Ro IV
  Inverness, 56, Ro V *Rocco*, 65; Ro
  VI/Beardmore Inflexible, 56; Ro VII
  Robbe I, 65; Ro VIII *Roland I*, 47,
  55; Ro XII (*Roka*), 47; *Romar*, 55
Rumpler C.IV/VII, 23, 32, 81

RWD: -14 *Czapla*, 81, 179; -25 *Sokol*,
  178
Savoia S.81, 124, 133
Short: Stirling, 84, 109; Sunderland,
  223
Siemens-Schuckert D.III, 19; D.IV,
  19, 20
Supermarine: Spitfire, 84–5, 164, 200,
  209, 212, 221, 239, 258–9; Walrus,
  230
Tupolev: ANT-9, 48; ANT-21/29/39,
  82; ANT-40, 82; R-3 (ANT-3), 48;
  R-6/KR-6, 82; SB-2M-100 *Katiuska*
  (SB)/Avia B 71, 82–3, 116, 125,
  127, 129, 134, 137, 141, 165; TB-1
  (ANT-4), 48; TB 3, 84, 91
Udet U 11 *Kondor*, 55; U 12 *Flamingo*,
  57–8, 80
UFAG C.I, 25
Vought V-156/SB2U Vindicator, 165,
  226, 242, 254
Westland Lysander, 81
Vickers: Wellseley, 164; Wellington,
  190, 221, 227
Zeppelin (Staaken): R.VIII, 19;
  R.XIVa, 32
Airfields:
Aalborg (Denmark), 219–20, 221, 227
Adlershof-Johannisthal (Germany),
  21–2
Alcala de Henares (Spain), 126
Almazan (Spain), 121
Amsterdam-Schipol (Netherlands),
  241
Asch-Eger(Czechoslovakia), 169
Badajoz (Spain), 121
Banak (Norway), 234
Bardufoss (Norway), 234–5
Berlin-Döberitz (Germany), 24, 121
Berlin-Staaken (Germany), 57–8
Berlin-Tempelhof (Germany), 113
Böblingen (Germany), 24, 30, 57
Bodø (Norway), 234
Boscombe Down (Britain), 215
Boulogne-Alprecht (France), 242–3
Brusthem (Belgium), 241
Bruxelles-Evère (Belgium), 152, 258
Le Bourget (France), 91
Caceres (Spain), 124
Calais-Marck (France), 243
Cambrai-Niergnies (France), 243
La Cenia (Spain), 138, 142
Conde Vraux (France), 247, 251
Le Coulot (France), 266
Cracow-Rakowice (Poland), 180
Dedelsdorf, 194
De Kooy (Netherlands), 241
Dijon-Longvic (France), 91, 243
Dôle-Taraux (France), 243
Ecury-sur-Coole (France), 251
Esbjerg (Denmark), 222
Farnborough (Britain), 64
Fischamend (Austria), 40
Gatow (Germany), 129
Getafe (Spain), 123
Giebelstadt (Germany), 251
Gotha (Germany), 121
Graz-Thalerhof (Austria), 25, 40–1,
  80
Haag-Ockenburgh (Netherlands), 244
Haag-Valkenburg (Netherlands), 244
Haag-Ypenburg (Netherlands), 241,
  244
Hörnum/Sylt (Germany), 69, 213, 218
Hutniki (Poland), 180
Jever (Germany), 191

Jerez de la Frontera (Spain), 122
Kiel-Holtenau (Germany), 59
Kitzingen (Germany), 115
Klagenfurt-Annabichl (Austria), 41
Köln-Butzweilerhof (Germany), 115
Kottbus (Germany), 21
Kristiansand-Kjevik, 226
Landfeld (Germany), 125, 251
Lechfeld (Germany), 241
Lippstadt (Germany), 207
List/Sylt (Germany), 58, 213
Loddenheide (Germany), 195
Lübeck-Blankensee (Germany), 24
Lyon-Bron (France), 91, 263
Malmaison (France), 251
Mannheim (Germany), 18
Merseburg (Germany), 121
Metz-Frescarty (France), 198
Mildenhall (Britain), 268
Monjos (Spain), 142
Nancy (France), 91
Neerhespen (Belgium), 241
Neuhausen (Germany), 188
Neukirchen-bei-Ansbach (Germany),
  121
Norderney (Germany), 59, 216
Oslo-Fornebu (Norway), 220, 222–3,
  226, 229, 232
Oslo-Kjeller (Norway), 223
Paderborn (Germany), 30
Paris-Orly (France), 266
Pate (Spain), 142
Pau (France), 91
Peterfeld (Latvia), 23
Poznan-Lawica (Poland), 21
Rechlin am Müritzsee (Germany), 79
Reims-Pommery (France), 251
Rhein-Main (Germany), 251
Roth-Kiliansdorf (Germany), 239
Rotterdam-Waalhaven (Netherlands),
  243
Rouvres (France), 198
Roye-Amy (France),
Sabadell (Spain), 143
Schaffen (Belgium), 241
Schleissheim (Germany), 24, 57–8,
  111
Schkeuditz (Germany), 57
Serrania de Ronda (Spain), 125
Setnesmoen (Norway), 230–1
Skånland (Norway), 234
Stavanger-Sola (Norway), 219, 223,
  226–30
Tablada (Spain), 122
Talavera (Spain), 124
Thionville (France), 198
Tellancourt (France), 17
Trondheim-Vaernes (Norway), 226–7,
  231–2
Vaerlose (Denmark), 222
Vilajuiga (Spain), 142
Vitoria (Spain), 129
Westerland (Germany), 230
Wien-Aspern (Austria), 25, 40–1, 161
Wiener Neustadt (Austria), 25, 40,
  161
Wilhelmshaven-Rüstringen (Ger-
  many), 60
Wright-Patterson (USA), 64
Zipser Neudorf (Slovakia), 178
Airlines:
Compagnie Franco-Roumaine, 54
Deruluft, 47, 49, 53
Deutsche Luftreederei, 30–2
Deutscher Aero Lloyd, 52
Dobrolet, 49

Imperial Airways, 52
Junkers Luftverkehr, 52
Lufthansa, 46–8, 52–3, 60–1, 65, 73, 94, 96–7, 106, 110, 112, 120–2, 125, 147, 160, 163, 219–20
ÖLAG, 41, 80
Allied Control Commission, 90
Allied Aviation Guarantee Committee, 54
Akaflieg, 58
Åndalsnes, 228–31
Andalusia, 120, 123
Anschluss Crisis, 138, 199, 160–2, 164, 167
Armée de l'Air, 90–1, 109, 113–16, 148, 159, 165, 166, 198, 202–3, 239–40, 250, 264–5, 268
Austria, 25–6, 40, 80, 91, 111, 149, 165, 191, 207
Aviation Militaire Belge (Belgian Air Force), 241, 254, 267
Aviation Militaire Française (French Army Air Force), 70, 90

Barcelona, 122, 135, 139, 142–3
Basques, 120, 129–3
BCR (*Bombardement, Combate et Reconnaissance*), 82, 88, 91, 109, 116
Beer Hall *Putsch*, 41, 42, 57, 162, 171, 232
Belgian Army, 243
Belgium, 36, 40–1, 193–4, 196, 199–200, 240–1, 243, 245, 247, 251, 253–7, 264
Bergen, 219–20, 223, 227–8
Berlin, 20–3, 31, 34–9, 44–5, 70, 73–4, 92, 96, 103, 114, 121, 124, 127, 138–9, 141, 143, 146, 149, 161, 163, 173, 175, 189, 208–9, 219, 227, 239, 250, 255, 269
Bilbao, 129–30, 132
'Bomber A', 109, 156
Boulogne Conference, 31
British Expeditionary Force (BEF), 194, 240, 261
Bundesministerium für Heerwesen (*BMfHW*), 80

Carinhall, 147, 168–9
Cartagena, 123–4, 139, 144
Catalonia, 120, 124, 139, 149
Chain Home (CH), 165, 172, 199
Chain Home Low (CHL), 217
Civil aviation, German, 52–4
Comité Permanent de la Défense Nationale, 165
Czech Air Force, 163, 165
Czechoslovakia, 25–7, 40–1, 100, 149, 162–70, 175, 207
Czechoslovakian Crisis, 139, 141, 150, 155, 162–70

Darmstädter und Nationalbank (Danatbank), 74, 94
Dawes Plan, 39
DEM, 116, 199
Denmark, 32, 55, 191, 219–21
Dessau, 22–3, 37, 53, 103, 108
Deterrent air fleet (*Risiko Luftflotte*), 101
Deutsches Luftamt, 18
Deutschösterreichische Fliegertruppe, 25
DLV, 35–6, 65, 95, 110, 152, 158
Dunkirk (Dunkerque), 216, 235, 254, 258–61
DVS, 58, 59, 61, 69, 70, 74, 80, 97, 110, 112

'Dynamo', Operation, 259, 261

E–Stelle Rechlin des RDLI, 64, 95, 107, 110, 165
E–Stelle Travemünde des RDLI, 61, 65
Enigma, 209, 263
'Erkudista' (*Erkundungsflugzeug für die Divisionsnahaufklärungsstaffeln*), 64, 65
'Erkunigros' (*Erkundungsflugzeug für mittlere Höhen und grösste Entfernungen*), 64, 65

Fall 'Gelb', 191–3, 196, 200, 205–6, 216–20, 227, 231–2, 239–40, 268
Fall 'Grün', 167
Fall 'Rot', 259, 261, 264–5, 267
Fall 'Weiss', 175, 178
Fertigungs GmbH, 64, 103
Fili, 47–8, 103
Flak commands:
    Höheren Kommandeur der Festungs-flakartillerie III, 163
    Höheren Kommandeure der Flakartillerie, 150
    Lv–Zone West, 163
Flak formations:
    Flakbrigade VI, 218
    Flakgruppe 18, 179
    Flakkorps I, 194, 247, 261,
    Flakkorps II, 247–8, 261
    F/88, 124, 127, 132, 135, 139
    I/Flak Regiment 22,
Flakwaffe, 111–12, 170, 172, 250
Flughafenpolizeistellen, 40
Flugsport, 30, 35
Flugzeug-Beschaffungsprogramm 1934 (*Rheinlandprogramm*), 104
Flugzeugmeisterei, 153
France, 26, 36, 40, 100, 112, 120, 124, 154, 162, 164–5, 168, 171, 190–1, 195, 197–200, 243–4, 248, 251, 253–5, 264, 266, 268–9
Franco-Soviet Mutual Assistance Pact 1935, 115
Freikorps, 20, 24, 30, 36–7, 51, 91, 98, 142, 150, 171
    Commands and formations:
        Bayerisches Schützenkorps, 22;
        Freikorps Hülsen, 22; *General-kommando Oven*, 22; *Grenzschutz Ost*, 22; *Landeschützenkorps*, 21;
        1.Garde Reserve Division, 23;
        III.Marinebrigade, 22, 30
French Army: 86–7, 253
    1er Armée, 259
    2ème Armée, 240, 253
    4e Armée, 265
    7e Armée, 240, 248, 250
    9e Armée, 240, 251
    Armée de Rhin, 38
    6e Bataillon de Chasseurs Alpins, 234
    Corps de Cavallerie, 250
    1er Groupe des Armées, 240, 257
    2ème Groupe des Armées, 251, 261
    3ème Groupe des Armées, 261–2, 267
    4e Groupe des Armées, 261
    Théatre d'Opérations de Nord Est (*TONE*), 240
Freya, see Radar

German Army Formations:
    Great War: *2.Armee*, 19; *4.Armee*, 24; *5.Armee*, 19; *7.Armee*, 21, 99; Eastern Field Army, 18; *112.Infanterie Regiment*, 96; Western Field Army, 17

Wehrmacht: *1.Armee*, 266; *2.Armee*, 167; *3.Armee*, 175, 182, 184; *4.Armee*, 175, 183, 185, 192–4, 209, 254; *6.Armee*, 192, 196, 209, 245, 247, 253; *8.Armee*, 160, 167, 175, 185–6; *9.Armee*, 262, 266; *10.Armee*, 167, 175–6, 192; *12.Armee*, 167, 192–4, 253; *14.Armee*, 167, 175, 186; *16.Armee*, 192; *18.Armee*, 193, 197, 249, 259; *XIV.Armee Korps (Motorisiert)*, 175; *XV.Armee Korps (Motorisiert)*, 175, 183, 209, 253; *XVI.Armee Korps (Motorisiert)*, 175, 182–3, 209, 250; *XIX.Armee Korps (Motorisiert)*, 193, 244, 248, 251–2; *XXI.Armee Korps*, 219; *XXXIX.Armee Korps (Motorisiert)*, 249; *XLI.Armee Korps (Motorisiert)*, 251; *3.Gebirgsjäger Division*, 232; *138.Gebirgsjäger Regiment*, 228; *Grenzschutz Abschnittkommando 1*, 182, 219; *Gruppe XXI*, 219; *Heeresgruppe A*, 192, 196, 256–7, 261; *Heeresgruppe B*, 192, 249, 261; *Heeresgruppe C*, 189, 192, 261; *Heeresgruppe Nord*, 175, 192; *Heeresgruppe Süd*, 175, 192; *Höheres Kommando XXXI*, 219, 222; *22.Infanterie Division (Luftlande)*, 244; *34.Infanterie Division*, 243; *181.Infanterie Division*, 228; *196.Infanterie Division*, 228; *2.Leicht Division*, 185; *1.Panzer Division*, 183–4, 253; *2.Panzer Division*, 254; *7.Panzer Division*, 258; *9.Panzer Division*, 244; *Panzergruppe Guderian*, 261; *Panzergruppe Kleist*, 196–7, 261–2; *III./Regiment Gross Deutschland*, 244
German Navy, 25, 45, 59, 65–66, 72, 74–6, 79, 97–9, 117, 138, 146, 151, 158, 171–2, 191–2, 212, 214, 216
Gesellschaft zur Förderung gewerblicher Unternehmungen (GEFU), 38
Gestapo, 113, 125, 152
Goerz-Vizier 219 bomb sight, 107
Great Britain, 55, 171–2, 189, 197, 199, 212, 224, 239, 255, 268–9
Grebbe Line, 245
Guernica, 130–1, 249
GU–VVS, 78–9

Haerens Flyvevåben (Norwegian Army Air Corps), 221, 226
Heimatschutzjäger, 110
'Heitag' (*Heimatjagdeinsitzer*), 64, 65, 110
Hispano-Marroqui de Transportes SL (HISMA), 121–2
Hungarian Air Force, 32, 162
Hungary, 25, 41, 187

Idflieg, 19, 23, 30, 52, 149, 153
Inter-Allied Aeronautical Commission of Control (IAACC), 27, 29, 32–4, 41
Italy, 25, 36, 55, 91, 121,122, 125, 134, 161, 168, 175, 191

Japan, 41, 55
Japanese Navy, 33, 213

Kellogg–Briand Pact, 72
Knickebein, 113, 215
Kogenluft, 71, 149
Königlich Bayerischen Fliegertruppen, 51
Königsberg, 21, 28–29, 47, 57, 73, 185

Kristiansand, 219, 221, 223, 227, 229
KuK Luftfahrtruppen/KuK Seefliegerkommando, 25

Latvia, 22, 47, 187
Latvian Air Force, 23
LdV 16, 100, 117
Lieferplänen Nr 1–14, 104–105, 109, 150,
153, 155, 169
Lipetsk, 40, 45–7, 49–50, 59, 64–5,
69–71, 74, 78–9, 152
Lithuania, 23, 36, 47, 100, 173
London Ultimatum, 31–2
Lotfe (Lotfernrohr) 7/7D bomb sight,
107
Luchtvaartafdeling (LVA)/Marine
Luchtvaartdienst (MLD) (Dutch Air
Forces), 241, 248, 267
Luftdienst GmbH, 61
Luftfahrt GmbH/Deutsche Luftfahrt
GmbH, 59
Luftgaue:
  I (Königsberg), 177
  III (Berlin), 163, 177
  IV (Dresden), 163, 177
  V (Stuttgart), 117
  VI (Münster), 117
  VII (Munich), 250
  VIII (Munich), 163, 177, 181
  XI (Hannover), 197
  XII (Wiesbaden), 117, 197, 200, 209,
  250
  XIII (Nuremberg), 250
  XVII (Vienna), 162, 190
  Belgien–Nord Frankreich (Brussels),
  151–2
  zbV 16, 177, 254
  Luftgaustab zbV 200, 221
  Luftgaustab zbV 300, 221
Luftpolizei Abteilungen, 31
Luftstreitkräfte and Armistice, 17–18, 28,
45, 49, 106, 113, 150
Luftüberwachungs Abteilungen, 31
Luftwaffe:
  Accidents, 155, 159, 206
  Air defences, 89, 163, 172, 190–1,
  197
  Co-operation with Army, 100, 183–4,
  189, 230–1, 237, 248, 254
  Expansion, 112, 117
  Groundcrews ('Black Men'), 129,
  135–6, 239, 245, 256, 264, 266
  Heavy Bomber requirement, 108
  Heavy Dive Bomber requirement,
  107–8
  Heavy Fighter requirement, 109
  Höhere Fliegerkommandeure, 112, 150
  Inspectorates: Generalinspekteur der
  Luftwaffe, 159; Luftwaffeninspektion
  der Jagdflieger (Fighters), 108, 111,
  151; Luftwaffeninspektion der
  Kampfflieger (Bombers), 177;
  Luftwaffeninspektion der Kampf-,
  Sturzkampf- und Aufklärungsflieger
  (Bombers, Dive Bombers and
  Reconnaissance), 177; Luftwaffen-
  inspektion der Luftlande- und
  Fallschirmtruppe (Airborne and
  Paratroops), 163; Luftwaffen-
  inspektion der Schulen der Luftwaffe,
  110; Luftwaffeninspektion der
  Sturzkampfflieger (Dive Bombers),
  253
  Intelligence, 171, 176, 190, 198, 221
  Light Dive Bomber requirement,
  107–8

Light Fighter requirement, 110
Losses, 143, 180, 188–9, 192, 202,
236, 267
Medium Bomber requirement, 106–7
Personnel, 110–11, 151, 157, 166,
170, 206, 257
Reconnaissance, 192, 197–200, 248,
264
Staffs and administration: Allgemeines
Luftamt (LB), 97; Chef des
Ausbildungs, 149, 159; Chef der
Luftwehr, 148; Führer der Luftstreit-
kräfte (FdL), 102, 112, 171, 213,
215, 217, 248; General der Luftwaffe
beim ObdM, 216; Generalluftzeug-
meister, 153, 235; Kommandeur der
Heeresflieger, 102; Kommando der
Blindflugschule der Luftwaffe, 220;
Kommando der Fliegerschulen, 158;
Luftkommandoamt (LA), 97,98 147;
Luftschutz Amt (LS-Amt), 95–96;
Lufttransportführer, 220; Luftwaffen-
führungstab, 97, 171; Luftwaffe
Generalstab, 146–149, 151, 156;
Personnelamt, 148; Sonderstab W,
121; Staatssekretär der Luftfahrt/
Luftwaffe, 97, 147–149, 153;
Technisches Amt (LC), 97, 108, 148,
151–153, 155
Störkampf/Nachtschlachtstaffeln, 138
Strength, 147, 164, 172, 177, 189,
200, 220, 232, 239
Supplies, 166, 184, 204–5, 256
Training, 110, 158–9, 207
Training establishments: LKS Dresden,
146; Luftkriegsakademie, 111, 141;
Luftkriegsschulen (LKS), 111, 156;
Lufttechnischen Akademie, 146;
'Reichsbahnstrecken' (RB-strecken),
110; Streckenschule Berlin, 111

Maastricht, 193–5, 243, 247, 252
Madrid, 120–2, 124–9, 133, 135–7, 143
Maginot Line, 115, 196, 201, 251, 261,
266
Manufacturers of aircraft and aero-
engines:
  AB Flugindustri, 48, 60
  AEG, 31, 35
  Albatros, 36–7, 47, 77
  Arado, 77, 113
  Argus, 56, 77
  ATG, 103
  Aviatik, 35
  Benz, 54, 56
  BFW (Bayerische Flugzeugwerk), 55,
  97, 102, 104, 152, 162
  BMW, 39, 53, 77, 103
  Bücker, 39
  Caspar, 33
  Central Aircraft Workshop (Poland),
  21
  Daimler, 54, 56
  Daimler-Benz, 56, 77, 103, 157
  De Havilland, 55
  Dornier, 36, 77, 113
  Focke Wulf, 77
  Fokker, 35, 39, 46
  Gotha, 103
  Halberstadt, 35
  Hannover, 35
  Heinkel, 33, 36, 39, 53, 76–7, 102,
  104, 113
  Henschel, 103
  Hirtenberger Patronen, 112
  Junkers, 22, 36–7, 47–8, 52, 55, 76–7,

103–4, 129
Klemm, 55
LVG, 35
MIAG, 103
Potez, 254
PZL, 181
Rohrbach, 55–6, 77
Rumpler, 35
Russo-Baltiysky, 47
Siebel, 103
Siemens, 56
Siemens und Halske, 103
Siemens Schuckert, 35
Svenksa Aero, 33, 39
Udet, 42
Weser Flugzeugbau/Blohm und Voss,
102, 103
'Market Garden', Operation, 244
Marinegruppen Kommando Ost, 218
Marinens Flyvevåben (Norwegian Naval
Air Corps), 221
Moscow, 36, 44–5, 47, 74, 165
Munich Agreement 1938, 141, 169,
170–1

'Najuku' (Nachtjagd und Erkundungs-
flugzeug), 64, 65, 78
Narvik, 218–19, 221, 228, 230, 232–6
National Sozialistische Fliegerkorps
(NSFK), 158
  Gruppe 17, 162
Netherlands, The, 32, 55, 192–3,
196–7, 241, 244–5, 247–8, 256
Norway, 41, 193, 206, 218–20, 222,
226–7, 230–1, 234, 249, 256, 260
Norwegian Navy, 79

ObdL, 149, 176–7, 180, 182, 186–7,
193–4, 196, 219, 231–2, 253, 262,
266
OHL, 99
OKH, 150, 193, 198, 219, 257, 261
OKW, 51, 150, 167–8, 178, 187, 189,
190, 192–4, 198, 212, 216, 218–19,
232, 249, 261, 265, 268
Oslo, 218–20, 222–3, 227–30, 232, 236
Oslo Report, 113
Österreichische Flugpolizei, 40
Österreichische Luftstreitkräfte (LStrKr),
111–12, 159–62
  Abteilung Luftfahrwesen, 80
  Abteilung L, 80
  Kommando der Luftstreitkräfte (KoLu),
  112
  Lehrabteilung II (LA II), 111
  Lehrabteilung III (LA III), 111
  Luftschutzkommando, 112

Paris, 38, 54, 100, 115–16, 165, 261–3,
265–6
Paris Agreements:
  1921, 34
  1926, 54, 58, 64–5
  1927, 79–80
Planning:
  Luftwaffe: in 1933, 100–2; in 1938,
  150; for anti-shipping operations,
  212–13, 216, 217; for Fall 'Gelb',
  192–7, 208–9; for Fall 'Rot', 261–2;
  for Fall 'Weiss', 176–8; for
  'Weserübung', 220–1; Studie Blau,
  171; Studie Grün, 171; Studie Rot,
  171
  Wehrmacht and war games:
  'Generalstabsreise 1939', 172 ;
  'Generalstabreise', 164;

'Generalstabsübung Schlesien', 175;
'Planstudie 39', 167, 171; 'Planstudie
1938', 163; 'Planstudie Grün', 163;
Studie Nord, 218
Poland, 18, 26–8, 36, 38, 40–1, 60,
171–2, 175–6, 186–9, 192, 197, 260
Polish Army, 175–6, 178, 182–4, 185
Polish Corridor, 175–6, 182
Polish Navy, 176, 182
Polizei-Fliegerstaffeln, 28, 30
Polizei-Flugstaffeln, 40
Polskie Lotnictwo Wojskowe/PLW (Polish
Air Force), 176–81, 187–8

Radar:
British: 116; GM, 199; Type 1, 172,
199; Type 2, 217
German: 172, 227; Freya, 172, 197;
Würzburg, 172, 199, 229
Regia Aeronautica (Italian Air Force), 65,
111, 144, 263
Reichsamt für Luftfahrt und Kraftfahr-
wesen, 35
Reichsluftamt, 19, 35
Reichsluftfahrtkommissariat, 94–95
Reichsluftministerium (RLM), 74–5, 94,
96–8, 103–4, 107, 113–14, 117, 121,
134, 138, 144, 147–8, 150, 153, 155,
168, 219
Reichswehrministerium (RWM), 22, 30,
32, 35, 37, 44, 50–1, 53, 57, 60–4,
71–2, 74–5, 77, 95, 103, 110, 269
Reichsheer:
A7L (Aviation Department in Prussian
War Ministry), 19, 30, 32
In 1 (L), 51, 101
Organizations: Heereswaffenamt, 48,
50–1, 56–7, 62, 64–5, 76, 78, 95,
106, 108; In 2 (Infantry Inspectorate), 72; In 4 (Artillery Inspectorate),
72; In 7 (Communications
Inspectorate), 72; Inspektion der
Waffenschulen und der Luftwaffe, 51;
Inspektion für Waffen und Gerät, 28,
50; T1, 50–1, 71–3, 76; T2, 51,
71–3; T3, 28, 50; Truppenamt, 28,
37, 49, 51, 63, 95, 99–100; Wa Prw
8, 52
Plans for air force in 1919, 25; in
1920, 28; in 1925, 60; in 1927, 63;
in 1930, 72; in 1932, 73, 74; in
1933, 101
Referenten zbV, 28–9, 39, 51–2, 57
Reklamestaffeln, 69
TA(L)/Luftschutzreferat, 28, 49–51
T2 III (L), 49
T2V (L), 50–51
Reichswehr formations and commands:
Artillerieführer V, 102
4.Artillerie Regiment, 98
9.Batterie/Artillerie Regiment Nr 1, 64
13.Brigade, 30
2.Fahrabteilung, 64
Generalkommando Lüttwitz, 31
6.Infanterie Regiment, 95

13.Infanterie Regiment, 136
11.Kavallerie Regiment, 64
Wehrkreis I, 57, 69, 95
Wehrkreis II, 57, 69
Wehrkreis III, 57
Wehrkreis IV, 57
Wehrkreis V, 57
Wehrkreis VI, 30, 57
Wehrkreis VII, 52, 57, 69
Reichsmarine:
A 11 I, 29
Flottenabteilung Referent für Seeflug-
wesen, 29
Gruppe BSx, 59
Gruppe LS, 59, 75
Organizations: Allgemeines Marineamt,
59; Marinekommandoamt, 59, 75–6;
Marineleitung, 102; Seetransport-
abteilung, 61; Stationskommando
Nordsee, 29; Stationskommando
Ostsee, 29
Reichsluftamt, 19
Reichsminister für die Luftfahrt, 96
Ring Deutscher Flieger, 29, 39
RKK–VVF, 36, 44–5, 48
Romania, 25, 40, 154, 186–7
'Rote Kapelle' (spy ring), 113
Rotterdam, 132, 197, 244–5, 249–50
Royal Air Force (RAF), 33, 35, 71, 83,
87, 89–90, 94, 105, 113–16, 147–8,
151, 157–9, 164, 166, 171, 198, 202,
204, 215, 218, 224–5, 231, 234,
239–42, 247–8, 252, 254, 259–60,
265, 268
Royal Navy, 90, 171, 190, 212, 219–20,
224–6, 230, 236–7
Ruhr Fund, 38, 39
Russo–Finnish War (1939–40), 218–19
Russo–German Non-Aggression Pact
(1939), 175, 187

Schnellbomber, 109
Schweizerische Flugwaffe (Swiss Air
Force), 263
Seeflugzeug–Erprobungsstelle (SES),
Travemünde, 61
Seville, 120–3, 125, 129, 144
Sicherheitspolizei (Sipo), 29–31
Slovak Air Force (Slovenske vzdusné
zbrane), 170
Slovak Army, 178
Slovakia, 170, 175, 178, 187
Soldiers' Councils, 18–19
Sonderfall 'Otto', 160
Sondergruppe R, 37
Soviet Union, 32, 34, 36–7, 40, 46–9,
55, 115, 125, 151, 154, 156, 165, 218
Spanish Air Force:
Nationalist, 123, 129, 133
Republican, 137–8, 140–1
Spanish Army, 120, 123, 129, 133,
137–8, 140
Spanish Navy, 129
Spain, 55, 117, 120–145, 149, 154
Sportflug GmbH, 57–9, 96

Staatsamt für Heereswesen, 40
Stavanger, 219–20, 223, 226–7, 229,
231, 233
Sudetenland, 162, 168–9
Sweden, 36, 41, 55, 79, 91, 154, 228,
232
Swedish Air Force, 167
Switzerland, 32, 36, 79–80, 161, 190
Synthetic fuel, 154

Transport Ministry (Reichsverkehr-
ministerium), 52, 59, 74–7, 94
Trondheim, 219, 223, 226–33, 236
TsAGI, 47
Turkey, 27, 48, 55

Übung 'Freudenthal', 163
Unternehmen:
'Döberitz', 143
'Feuerzauber', 121, 124
'Guido', 123–124
'Neptun', 144
'Niwi', 244
'Paula', 263
'Schulung', 163
'Tiger', 266
'Wasserkante', 176, 181, 186
'Weserübung', 101, 189–91, 208,
214–15, 217, 219–20, 227, 268
'Ural Bomber', 84, 108–9, 156
USAAC (US Army Air Corps), 70,
83–4, 89, 115–16, 158
Air Corps Tactical School, 89, 156
USAAF (US Army Air Force), 88–9

Valencia, 137–40, 144
Versailles Peace Treaty, 21, 26–30,
32–4, 36, 39–40, 53, 70–1, 94, 114,
168, 175
Vienna (Wien), 40–1, 79, 112, 161, 163,
169, 190
Volkswehr, 25
VVS–RKKA (Russian Air Force), 45,
47, 84, 91, 115–16, 144, 165

Warsaw (Warszawa), 36, 54, 132,
175–6, 178, 180–8, 259
Wehrflugorganization, 74
Wehrmacht, 99, 114, 124, 149–51,
165–6, 172, 175, 187, 253
West Wall, 163, 189, 197, 200–1, 261
Wienerbärn, 41, 218, 234
'Winterübung', 115
Winterübung 'Rügen', 124
World Disarmament Conference, 72–3,
114

X–Gerät, 172, 176, 185, 199, 269

Y–Gerät, 113, 172
Yugoslavia, 25, 55, 154

Zealand (Sjaelland), 219, 221
Zentrale Moskau (ZMo), 44

Index of Personalities

Note: Military ranks are the highest given in the text.

Adam, Generalleutnant Wilhelm, 72–3,
75
Adolph, Oberleutnant Walter, 248
Ahlefeld, Oberstleutnant Hans, 227
Albach, Heinrich, 208
Albert, Heinrich, 19

Aldinger, Hauptmann, 123
Alksnis, Yakov I., 47–8, 79, 91, 165
Andrae, Generalleutnant Alexander, 221
Andreas, Hauptmann Otto, 215
Andrée, Hauptmann, 20
Angerstein, Karl, 31

Aschenbrenner, Hauptmann Kurd, 269
d'Astier de la Vigerie, Général François,
198

Baier, Hauptmann, 69
Balbo , Marshal Italo, 111

Balthasar, *Oberleutnant* Wilhelm, 137
Bär, Heinz, 202
Baranov, Pyotr I., 47, 49
Barès, Joseph, 41, 90
Barratt , Air Marshal Arthur, 199, 239, 245, 252, 264
Battmer, Ernst, 153
Bauer, Gustav, 27
Baur de Betaz, *Oberst* Wilhelm, 232
Bäumer, Paul, 28
Beck, *General der Artillerie* Ludwig, 87, 99–100, 147, 164–5
Behrendt, *Oberst*, 176, 196
Beitzke, *Hauptmann*, 235
Berg, *Major* Ernst *Freiherr* von, 265
Best, Lieutenant General P. W., 244
Billotte, *Général* Gaston, 240, 243, 248
Blaskowitz, *General der Infanterie* Johannes, 185–6, 188
Blattner, *Hauptmann*, 182
Blomberg, *Generalfeldmarschall* Werner von, 50–51, 60–1, 72, 76, 95–6, 98, 101–2, 104, 108, 110, 114–15, 121, 123–4, 131, 134–5, 147–50
Bodenschatz, *Generalmajor* Karl, 18, 151, 170, 219
Boddem, *Leutnant* Peter, 134
Bock, *Generalfeldmarschall* Fedor von, 175, 177, 192–4, 196–7, 261–2, 265–6
Boenigk, Oskar *Freiherr* von, 21
Bogatsch, Rudolf, 23
Böhne, *Oberleutnant* Heinz, 235
Bohnstedt, *Oberst* Eberhardt, 95–8, 101–2
Bolle, Carl, 97, 113
Bongartz, Heinz, 28
Bonin, *Oberleutnant* Hubertus von, 143
Borries, *General* von, 38
Brack, *Sonderführer*, 235
Brandenburg, Ernst, 19, 35, 52–4, 57, 59, 62, 71–2, 75, 88, 94, 96
Brandis, *Oberleutnant Freiherr* von, 233
Bredow, *Oberstleutnant* Ferdinand von, 69
Brockdorff-Rantzau, Ulrich, 27
Bruch, *Generalmajor* Hermann, 102, 216
Brutzer, *Konteradmiral*, 75
Büchner, Franz, 22
Buckler, Julius, 28
Bülow, *Oberst Freiherr* von, 171
Bülowius, *Oberst* Alfred, 241

Canaris, *Admiral* Wilhelm, 124, 150
Caproni, Gianni, 88
Carls, *Admiral* Rolf, 218
Casimir, *Hauptmann* Artur von, 235
Chamberlain, Neville, 167–8, 170–1, 175
Chennault, Major Claire L., 89
Chiang Kai-shek, 57
Christ, *Hauptmann* Torsten, 140
Christiansen, Friedrich, 21–2, 33, 65, 110, 112, 158–9
Christian X , King, 222
Churchill, Winston, 148, 218, 247, 255, 268
Ciano, Count Galeazzo, 172
Coeler, *Generalmajor* Joachim, 59, 171, 177, 216–17
Conrad, *Oberst* Gerhard, 244
Cramon-Taubadel, *Oberst* Hans-Jürgen von, 206
Creydt, *Hauptmann*, 22
Cuno, Wilhelm, 38–9
Curtius, Julius, 71

Davila, *General* Fidel, 137
Deichmann, *Hauptmann* Paul, 71, 99, 117, 148, 156
Dessloch, Otto, 21, 71
Dietl, *Generalmajor* Eduard, 232–3, 235
Dietrich, Richard, 39
Dilley, *Oberleutnant* Bruno, 179
Dinort, *Major* Oscar, 71, 115, 179, 182, 258
Dollfus, Engelbert, 111–12
Dornier, Claude, 102
Dörning, *Generalmajor* Hans von, 191, 247
Dörr, Gustav, 28
Douhet, Giulio, 82, 88–90, 101, 143, 234, 240
Dowding, Air Chief Marshal Sir Hugh, 148, 204, 247, 255
Drewes, *Oberstleutnant* Martin, 222–3
Duval, Maurice, 87–88

Eberhardt, *Oberleutnant* Kraft, 122–3, 125–6
Eberhardt, *Generalleutnant* Walter von, 23
Ebert, Friedrich, 18, 20, 39
Ellguth, *Hauptmann* Dietrich *Graf* von Pfeil und Klein, 202
Enver Pasha, 36
Euler, August, 19, 31–2, 35

Faber, *Kapitänleutnant*, 29, 33, 59
Falcke, *Hauptmann* Wolfgang, 179, 222, 227
Falkenhorst, *General* Nikolaus von, 219, 232–3
Faupel, Wilhelm von, 134
Felmy, *General der Flieger* Hellmuth, 32, 49–50, 62, 71–4, 78, 83, 95, 100–1, 149, 151, 167–8, 171, 181, 189–90, 192–3, 195
Fiebig, *Oberst* Martin, 186, 241
Fink, *Oberst* Johannes, 189
Fischer, *Oberst* Veit, 37
Fisser, *Oberst Dr* Johann-Volkmar, 206, 264
Fokker, Anthony, 32, 39
Förster, *Hauptmann*, 232, 233
Förster, *Oberst* Helmuth, 155, 177, 185, 189
Framm, *Oberleutnant* Gert, 248
Franco, *General* Francisco, 120–5, 128–30, 133–4, 136–8, 140, 142–3
Fritsch, *General der Artillerie* Werner von, 115, 146, 149
Frankenberger, *Unteroffizier* Arno, 198
Fuchs, *Oberst* Robert, 126, 133, 230, 232–3, 236

Gablenz, *Oberstleutnant* Karl August *Freiherr* von, 220, 222–3, 227, 233, 235–6
Galland, *Hauptmann* Adolf, 111, 128, 135, 138–9, 248, 265
Gallera, *Leutnant* Hans-Peter von, 126
Garcia Morato, *Capitán* Joaquin, 123
Garski, *Oberstleutnant* Eugen, 244
Geisler, *General der Flieger* Hans Ferdinand, 59, 171, 212, 220, 222, 227–8, 230, 232, 269
Gentzen, *Hauptmann* Hannes, 179, 202, 257
Georges, *Général* Jacques, 240, 245
Gessler, Otto, 30, 38, 44, 60–1
Geyer, *Oberst* H., 71
Gilsa, *Leutnant* Kurt von, 127, 129

Gnys, Second Lieutenant Wladyslav, 179
Gockel, *Leutnant*, 132
Göhre, Paul, 18
Goebbels, Josef, 175, 212, 243
Goldwyn, Samuel, 48
Gollob, Gordon, 179
Goltz, General Rüdiger *Graf* von der, 22–3, 38, 96, 219
Göring, Carin, 42, 92
Göring, *Reichsmarschall* Hermann, 17, 21, 24, 41–2, 53, 91–2, 94–9, 101–2, 104, 108–14, 121, 124, 128, 134, 146–56, 159, 161, 163, 167–71, 175, 178, 180, 184–6, 188, 190, 192–3, 195–6, 198, 209, 212–14, 216–17, 219–20, 227, 231, 236, 241, 253, 258–62, 267–9
Göring, Emmy, 97
Gossage, Group Captain E. L., 63, 69–70
Gotz, Horst, 234
Grabmann, *Major* Walter, 141–2, 179–81
Grauert, *Generaloberst* Ulrich, 21–2, 63, 71, 163, 176–7, 185–7, 193, 251, 253–5, 257–9, 265, 269
Greim, *General der Flieger* Robert *Ritter* von, 24, 41, 57, 70, 110, 148, 152–3, 251, 255, 269
Groener, Wilhelm, 27, 38, 57, 61, 71–2
Gronau, *Oberleutnant zur See* Wolfgang von, 59
Guderian, *General der Panzertruppe* Heinz, 193, 248, 251–6, 261, 266
Guerard, Theodor von, 71
Günter, Siegfried, 35
Günter, Walter, 35

Haakon VII, King, 220
Haas, *Leutnant* Hermann, 137
Hácha (Czech), Emil, 170
Hackl, Anton, 202
Haehnelt, *Generalmajor* Wilhelm, 19–20, 30–1, 97
Halder, *Generalmajor* Franz, 147, 165, 249, 257, 265
Hammerstein-Equord, *Generalleutnant* Kurt *Freiherr* von, 51, 63, 71–74, 95, 101–2, 110
Handrick, *Major* Gotthard, 133, 135, 141
Hansen, *Oberleutnant* Werner, 222–3
Harder, *Oberleutnant* Harro, 135
Harding, *Leutnant*, 30
Härle, *Major* Fritz, 141, 143
Harlinghausen, *Major* Martin, 138–9, 168, 215, 223, 231–2, 235–236
Harris, Group Captain Arthur, 116
Hartmann, *Oberst*, 44
Hasse, *General* Otto, 37, 44, 50, 60
Hedderich, *Leutnant* Wenner, 243
Hefele, *Major*, 138
Heinkel, Ernst, 33, 55, 102, 104
Held, Alfred, 190
Heller, Colonel Wladyslaw, 181
Henke, *Flugkapitän* Alfred, 120–3, 125
Hepe, *Leutnant* Hans-Jürgen, 127
Hering, *Hauptmann* Hubertus, 138
Herring, Group Captain J. H., 113
Hermann, *Leutnant* Hans-Joachim ('Hajo'), 121–2, 155
Hess, Rudolf, 20, 41, 94, 97, 120, 152
Heye, *Generaloberst* Wilhelm, 44, 60–1, 63, 72
Hilferding, Rudolf, 44

Hindenburg, Paul, 60, 92, 99
Himer, *Generalmajor* Kurt, 222
Hippel, *Oberstleutnant* Walter von, 252
Hitler, Adolf, 41–2, 58, 73, 76, 89, 92,
  94–6, 101, 104, 108, 110, 114–15,
  120–4, 134–5, 147, 149–1, 154–5,
  160–3, 165, 168–72, 175–8, 186,
  188–90, 192–7, 209, 212, 214,
  216–20, 232, 240, 250, 253, 258,
  260–1, 267, 269
Hoepner, *General der Panzertruppe* Erich
  250
Hoeppner, *Generalleutnant* Ernst von,
  18, 24, 29, 149
Hoffmann von Waldau, *Oberst* Otto, 71,
  151, 173, 250, 261, 263
Höhne, *Oberstleutnant* Otto, 262
Holle, *Oberstleutnant* Alexander, 124,
  127–8
Höness, *Unteroffizier* Guido, 133
Hönmanns, *Major* Erich, 195
Hoth, *General der Panzertruppe*
  Hermann, 253
Houwald, *Oberleutnant* Freiherr von, 69
Hozzel, *Hauptmann* Paul-Werner, 182,
  231

Ibel, *Oberst* Max-Josef, 248
Ihlefeld, Herbert, 145
Inskip, Sir Thomas, 165

Jacobs, Josef Carl, 23, 28
Jäger, *Hauptmann*, 228, 237
Jeschke, Oberfeldwebel Martin, 230
Jeschonnek, *General der Flieger* Hans, 21,
  71, 75, 101, 146–8, 150–2, 155–6,
  162, 164–5, 167, 169, 172, 175, 186,
  193, 195, 216, 218, 250, 258–9, 261,
  269
Jeschonnek, *Hauptmann* Paul, 52, 75,
  150
Jodl, *Generaloberst* Alfred, 150, 167,
  193–6, 219
John, *Oberleutnant*, 267
Jordan, *Oberleutnant*, 71
Junck, *Oberst* Werner, 21, 264
Junkers, Hugo, 37, 48, 52, 102

Kammhuber, *Oberst* Joseph, 51, 73,
  150–1, 168, 195, 206, 264, 267
Kapp, Wolfgang, 30
Karlewski, *General der Flieger* Erich, 147,
  149
Kastner, *Hauptmann*, 30
Kaupisch, *General der Infanterie*
  Leonhard, 102, 149, 182, 219
Keitel, *Generaloberst* Wilhelm, 51, 71,
  150, 167, 194, 219
Keller, *Generaloberst* Alfred, 58, 71, 105,
  167, 194, 219
Kenney, Captain George C., 89
Kessel, *Oberleutnant* Hans-Detlef von,
  135
Kesselring, *Generalfeldmarschall* Albert,
  51, 98, 103, 127, 134, 147–9, 151,
  156, 160, 163, 167, 171, 175, 176,
  177, 178, 182, 184–5, 187, 195, 197,
  202, 208, 227, 239, 2445, 250, 254,
  258, 261–2, 265, 267–9
Kessler, *Oberst* Ulrich, 59, 97, 168, 182
Kindélan, General Alfredo, 129
Kitzinger, *Generalleutnant* Karl, 163, 166
Kleffel, Walter, 152
Klein, *Generalmajor* Hans, 191, 200–1,
  206
Kleist, *General der Panzertruppe* Ewald

von, 196, 243, 245, 250–8, 261–2,
  266
Kluge, *Generalmajor* Günther, 185, 254
Klümper, *Oberleutnant*, 138
Knauss, *Oberst* Robert, 101, 219
Kneiding, *Unteroffizier* Kurt, 126
Knüppel, *Hauptmann* Herwig, 123, 127,
  254
Koch, *Hauptmann* Walter, 194, 243
Köhl, *Hauptmann* Hermann, 54
Kolbitz, *Leutnant* Oskar, 126
Koller, Karl, 31
Koppenberg, Heinrich, 103, 155–6
Korten, *Oberstleutnant* Günther, 42, 71,
  162, 186–7
Könnecke, Otto, 28
Kopp, Viktor, 36–8
Krahl, *Oberstleutnant*, 69
Krassin, Leonid, 37–8
Kreipe, *Generalleutnant* Werner, 42, 71,
  159
Krocker, *Hauptmann*, 21
Küchler, *General* Georg von, 193, 249
Kühl, *General der Flieger* Bernard, 102,
  149, 159
Kusserow, *Major* Ernst, 230

Lackner, *Oberst* Walter, 249, 254
Lahs, *Admiral* Rudolf, 59, 96
Lawrence, T. E., 36
Lebedev, P. P., 44, 47
Leeb, *Generaloberst* Wilhelm *Ritter* von,
  189, 192, 261
Leie, Erich, 202
Leigh-Mallory, Wing Commander
  Trafford, 87
Leizaola, Jesús Maria, 132
Lent, *Leutnant* Helmuth, 179–80, 222–3
Lessing, *Major*, 220
Lettow-Vorbeck, *Generalmajor* Paul von,
  22
LiethThomsen, *General der Flieger*
  Hermann von der, 19, 24–5, 27, 30–1,
  44, 50, 94
Lindbergh , Charles A., 55, 164
Loeb, *Oberst* Friedrich, 151–2
Loerzer, *General der Flieger* Bruno, 21,
  24, 28, 95–6, 110, 152, 176–7, 182,
  184–6, 251–3, 257, 259, 265, 269
Lohmann, *Kapitän zur See* Günther,
  59–61
Löhr, *General der Flieger* Alexander, 80,
  112, 159, 161–3, 167, 175, 177–9,
  183–4, 186
Lonsdale, Lieutenant Commander
  Rupert, 231
Ludendorff, Erich, 19, 99
Ludewig, *Leutnant* Wolfgang, 244
Ludlow-Hewitt, Air Chief Marshal Sir
  Edgar, 148, 164
Lüttwitz, *General* Walter von, 31
Lützow, *Oberleutnant* Günther, 128, 145
Lux, Albert, 20

MacArthur, General Douglas, 89
Mahnke, Alfred, 31
Mahrenholtz, *Hauptmann* Siegfried, 226
Maltzahn, *Hauptmann* Günther *Freiherr*
  von, 202
Manstein, *Generalleutnant* Erich von , 98,
  193
Martini, *Generalmajor* Wolfgang, 71, 98,
  172, 264
Massenbach, *Major* Dietrich *Freiherr*
  von, 269
Massow, *Oberst* Gerd von, 209, 245,

247, 251–2, 261, 264
Masterman, Air Commodore E. A. D.,
  29, 33–4
Medwecki, Captain, 179
Mehnert, *Major* Karl, 133, 135, 141
Mehrens, *Leutnant zur See* Günther, 231
Meister, *Oberstleutnant* Rudolf, 21, 71,
  159
Menckhoff, Karl, 28
Merhart, *Hauptmann* Hubertus von,
  126–7
Messerschmitt, Willi, 97, 102, 104, 110,
  152
Methfessel, *Leutnant* Werner, 179
Mezhenikov, Sergei, 78
Michalski, Gerhard, 202
Miklas, Wilhelm, 161
Milch, Anton, 97
Milch, *Generalfeldmarschall* Erhard, 21,
  24, 29, 53, 60, 62, 74, 94–9, 101–6,
  110–14, 121, 127, 146–53, 155, 161,
  163, 165, 167, 169–71, 184, 195, 227,
  231, 236, 256, 261, 264, 268–9
Mitchell, Brigadier General William,
  88–9
Mittelberger, *Generalmajor* Hilmar *Ritter*
  von, 51, 71–3, 75–8, 95
Mix, *Hauptmann* Erich, 202
Modrow, Hans, 234
Mola, *General* Emilio, 120, 123, 130–2
Mölders, *Hauptmann* Werner, 139–40,
  143, 145, 201–2, 265, 267
Moreau, *Hauptmann* Rudolf *Freiherr* von
  122–5, 127–8, 131, 155
Morzik, Friedrich ('Fritz'), 49, 244
Mratzek, *Unteroffizier* Ernst, 129
Muff, *General* Wolfgang, 162
Müller, Heinrich, 112
Müller, *Major*, 45
Müncheberg, Joachim, 202
Mussolini, Benito, 80, 111, 139, 178

Neubert, *Oberleutnant* Frank, 179
Newall, Marshal of the Royal Air Force
  Sir Cyril, 164, 204
Niedermayer, *Major* Oskar *Ritter* von,
  37–8, 44
Nielsen, Andreas, 42
Niepmann, *Hauptmann*, 69
Nirminger, *Unteroffizier* Hans, 143
Normann, *Hauptmann* von, 198

Oesau, *Leutnant* Walter, 140, 145
Oppeln-Bronikowski, *Major* von, 24
Orgaz, *General* Luis, 120
Osterkamp, *Oberstleutnant* Theo, 20, 60,
  70
Oxland, Group Captain R. D., 157

Park, Air Vice Marshal Sir Keith, 259,
  261
Papen, Franz von, 96
Paulus, *Oberst* Friedrich, 176
Pawlikowski, Colonel Stefan, 180
Pellengahr, *Generalleutnant* Richard, 228,
  230
Peltz, *Oberleutnant* Dieter, 177
Pennès, *Général* Roger, 198
Pétain, Marshal Philippe, 87, 266
Petersen, *Major* Edgar, 213
Pflugbeil, *General der Flieger* Kurt, 23,
  64, 69, 71, 111, 177, 184, 254
Picasso, Pablo, 132
Pilsudski, Marshal Josef, 178
Pitcairn, *Oberleutnant* Douglas, 135
Ploch, *Hauptmann* August, 152

INDEX 319

Plocher, *Major* Hermann, 71, 136, 138–40, 142

Poincaré, Raymond, 38

Pohle, *Hauptmann* Helmut, 212

Poindron, *Général*, 41

Prada, *Coronel* Adolfo, 134

Price, Ward, 114

Putzier, *Generalmajor* Richard, 194–5, 243, 245, 248–9

Quisling, Vidkun, 218

Radek, Karl, 36, 38

Raeder, *Grossadmiral* Erich, 61, 66, 75, 102, 146, 216–19, 221

Reichenau, *General der Infanterie* Walther von, 95, 176–85, 192–4, 196, 245, 250–1, 253, 259

Reinberger, *Major* Helmut, 195

Reinecke, *Hauptmann* Günther, 231

Reuss, Franz, 31

Rheinhardt, *Generalleutnant* Georg-Hans, 251–2

Rhoden, *Oberstleutnant* Hans-Detlef Herhudt von, 241

Ribbentrop, Joachim, 172, 175

Richthofen, Manfred von, 17, 106, 152

Richthofen, Lothar von, 17, 28, 35

Richthofen, *Generalfeldmarschall* Wolfram von, 17, 71, 105–9, 125, 127–31, 135–6, 142–4, 150–1, 176–9, 183–9, 192–4, 196, 245, 248, 250–1, 253–5, 257–9, 264–6, 268–9

Rieckhoff, Herbert, 31

Ritter, *Generalmajor* Hans, 216

Rohrbach, Adolph, 32, 102

Rommel, *Generalmajor* Erwin, 136, 258

Rómmel, General Juliusz, 188

Rorand, *Colonel*, 32

Rosengol'ts, A. P., 44, 47, 165

Rotberg, *Oberleutnant* von, 197

Rowehl, *Oberstleutnant* Theodor, 163, 171

Rüdel, *General der Flieger* Otto Günther, 148–9

Ruge, Colonel Otto, 233, 235

Rumbold, Sir Horace, 69

Rundstedt, *Generaloberst* Gerd von, 175, 177, 185, 192–4, 196, 198, 257, 261, 265–6

Salazar, Antonio de Oliveira, 123

Sachsenberg, Gotthard, 21–3, 53, 62–3, 77, 80

Sanjurjo, *General* José, 120

Schatz, *Oberleutnant*, 20

Scheele, *Major* Alexander von, 121–5, 129

Scheidemann, Philipp, 48

Scheneback, *Freiherr* von, 45

Schlageter, Leo, 38

Schleich, Eduard *Ritter* von, 24, 28, 206

Schleicher, Kurt von, 20, 37, 51, 73, 75–6, 92

Schmale, *Oberfeldwebel* Willi, 191

Schmid, *Oberstleutnant* Josef, 23, 146, 171–2, 190, 260, 268

Schmidt, *Oberleutnant* Hermann, 199

Schmidt, *Oberleutnant*, (*Fallschirmjäger*), 228

Schmidt, *Generalleutnant* Rudolf, 244, 249

Schober, Johann, 80, 112

Schöndorff, *Hauptmann*, 47

Schoenebeck, Carl-August von, 21, 23

Schröder, Joachim von, 21

Schubert, Carl von, 49

Schulze, *Unteroffizier* Helmut, 122–3

Schulze-Boysen, Harro, 113, 125

Schulze-Heyn, *Oberleutnant*, 69

Schumacher, *Major* Carl, 190–1

Schuschnigg, Kurt von, 111, 160

Schwartzkopff, *Oberst* Günther, 253

Seeckt, *Generaloberst* Hans von, 25, 27–30, 35–9, 44, 48–9, 57, 60

Seidel, *Generalmajor* Hans-Georg von, 173, 256

Seidel, *Leutnant* Paul, 243

Seidemann, *General der Flieger* Hans, 71, 142, 194, 196, 252, 254, 258

Seiler, *Oberfeldwebel* Reinhard, 137

Sellhorn, Heinz, 207

Seyss-Inquart, Arthur, 160–1

Seywald, *Oberst*, 184, 186

Siburg, *Generalmajor* Hans, 59, 183

Siebel, Friedrich ('Fritz'), 37, 57

Siebert, *Hauptmann* Anton, 25

Siegert, *Major* Wilhelm, 19, 29

Simon, Sir John, 114

Slessor, Group Captain John, 167

Smidt, *Hauptmann*, 139

Smith, Captain Truman, 164

Smut, Jan, 35, 88

Speidel, *Oberst* Hans, 71, 136, 163, 188, 195, 202, 250

Sperling, *Rittmeister*, 69

Sperrle, *Generalfeldmarschall* Hugo, 21, 51, 71, 102, 105, 124–31, 133–5, 149, 151, 160–1, 163, 167, 169, 171, 189, 192–4, 196, 198, 209, 239, 241, 250–2, 254, 257–8, 261, 266, 269

Spielvogel, *Major* Werner, 181, 184

Spiller, *Hauptmann* Eberhard, 220, 223

Sponeck, *Generalleutnant* Hans Graf von, 194, 244–5, 250

Stahr, *Major*, 45–6

Stahl, *Oberst*, 191

Stahl, Peter, 207

Stalin, Josef, 47, 91, 125

Steffen, *Hauptmann*, 22

Stein, *Hauptmann* Gert, 216

Steinhoff, Johannes, 202

Stinnes, Hugo, 39

Stitz, Max, 202

Stoeckel, *Oberst* Alois, 206

Stollbrock, *Oberstleutnant* Joachim, 199, 215

Stresemann, Gustav, 39–41, 44, 49, 61, 66, 69

Student, *Generalmajor* Kurt, 28, 35, 39, 44, 50, 52, 64, 71, 163, 169, 194, 244–5, 249

Stülpnagel, *General der Flieger* Otto, 146, 149

Stumpff, *Generaloberst* Hans-Jurgen, 98, 148, 165, 167, 169, 231–2, 235–6, 269

Stutterheim, *Generalmajor* Wolf von, 186, 251, 261, 266

Süssmann, *Generalleutnant* Wilhelm, 206, 221, 232, 249

Sykes, Sir Frederick, 164

Tank, Kurt, 65

Têtu, *Général* Marcel, 198, 240, 245, 247

Thomsen, *see* Lieth-Thomsen

Tirpitz, *Grossadmiral* Alfred von, 101, 113

Thoma, *Oberst* Wilhelm *Ritter* von, 265

Thuy, Emil, 47

Trautloft, *Leutnant* Hans, 121, 123

Trenchard, Marshal of the Royal Air Force Sir Hugh, 88, 90, 99

Troitsch, Hans, 190

Trotsky, Leon, 36, 44, 47

Tschunke, *Major* Fritz, 38

Tupolev, Andrei N., 48, 82, 91, 165

Udet, *Generaloberst* Ernst, 24, 28, 42, 52, 55, 57, 76, 107–10, 127, 143, 148–9, 151–3, 155, 157, 166, 169–70, 194, 204–5, 227, 235, 265, 269

Ursinus, Oskar, 35

Vachell, Group Captain J. L., 164

Varela, *General* José Enrique, 133

Velardi, General Vicenzo, 129

Veltjens, Joseph, 28, 123

Vierling, *Oberst*, 161

Vigón, *Coronel* Juan, 131

Vogt, *Hauptmann* Wilhelm, 28

Vuillemin, *Général* Joseph, 82, 164–5, 167, 190, 203, 240, 243, 247, 255, 262, 265–6

Volkmann, *General der Infanterie* Helmuth, 50, 52, 64–5, 71, 77–8, 134–41

Wagner, *Hauptmann* Richard, 222

Wandel, *Leutnant* Joachim, 132

Warlimont, *Oberstleutnant* Walter, 123–4, 134, 156, 218–19

Watson-Watt, Robert, 200

Wechmar, *Oberstleutnant* Karl *Freiherr* von, 235

Wegener, *Vizeadmiral* Wolfgang, 218

Weiss, *Hauptmann* Otto, 245, 258

Weiss, *Oberst* Wilhelm, 228

Weissmann, *Generalmajor*, 200

Wenniger, *Generalleutnant* Ralf, 76, 95, 97, 171, 196

Wessel, Horst, 114

Wetzell, *General* Wilhelm, 149

Wever, *Generalleutnant* Walther, 75, 84, 87, 98, 100, 108–9, 114–15, 146–7, 150–1, 269

Weygand, Marshal Maxime, 91

Wiart, Major General Sir Adrian Carton de, 229

Wick, *Major* Helmuth, 158, 202

Wilberg, *General der Flieger* Helmuth, 24–5, 27–8, 30, 32, 34–5, 49, 51, 59, 71, 97, 121, 123–4, 127, 149, 156

Wilcke, Wolf-Dietrich, 202

Wild, Wolfgang von, 23

Wimmer, Wilhelm, 52, 57, 71, 78, 80, 89, 97, 105–6, 108–9, 151–3, 155, 163, 177, 182, 185

Winkler, Max, 129

Wirth, Chancellor, 37–8

Witt, *Major* Hugo, 206

Witzig, *Oberleutnant* Rudolf, 243

Wolff, *Major* Karl-Heinz, 138, 189

Wolff, *Generalmajor* Ludwig, 160–1

Woytasch, *Generalmajor* Kurt, 228, 230–1

Wühlisch, *Generalmajor* Heinz-Hellmuth von, 51, 151, 195

Wurm, *Major*, 194

Wurmheller, Josef, 202

Yllam, Julius, 25–6, 80, 162

Zajac, Brigadier Jozef, 177–8

Zander, *Konteradmiral* Konrad, 59, 76, 102, 149

Zech, *Unteroffizier* Herbert, 122–3